WATER RESOURCES SYSTEMS MANAGEMENT TOOLS

WATER RESOURCES SYSTEMS MANAGEMENT TOOLS

Larry W. Mays, Ph.D., P.E., P.H.

Editor-in-Chief

Department of Civil and Environmental Engineering
Arizona State University
Tempe, Arizona

McGraw-Hill

New York Chicago San Francisco Lisbon London
Madrid Mexico City Milan New Delhi San Juan
Seoul Singapore Sydney Toronto

The McGraw-Hill Companies

Cataloging-in-Publication Data is on file with the Library of Congress.

1 2 3 4 5 6 7 8 9 0 DOC/DOC 0 1 0 9 8 7 6 5 4

ISBN 0-07-144381-9

The sponsoring editor for this book was Larry S. Hager, the editing supervisor was David E. Fogarty, and the production supervisor was Pamela A. Pelton. This book was set in the HB1 design in Times Roman by Joanne Morbit and Paul Scozzari of McGraw-Hill Professional's Hightstown, N.J. composition unit. The art director for the cover was Handel Low.

Printed and bound by RR Donnelley.

This book is printed on recycled, acid-free paper containing a minimum of 50% recycled, deinked paper.

CONTENTS

CONTRIBUTORS

Messele Z. Ejeta *California Department of Water Resources, Sacramento, California* (CHAP. 3)

Fred E. Goldman *Kennedy/Jenks Consultants, Phoenix, Arizona* (CHAP. 5)

Larry W. Mays *Department Of Civil And Environmental Engineering, Arizona State University, Tempe, Arizona* (CHAPS. 1, 3, 5, 7, 8)

John W. Nicklow *Southern Illinois University, Carbondale, Illinois* (CHAP. 4)

Sukru Ozger *Carollo Engineers, Phoenix, Arizona* (CHAP. 7)

Richard L. Skaggs *Pacific Northwest National Laboratory, Richland, Washington* (CHAP. 8)

Yeou-Koung Tung *Department of Civil Engineering, Hong Kong University of Science and Technology, Kowloon, Hong Kong* (CHAPS. 1, 2)

B. T. Wahlin *WEST Consultants, Inc., Tempe, Arizona* (CHAP. 6)

PREFACE

This book is the third book in a series of books dealing with management tools for water. The titles of the first two books in this series are *Urban Water Supply Management Tools* and *Urban Stormwater Management Tools,* both published by McGraw-Hill in 2004. This third book, first and foremost, is intended to be a reference book for those wishing to expand their knowledge of state-of-the-art techniques for the management of various types of water resources systems. This book focuses upon the use of optimization techniques, in many cases interfaced with simulation models, to operate water systems such as river-reservoir systems, water distribution systems, and irrigation canals. Also this book looks at the use of optimization techniques for remediation design of groundwater systems and the operation and capacity expansion of regional water supply systems. A second major focus is the use of uncertainty and reliability analysis for the analysis of water resources systems. This book will be of value to engineers, managers, operators, and analysts involved with the various aspects of analysis and operation of various water resource systems. Another use of this book will be as a text for graduate level courses in water resource systems analysis.

Preparation of this book has been a special treat for me because all of the authors (Drs. Messele Z. Ejeta, Fred Goldman, John Nicklow, Sukru Ozger, Richard Skaggs, Y.K. Tung, and Brian Whalin) are former Ph.D. students and close friends of mine. Each of these chapter authors has become a leading expert in the field of water resource systems. The authors were chosen, not only because they are former students of mine, but because of their proven knowledge in the specific area of their contribution. Chapters 1 and 2 are updated versions, respectively, of Chapters 6 and 7 of the *Water Resources Handbook,* published in 1996 by McGraw-Hill, for which I was also the Editor-in-Chief.

One of the quotes that I used in the *Water Resources Handbook* was by Chief Seattle, chief of the Suquamish tribe, who lived across Puget Sound from the site of the city that later arose in Seattle's name, "Man did not weave the web of life, he is merely a strand in it. Whatever he does, to the web, he does to himself." Humans have obviously done some fairly major changes to affect this web that will have serious consequences for the future. Among these changes on Earth are those that affect our water resources and environment. Humans have created some very interesting and challenging water resource problems that will challenge us for many decades and centuries into the future. These include the challenges that we face from global climate change, the challenges that we face for sustainability, the challenges related to water supply systems security from terrorist activity, just to name a few. Hopefully the types of methodologies discussed in this book will at least fulfill some small advancement in the solution of our future water resources challenges.

During the past 28 years of my academic career as a professor, I have received help and encouragement from so many people that it is not possible to name them all. These people represent a wide range of universities, research institutions, government agencies, and professions around the world. To all of you I express my deepest and sincerest thanks.

Each book that I have developed has been a part of my lifelong journey in water resources and this book certainly is no exception. I have gained more from my experiences in developing books than can ever be measured in words. To develop a book such as this one, with all the authors being former Ph.D. students has been a very special experience. All of these authors, I consider a part of my extended family, which includes all of my former Ph.D. students. They have represented a very valuable part of my life.

I must acknowledge my three children, Travis, Elyssa, and Tyler for their love and willingness to enjoy the water-related sports with me, especially at our second home in Pagosa Springs, Colorado. Probably they will never read any part of this book, but they do represent the most valuable part of my life.

This book has been a part of my personal journey in life to learn as much as possible about water and to use this knowledge in my teaching, research, and writing. I hope that you will be able to use this book in your own journey of learning about water. As I continue my efforts in the study and photography of ancient water structures, especially those built by the Romans, I am placing many of my photographs on my web site (www.public.asu.edu/~lwmays/). Take a look, these may be of interest to some of you readers.

I dedicate this book to humanity and human welfare.

Larry W. Mays
Scottsdale, Arizona
Pagosa Springs, Colorado

WATER RESOURCES SYSTEMS MANAGEMENT TOOLS

CHAPTER 1
SYSTEMS ANALYSIS

Larry W. Mays
Department of Civil and Environmental Engineering
Arizona State University
Tempe, Arizona

Yeou-Koung Tung
Department of Civil Engineering
Hong Kong University of Science & Technology
Kowloon, Hong Kong

1.1 SYSTEMS CONCEPT

1.1.1 What Is Optimization?

The major types of water resources problems that must be solved for various hydrosystems include (Buras, 1972):

1. Determination of the optimal scale of project development
2. Determination of the optimal dimensions of system components
3. Determination of the optimal operation of the system

Water resource problems deal with both design and analysis. *Analysis* is concerned with determining the behavior of an existing system or a trial system that is being designed. In many cases, study of the system behavior is to determine operation of the system or the response of a system under specified inputs. The *design problem* is to determine the sizes of system components. As an example, the design of a reservoir system determines the size and location of reservoirs. Analysis of a reservoir system is the process of determining operation policies for the reservoir system. In other words, a design is formulated and followed by an analysis to see if it performs according to specifications. If a design satisfies the specifications, then an acceptable design is found. New designs can be formulated and then analyzed.

Conventional procedures for design and analysis are iterative. The effectiveness of conventional procedures is dependent upon an engineer's intuition, experience, skill, and knowledge of the hydrosystem under investigation. Therefore, conventional procedures are highly related to the human element which could lead to inefficient design and analysis of complex systems. Conventional procedures are typically based on simulation models in a trial and error process. Sometimes a simulation model is applied iteratively in an attempt to arrive at an optimal solution.

Optimization procedures eliminate the trial and error process of changing a design and resimulating with each new design change. Instead, an optimization model automatically searches for the optimum design parameters. An optimization procedure has mathematical expressions that describe the system and its response to the system inputs for various design parameters. These mathematical expressions include constraints to define the limits of the design variables, and objective functions for evaluating system performance.

An advantage of the conventional process is that the engineer's experience and intuition are used in making changes in the system or to make additional specifications. But the conventional procedure can lead to nonoptimal, or uneconomical, designs and operation policies. Also, the conventional procedure can be very time consuming and labor intensive. An optimization procedure requires one to explicitly identify the design variables, the objective function or measure of system performance to be optimized, and the constraints for the system. In contrast to the conventional procedure, the optimization procedure is more organized, using appropriate mathematical approaches to select the decisions. However, optimization models generally can accommodate fewer mathematical and probabilistic complications.

An optimization problem may be formulated in a general framework with an objective function to

$$\text{Optimize } f(x) \tag{1.1}$$

subject to constraints

$$g(x) = 0 \tag{1.2}$$

and bound constraints on the decision variables

$$\underline{x} < x < \overline{x} \tag{1.3}$$

where x is a vector of n decision variables $(x_1, x_2, \ldots, x_n)$, $\mathbf{g}(x)$ is a vector of m equations called constraints, and $\underline{x}$ and $\overline{x}$ represent vectors of the lower and upper bounds, respectively, on the decision variables. In general, the objective equation (1.1) is to be maximized or minimized. Maximizing $f(x)$ is equivalent to minimizing $-f(x)$ or vice versa.

Every optimization problem has two essential parts: the objective function and the set of constraints. The *objective function* describes the performance criteria of the system. *Constraints* describe the system or process that is being designed or analyzed, which can be of two forms: equality constraints and inequality constraints. A *feasible solution* to an optimization problem is a set of values of the decision variables that simultaneously satisfies all constraints. The *feasible region* is the region of feasible solutions defined by the constraints. An *optimal solution* is a set of values of the decision variables that satisfies the constraints and provides an optimal value of the objective function.

Depending upon the nature of the objective function and the constraints, an optimization problem can be classified as: (1) linear vs. nonlinear, (2) deterministic vs. probabilistic, (3) static vs. dynamic, (4) continuous vs. discrete, and (5) lumped parameter vs. distributed parameter. *Linear programming* problems consist of a linear objective function where all constraints are linear. *Nonlinear programming* problems have nonlinear constraints and/or the objective function. *Deterministic* problems consist of parameters that can be assigned fixed values, whereas *probabilistic* problems consist of parameters that are considered as random variables. *Static* problems do not explicitly consider the variable time aspect, whereas *dynamic* problems do consider variable time. Static problems are referred to as *mathematical programming* problems, and dynamic problems are often referred to as *optimal control* problems, which involve difference or differential equations. *Continuous* problems have variables that can take on continuous values, whereas with *discrete* problems, the variables must take on discrete values. Typically, discrete problems are posed as *integer programming* problems, in which the variables must be integer values. A *lumped* problem considers the parameters and variables to be homogeneous throughout the system, whereas *distributed* problems must take into account detailed variations of system behavior from one location to another.

The method of optimization used depends on: (1) the type of objective function, (2) the type of constraints, and (3) the number of decision variables. Table 1.1 lists some general steps to solve an optimization problem. Some problems may not require that the engineer follow the steps in the exact order, but each step should be considered in the process. The overall objective in optimization is to determine a set of values of the decision variables that satisfies the constraints and provides the optimal response to the objective function.

1.1.2 Single-Objective versus Multiple-Objective Optimization

The solutions to a growing number of problems facing water resource professionals today are becoming more complex. In most water resource problems, the decision-making process is cultivated by the desire to achieve several goals simultaneously, and many of them could be *noncommensurate* and conflicting. In such circumstances, improvement of some objectives cannot be obtained without

TABLE 1.1 Six Steps Used to Solve Optimization Problems

1. Analyze the process itself so that the process variables and specific characteristics of interest are defined, i.e., make a list of all of the variables.
2. Determine the criterion for optimization and specify the objective function in terms of the above variables together with coefficients. This step provides the performance model (sometimes called the economic model when appropriate).
3. Develop, via mathematical expressions, a valid process or equipment model that relates the input-output variables of the process and associated coefficients. Include both equality and inequality constraints. Use well-known physical principles (mass balances, energy balances), empirical relations, implicit concepts, and external restrictions. Identify the independent and dependent variables to get the number of degrees of freedom.
4. If the problem formulation is too large in scope:
   a. Break it up into manageable parts, or
   b. Simplify the objective function and model.
5. Apply a suitable optimization technique to the mathematical statement of the problem.
6. Check the answers, and examine the sensitivity of the result to changes in the problem coefficients and the assumptions.

Source: Edgar and Himmelblau, 1988.

the sacrifice of others. Therefore, the ideological theme of optimality in the single-objective context is no longer appropriate. Instead, the goal of optimality in the single-objective framework is replaced by the concept of *noninferiority* in the multiple-objective analysis. More discussion of the multiobjective problem is given in Sec. 1.7.

1.2 LINEAR PROGRAMMING (LP)

Linear programming models have been extensively applied to optimal resource allocation problems. As the name implies, LP models require that both the objective function and constraints be linear functions of the decision variables. The general form of an LP model can be expressed as

$$\text{max (or min)} \qquad x_o = \sum_{j=1}^{n} c_j x_j \tag{1.4}$$

subject to

$$\sum_{j=1}^{n} a_{ij} x_j = b_i, \qquad \text{for} \qquad i = 1, 2, \ldots, m \tag{1.5}$$

$$x_j \geq 0, \qquad \text{for} \qquad j = 1, 2, \ldots, n \tag{1.6}$$

where the c_j's are the *objective function coefficients,* the a_{ij}'s are the *technological coefficients,* and the b_i's are the right-hand side (RHS) coefficients. In matrix form, an LP model can be concisely expressed as

$$\text{max (or min)} \qquad x_o = c^T x \tag{1.7}$$

subject to

$$Ax = b \tag{1.8}$$

$$x \geq 0 \tag{1.9}$$

where c is an n by 1 column vector of objective function coefficients, x is an n by 1 column vector of decision variables, A is an m by n matrix of technological coefficients, b is an m by 1 column vector of the RHS coefficients, and the superscript T represents the transpose of a matrix or a vector. Excellent textbooks on LP include: Gass (1985), Hillier and Lieberman (1990), Taha (1987), and Winston (1987).

Four basic assumptions are implicitly built into LP models:

1. *Proportionality Assumption.* This implies that the contribution of the jth decision variable to the system effectiveness $c_j x_j$ and its usage of the various resources $a_{ij} x_j$ are directly proportional to the value of the decision variable.

2. *Additivity Assumption.* This assumption means that, at a given level of activity $(x_1, x_2, \ldots, x_n)$, the total usage of resources and contribution to the overall system effectiveness are equal to the sum of the corresponding quantities generated by each activity conducted by itself.

3. *Divisibility Assumption.* Activity units can be divided into any fractional level, so that noninteger values for the decision variables are permissible.

4. *Deterministic Assumption.* All parameters in the model are known without uncertainty. The effect of parameter uncertainty on the result can be investigated by conducting sensitivity analysis.

1.2.1 Simplex Method

A well-known algorithm for solving an LP problem is the *simplex method* which consists of the following three steps:

1. *Initialization Step.* Start at a feasible extreme point.
2. *Iterative Step.* Move to a better adjacent feasible extreme point.
3. *Stopping Rule.* Stop iterations when the current feasible extreme point is better than all its adjacent extreme points.

In linear algebra, the system of equations is called *indeterminant* if there are more unknowns than equations. For problems with n unknowns and m equations, where $n > m$, there is no unique solution to the problem. The possible solutions to an indeterminant system of equations can be obtained by letting $(n - m)$ unknowns equal zero and solving for the remaining m unknowns. The solutions so obtained are called the *basic solutions.* In theory, there will be, at maximum, a total of $n!/(m!(n - m)!)$ basic solutions to such problems, if all exist. Geometrically, each basic solution represents the intersection point of each pair of constraint equations, an *extreme point.*

The $(n - m)$ decision variables set to zero are called the *nonbasic variables* and the remaining m decision variables, whose values are obtained by solving a system of m equations with m unknowns, are called the *basic variables.* The solution of a basic variable that is feasible is called the *basic feasible solution* corresponding to a feasible extreme point in the feasible space.

The simplex method solves an LP model by exploiting the feasible extreme points. The algorithm searches for the optimum of an LP model by abiding by two fundamental conditions: (1) the *optimality condition* and (2) the *feasibility condition.* The *optimality condition* ensures that no inferior solutions (relative to the current solution point) are ever encountered. The *feasibility condition* guarantees that, starting with a basic feasible solution, only basic feasible solutions are enumerated during the course of computation.

1.2.2 New Methods of Solution

The simplex algorithm searches the optimality of an LP problem by examining the feasible extreme points. The path of optimum seeking is made along the boundary of the feasible region. The algorithm, since its conception (Dantzig, 1963), has been widely used and considered as the most efficient approach for solving LP problems. Later, two methods that have an entirely different algorithmic philosophy for solving LP problems were developed (Khatchian, 1979; Karmarkar, 1984).

In contrast to the simplex algorithm, *Khatchian's ellipsoid method* and *Karmarkar's projective scaling method* seek the optimum solution to an LP problem by moving through the interior of the feasible region. A schematic diagram illustrating the algorithmic differences between the simplex and the two algorithms is shown in Fig. 1.1. Khatchian's ellipsoid method approximates the optimum solution of an LP problem by creating a sequence of ellipsoids (an ellipsoid is the multidimensional analog of an ellipse) that approach the optimal solution. Both Khatchian's method and Karmarkar's method have been shown to be *polynomial time algorithms*. This means that the time required to solve an LP problem of size n by the two methods would take at most αn^{β} where α and β are two positive numbers.

On the other hand, the simplex algorithm is an *exponential time algorithm* in solving LP problems. This implies that, in solving an LP problem of size n by the simplex algorithm, there exists a positive number γ such that for any problem involving n variables the simplex algorithm would find its solution in a time of at most $\gamma 2^n$. For a large enough n (with positive α, β, and γ), $\gamma 2^n > \alpha n^{\beta}$. This

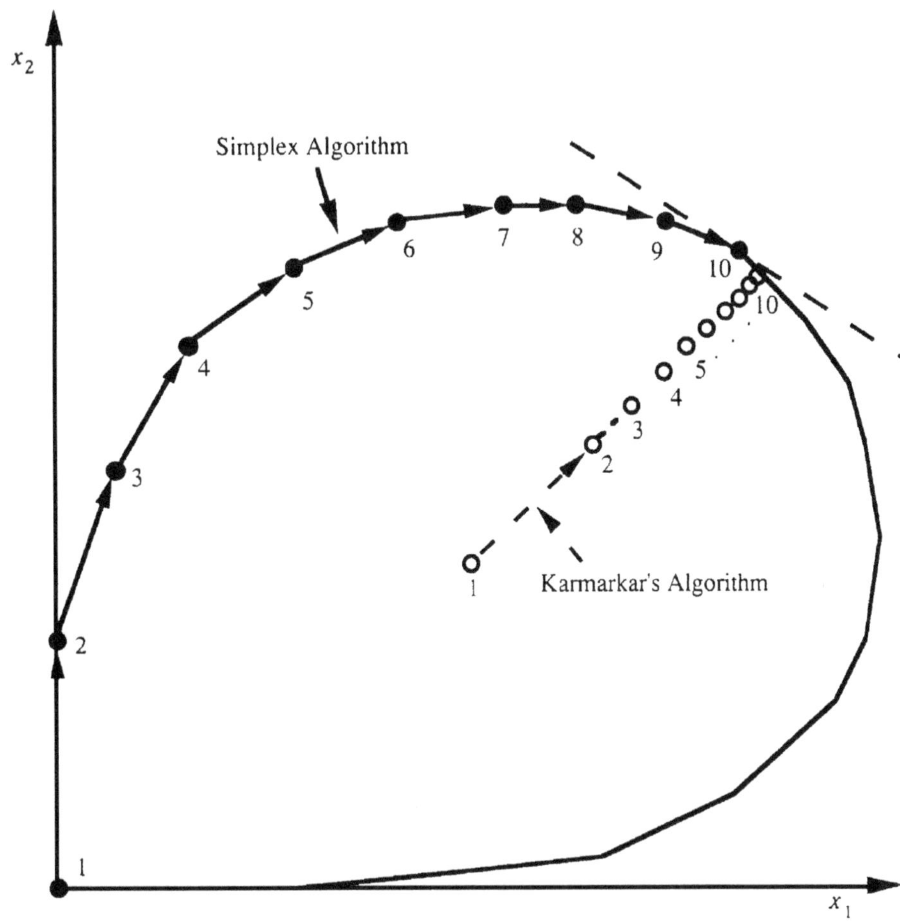

FIGURE 1.1 Search path for simplex algorithm and Karmarkar's algorithm. *(Mays and Tung, 1992.)*

means that, in theory, the polynomial time algorithms are superior to exponential time algorithms for large LP problems.

1.2.3 Integer Programming (IP)

Many water resource engineering problems involve decision variables that must be integer-valued. Typical examples are those representing a discrete number of objects, such as pumps and culvert barrels, and those representing quantification of codes, such as whether to accept or reject a particular water development alternative or alternatives. Mathematical model formulation for an IP problem is identical to Eqs. (1.7) to (1.9) for an *integer linear program* or to Eqs. (1.17a–c) for an *integer nonlinear program.* An integer program can be *mixed* or *pure,* depending on whether some or all its decision variables are restricted to be integer-valued.

Several algorithms have been developed for solving IP problems, and the ones that are mature in commercial application are those for integer linear programming problems. However, unlike the LP algorithms, the computation performance of IP algorithms is not uniformly efficient, and is erratic (Taha, 1987). This is especially true when the size of an IP problem gets larger. Reference books on IP optimization can be found elsewhere (Garfinkel and Nemhauser, 1972; Salkin, 1975; Taha, 1975; Nemhauser and Wolsey, 1988).

Methods for solving an IP problem can be categorized as *cutting methods* and *search methods.* The cutting methods, developed by Gomory (Taha, 1995), include the *fractional algorithm* for the pure integer programs and the *mixed algorithm* for mixed integer programs. In essence, the methods solve an IP problem by first treating all integer variables as if they are continuous variables. If all the integer variables are integer-valued, the optimal solutions are found. Otherwise, a sequence of secondary constraints is added systematically to the original problem until the resulting optimal solution satisfies the integer conditions. The added secondary constraints, in effect, cut off portions of the original feasible space that do not contain feasible integer points.

The search methods originate from the idea to enumerate all feasible integer solutions. By incorporating some efficient and smart tests, the required number of enumerations to reach the optimum can be reduced to a minimum. Among the search methods, the *branch-and-bound algorithm,* originally developed by Land and Doig (1960), is the most widely used. This algorithm, unlike the cutting methods, applies directly to both pure and mixed IP problems. It also starts solving the problem by ignoring the restriction of the integer. For those integer-constrained variables with fractional values, one of them is selected to branch the original problem into two subproblems that are subsequently solved. The branching process eliminates the feasible space that does not contain feasible integers. The process can be repeated until the integrality of all integer-constrained variables is satisfied. However, the process can be very time consuming and impractical. As a result, the concept of bounding is introduced based on the concept that, if the continuous optimum solution of a subproblem yields a worse objective function value than the one associated with the best available integer solution, it does not deserve further exploration. In this case the subproblem is said to be *fathomed* and may henceforth be deleted. The computational efficiency of the branch-and-bound algorithm depends on the selection of good branching variables and the sequence in which the subproblem is evaluated. Unfortunately, there is no definite optimal rule for this.

1.2.4 Network Methods

Many water resource planning and management problems are concerned with allocating water among various locations at different times. Some allocations are directional, as dictated by flow direction. Figure 1.2 is a typical example in which the system consists of several reservoirs connected by a network of channels. Problems of this nature can often be represented by a network involving nodes (reservoirs) and arcs (channels). Decision variables are the releases from the reservoirs at different times. Network problems can be categorized into one of four models: (1) network minimization model, (2) shortest-route model, (3) maximum-flow model, and (4) minimum-cost capacitated network model (Taha, 1987; Hillier and Lieberman, 1990).

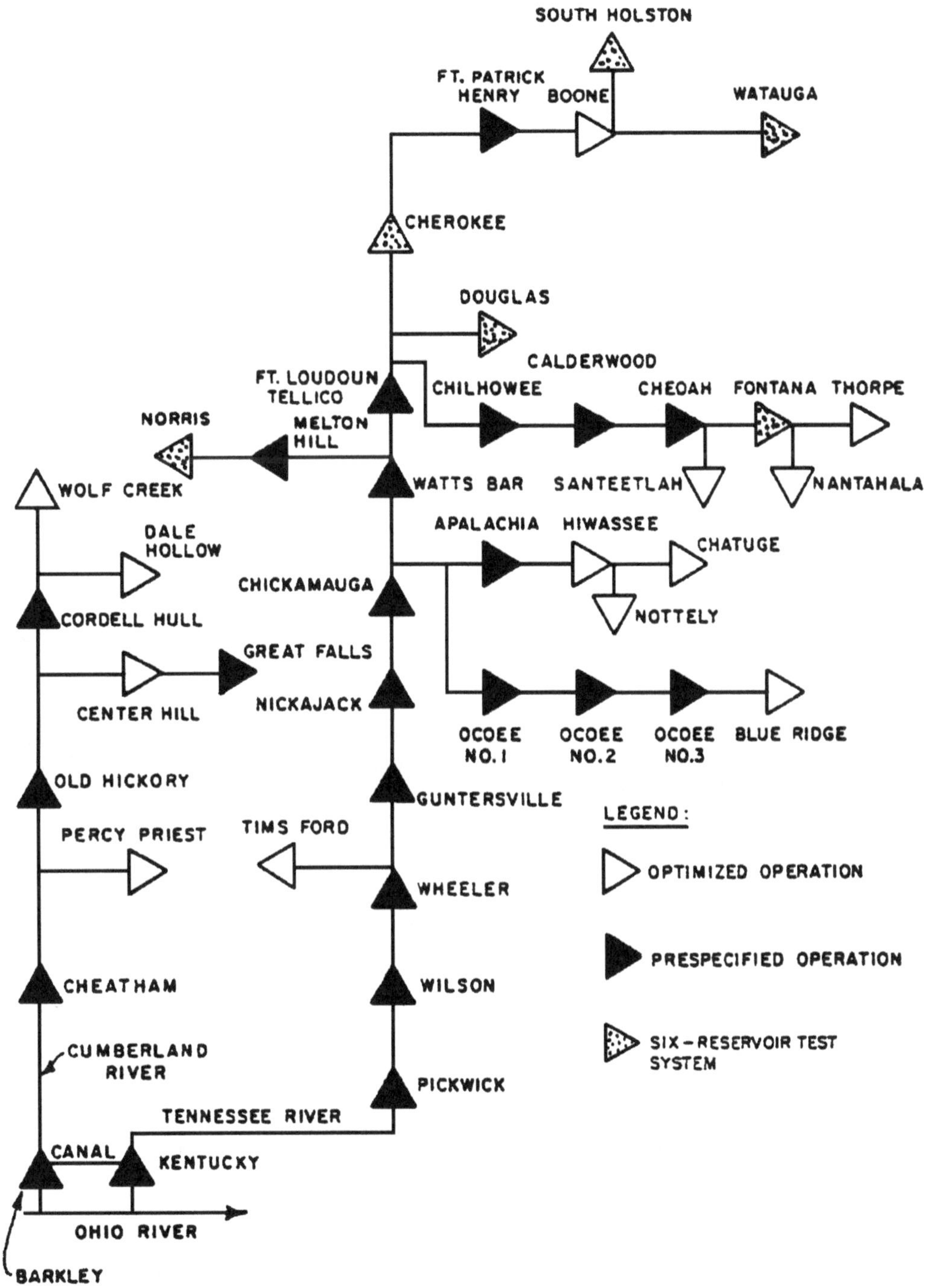

FIGURE 1.2 Schematic diagram of multireservoir system. *(Wunderlich, 1991.)*

Constraints in the mathematical programming model for a network problem are largely made up of the flow continuity equations. Therefore, the constraint matrix is generally very sparse, and its elements are either +1 or −1. Although most network problems can be formulated as LP problems, they are not solved by the LP algorithms. Instead, highly efficient algorithms have been developed to take advantage of the special structure of network optimization models (Jensen and Barnes, 1980). Nonlinearity in network optimization presents a challenge ahead.

1.3 NONLINEAR PROGRAMMING (NLP)

1.3.1 Unconstrained Nonlinear Optimization

This section describes the basic concepts of unconstrained nonlinear optimization, including the necessary and sufficient conditions of a local optimum. Further, unconstrained optimization techniques for univariate and multivariate problems are described. Understanding unconstrained optimization procedures is important because these techniques are the fundamental building blocks in many of the constrained nonlinear optimization algorithms.

The problem of *unconstrained* NLP can be stated as

$$\min_{x \in E^n} \; f(x) \tag{1.10}$$

in which x is a vector of n decision variables $x = (x_1, x_2, \ldots, x_n)^T$ defined over the entire Euclidean space E^n. Because the feasible region is infinitely extended without bound, the optimization problem does not contain any constraints.

Assume that $f(x)$ is a nonlinear function and twice differentiable; it could be convex, concave, or a mixture of the two over E^n. In the one-dimensional case, the objective function $f(x)$ could behave as in Fig. 1.3, consisting of peaks, valleys, and inflection points. The *necessary conditions* for a solution to Eq. (1.10) at an optimum x^* are: (1) the gradient vector $\nabla f(x^*) = \mathbf{0}$ and (2) the Hessian matrix $\nabla^2 f(x^*) = \mathbf{H}(x^*)$ is positive semidefinite. The *sufficient conditions* for an unconstrained minimum are: (1) $\nabla f(x^*) = \mathbf{0}$ and (2) $\nabla^2 f(x^*) = \mathbf{H}(x^*)$ is strictly positive definite.

In theory, the solution to Eq. (1.10) can be obtained by solving the following system of n nonlinear equations with n unknowns,

$$\nabla f(x^*) = \mathbf{0} \tag{1.11}$$

The approach has been viewed as indirect in the sense that it backs away from the original problem of minimizing $f(x)$. Furthermore, an iterative numerical procedure is required to solve the system of nonlinear equations which tends to be computationally inefficient.

By contrast, preference is given to those solution procedures that directly attack the problem of minimizing $f(x)$. Direct solution methods, during the course of iteration, generate a sequence of solution points in E^n that terminate or converge to a solution to Eq. (6.10). Such methods can be characterized as *search algorithms.*

In general, all search algorithms for unconstrained minimization consist of two basic steps: (1) determine the *search direction* along which the objective function value decreases; (2) a *line search* (or *one-dimensional search*) to obtain the optimum solution along the search direction determined by the first step. Mathematically, minimization for the line search can be stated as

$$\min_{\beta} \; f(x^0 + \beta d) \tag{1.12}$$

in which x^0 is the current solution point, d is the vector indicating the search direction, and β is a scalar, $0 < \beta < \infty$, representing the step size whose optimal value is to be determined. There are many search algorithms whose differences primarily lie in the way the search direction d is determined.

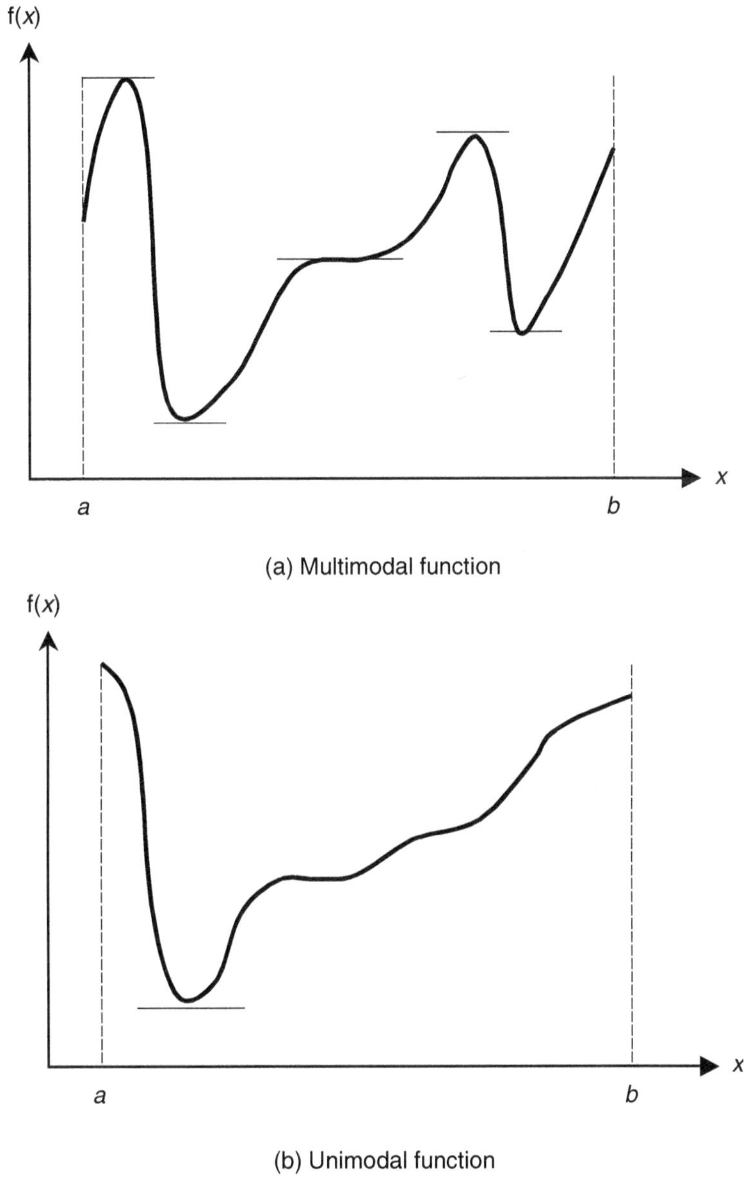

FIGURE 1.3 Shapes of one-dimensional objective function.

Due to the very nature of search algorithms, it is likely that different starting solutions might converge to different local minima. Hence, there is no guarantee of finding the global minimum by any search algorithm applied to solve Eq. (1.10) unless the objective function is a convex function over E^n.

To implement search algorithms, specification of convergence criteria or stopping rules is an important element that affects the performance of the algorithm and the accuracy of the solution. Several commonly used stopping rules in an optimum-seeking algorithm are

$$\|x^k - x^{k+1}\| < \epsilon_1 \tag{1.13a}$$

$$\frac{\|x^k - x^{k+1}\|}{\|x^k\|} < \epsilon_2 \tag{1.13b}$$

$$|f(x^k) - f(x^{k+1})| < \epsilon_3 \tag{1.13c}$$

$$\left| \frac{f(x^k) - f(x^{k+1})}{f(x^k)} \right| < \epsilon_4 \tag{1.13d}$$

in which superscript k is the index for iteration; ϵ represents the tolerance or accuracy requirement; $\|x\|$ is the length of the vector x; and $|\bullet|$ is the absolute value. The specification of the tolerance depends on the nature of the problem and on the accuracy requirement. Too small a value of ϵ (corresponding to a high accuracy requirement) could result in excessive iterations, wasting computer time. On the other hand, too large a value of ϵ could make the algorithm terminate prematurely at a nonoptimal solution.

Referring to Eq. (1.10), the solution of these types of problems can be stated in an algorithm involving the following basic steps:

Step 0 Select an initial starting point $x^{k=0} = (x_1^0, x_2^0, \ldots, x_n^0)$.

Step 1 Determine a search direction, d^k.

Step 2 Find a new point $x^{k+1} = x^k + \beta^k d^k$ where β^k is the step size, a scalar, which minimizes $f(x^k + \beta^k d^k)$.

Step 3 Check convergence criteria such as Eq. (1.13a–d) for termination; if not satisfied, set $k = k + 1$ and return to Step 1.

Various unconstrained multivariate methods differ in the way the search directions are determined. The recursive line search for an unconstrained minimization problem in Step 2 above is expressed as

$$x^{k+1} = x^k + \beta^k d^k \tag{1.14}$$

Table 1.2 lists the equations for determining the search direction for four basic groups of methods: descent methods, conjugate direction methods, quasi-Newton methods, and Newton's method. The simplest ones are the steepest descent methods, and the Newton methods are the most computationally intensive.

In the *steepest descent method* the search direction is $-\nabla f(x)$ because $\nabla f(x)$ points in the direction of the maximum rate of increase in objective function value, therefore a negative sign is associated with the gradient vector in Eq. (1.14) for a minimization problem. The recursive line search equation for the steepest descent method, then, is written as

$$x^{k+1} = x^k - \beta^k \nabla f(x^k) \tag{1.15}$$

Using *Newton's method,* the recursive equation for a line search is

$$x^{k+1} = x^k - \mathbf{H}^{-1}(x^k)\nabla f(x^k) \tag{1.16}$$

Although Newton's method converges faster than most other algorithms, the major disadvantage is that it requires inverting the Hessian matrix in each iteration, which is a cumbersome task, especially when the number of decision variables is large.

The *conjugate direction methods* and *quasi-Newton methods* are intermediate between the steepest descent and Newton's methods. The conjugate direction methods are motivated by the need to accelerate the typically slow convergence of the steepest descent method. Conjugate direction methods, as can be seen in Table 1.2, define the search direction by utilizing the gradient vector of

TABLE 1.2 Computation of Search Directions*

Search Direction	Definition of Terms
Steepest Descent $d^{k+1} = -\nabla f(x^{k+1})$	
Conjugate gradient methods (1) Fletcher-Reeves $d^{k+1} = -\nabla f(x^{k+1}) + a_1 d^k$	$a_1 = \dfrac{\nabla^T f(x^{k+1})\nabla f(x^{k+1})}{\nabla^T f(x^k)\nabla f(x^k)}$ $d^0 = -\nabla f(x^0)$
(2) Polak-Ribiere $d^{k+1} = -\nabla f(x^{k+1}) + a_2 d^k$	$a_2 = \dfrac{\nabla^T f(x^{k+1})\mathbf{Y}^{k+1}}{\nabla^T f(x^k)\nabla f(x^k)}$ $\mathbf{Y}^{k+1} = \nabla f(x^{k+1}) - \nabla f(x^k)$ $d^0 = -\nabla f(x^0)$
(3) 1-step Broyden-Fletcher-Goldfard-Shanno (BFGS) $d^{k+1} = -\nabla f(x^{k+1}) + a_3(a_4\mathbf{S}^{k+1} + a\mathbf{Y}^k)$	$a_3 = \dfrac{1}{(\mathbf{S}^{k+1})^T\mathbf{Y}^{k+1}}$ $a_4 = -\left(1 + \dfrac{(\mathbf{Y}^{k+1})^T\mathbf{Y}^{k+1}}{(\mathbf{S}^{k+1})^T\mathbf{Y}^{k+1}}\right)(\mathbf{S}^{k+1})^T\nabla f(x^{k+1})$ $+(\mathbf{Y}^{k+1})^T\nabla f(x^{k+1})$ $a_5 = (\mathbf{S}^{k+1})^T\nabla f(x^k)$ $\mathbf{S}^{k+1} = x^{k+1} - x^k$ $d^0 = -\nabla f(x^0)$
Quasi-Newton methods (1) Davidon-Fletcher-Powell (DFP) Method (variable metric method)† $d^{k+1} = \mathbf{G}^{k+1}\nabla f(x^{k+1})$	$\mathbf{G}^{k+1} = \mathbf{G}^k + \dfrac{\mathbf{S}^k(\mathbf{S}^k)^T}{(\mathbf{S}^k)^T\mathbf{Y}^k} - \dfrac{\mathbf{G}^k\mathbf{Y}^k(\mathbf{G}^k\mathbf{Y}^k)^T}{(\mathbf{Y}^k)^T\mathbf{G}^k\mathbf{Y}^k}$
(2) Broyden-Fletcher-Goldfarb-Shanno (BFGS) method† $d^{k+1} = \mathbf{G}^{k+1}\nabla f(x^{k+1})$	$\mathbf{G}^{k+1} = \mathbf{G}^k + \left(\dfrac{1 + (\mathbf{Y}^k)^T\mathbf{G}^k\mathbf{Y}^k}{(\mathbf{Y}^k)^T\mathbf{S}^k}\right)\dfrac{\mathbf{S}^k(\mathbf{S}^k)^T}{(\mathbf{S}^k)^T\mathbf{Y}^k}$ $-\dfrac{\mathbf{Y}^k(\mathbf{S}^k)^T\mathbf{G}^k + \mathbf{G}^k\mathbf{S}^k(\mathbf{Y}^k)^T}{(\mathbf{Y}^k)^T\mathbf{S}^k}$
(3) Broyden Family $\mathbf{G}^{\phi} = (1 - \phi)\mathbf{G}^{DFP} + \phi\mathbf{G}^{BFGS}$	
Newton method $d^{k+1} = -\mathbf{H}^{-1}(x^k)\nabla f(x^k)$	

* Formulas for other search directions can be found in Luenberger (1984).

† $\mathbf{G}^{\phi} = \mathbf{I}$ (identity matrix).

Source: Mays and Tung, 1992.

the objective function of the current iteration and the information on the gradient and search direction of the previous iterations. The motivation of quasi-Newton methods is to avoid inverting the Hessian matrix as required by Newton's method. These methods use approximations to the inverse Hessian with a different form of approximation for the different quasi-Newton methods. Detailed descriptions and theoretical development can be found in textbooks such as Luenberger (1984); Fletcher (1980); Dennis and Schnable (1983); Gill, Murray, and Wright (1981); and Edgar and Himmelblau (1988).

1.3.2 Constrained Optimization: Optimality Conditions

Consider the general nonlinear programming (NLP) problem with the nonlinear objective:

$$\text{Minimize } f(x) \tag{1.17a}$$

subject to

$$g_i(x) = 0 \qquad i = 1, \ldots, m \tag{1.17b}$$

and

$$\underline{x}_j \le x_j \le \overline{x}_j \qquad j = 1, 2, \ldots, n \tag{1.17c}$$

in which Eq. (1.17c) is a bounding constraint for the jth decision variable x_j with $\underline{x}_j$ and $\overline{x}_j$ being the lower and upper bounds, respectively.

Unlike an unconstrained problem, the feasible space of a constrained optimization problem is not infinitely extended. As a result, the solution that satisfies the optimality condition of the unconstrained optimization problem is not guaranteed to be feasible in constrained problems. In other words, a local optimum for a constrained problem might be located on the boundary or a corner of the feasible space at which the gradient vector is not equal to zero. Therefore, modifications to the optimality conditions for unconstrained NLP problems are necessary.

The most important theoretical results for the nonlinear constrained optimization are the *Kuhn-Tucker conditions.* These conditions must be satisfied at any constrained optimum, local or global, of any LP and NLP problems. They form the basis for the development of many computational algorithms.

Without losing generality, consider an NLP problem stated by Eqs. (1.17a–b) with no bounding constraints. Note that constraint Eq. (1.17b) is all equality constraints. Under this condition, the *Lagrange multiplier method* can be used to convert a constrained NLP problem into an unconstrained one by developing an augmented objective function, called the *Lagrangian function* $L(x, \lambda)$ as

$$L(x, \lambda) = f(x) + \lambda^T \mathbf{g}(x) \tag{1.18}$$

in which λ is the vector of *Lagrange multipliers* and $\mathbf{g}(x)$ is a vector of the constraint equations. Algebraically, Eq. (1.18) can be written

$$L(x_1, \ldots, x_n; \lambda_1, \ldots, \lambda_m) = f(x_1, \ldots, x_n) + \sum_{i=1}^{m} \lambda_i g_i(x_1, \ldots, x_n) \tag{1.19}$$

For a minimization problem, $L(x, \lambda)$ is the objective function, $m + n$ variables, to be minimized. The necessary and sufficient conditions for the optimal solution x^* are:

1. $f(x^*)$ is convex and $g(x^*)$ is convex in the vicinity of x^*

2. $$\frac{\partial L(x^*)}{\partial x_j} = \frac{\partial f}{\partial x_j} + \sum_{i=1}^{m} \lambda_i \frac{\partial g_i}{\partial x_j} = 0 \qquad j = 1, \ldots, n \tag{1.20a}$$

3. $$\frac{\partial L}{\partial \lambda_i} = g_i(x) = 0 \qquad i = 1, \ldots, m \tag{1.20b}$$

4. λ_i is unrestricted-in-sign $\quad i = 1, \ldots, m$ (1.20c)

Solving Eqs. (1.20a) and (1.20b) simultaneously provides the optimal solution.

Lagrange multipliers have an important interpretation in optimization problems. For a given constraint, the corresponding multiplier indicates the sensitivity of the optimal objective function value due to a unit change in the right-hand side of the constraint; that is,

$$\left.\frac{\partial f}{\partial b_i}\right|_{x = x^*} = \lambda_i$$

The λ_i's are called *dual variables* or *shadow prices.*

Kuhn-Tucker Conditions. Equations (1.20a–b) form the optimality conditions for an optimization problem involving only equality constraints. The Lagrange multipliers associated with the equality constraints are unrestricted-in-sign. Using the Lagrange multiplier method, the optimality conditions for the following generalized NLP problem can be derived.

$$\text{Minimize } f(x)$$

subject to

$$g_i(x) = 0 \qquad i = 1, \ldots, m$$

and

$$\underline{x}_j \le x_j \le \overline{x}_j \qquad j = 1, \ldots, n$$

In terms of the Lagrangian method, the above nonlinear minimization problem can be written as

$$\min L = f(x) + \lambda^T \mathbf{g}(x) + \underline{\lambda}^T(x - \underline{x}) + \overline{\lambda}^T(x - \overline{x}) \tag{1.21}$$

in which λ, $\underline{\lambda}$, and $\overline{\lambda}$ are vectors of Lagrange multipliers corresponding to constraints $\mathbf{g}(x) = \mathbf{0}$, $\underline{x} - x \le \mathbf{0}$, and $x - \overline{x} \le \mathbf{0}$, respectively. The Kuhn-Tucker conditions for the optimality of the above problem are

$$\nabla_x L = \nabla_x f + \lambda^T \nabla_x \mathbf{g} - \underline{\lambda} + \overline{\lambda} = 0 \tag{1.22a}$$

$$g_i(x) = 0 \qquad i = 1, 2, \ldots, m \tag{1.22b}$$

$$\underline{\lambda}_j(\underline{x}_j - x_j) = \overline{\lambda}_j(x_j - \overline{x}_j) = 0 \qquad j = 1, 2, \ldots, n \tag{1.22c}$$

The m basic variables in theory can be expressed in terms of the $n - m$ nonbasic variables as $x_B(x_N)$. Assume that constraints $\mathbf{g}(x) = \mathbf{0}$ are differentiable and the m *basis matrix* $\mathbf{B}$ can be obtained as

$$\mathbf{B} = \left[\frac{\partial \mathbf{g}(x)}{\partial x_B}\right]$$

which is nonsingular such that there exists a unique solution of $x_B(x_N)$. *Nonsingular* means that the determinant of the basis matrix $\mathbf{B}$ is not equal to 0.

The objective called a *reduced objective* can be expressed in terms of the nonbasic variables as

$$F(x_N) = f(x_B(x_N), x_N) \tag{1.23}$$

The original NLP problem is transformed into the following *reduced problem*

$$\text{Minimize } F(x_N) \tag{1.24a}$$

subject to

$$\underline{x}_N \le x_N \le \overline{x}_N \tag{1.24b}$$

which can be solved by an unconstrained minimization technique with slight modification to account for the bounds on nonbasic variables. *Generalized reduced gradient algorithms,* therefore, solve the original problem (1.17a–c) by solving a sequence of reduced problems (1.24a–b) using unconstrained minimization algorithms.

Generalized Reduced Gradient Methods. Consider solving the reduced problem (1.24a–b) starting from an initial feasible point x^0. To evaluate $F(x_N)$ by Eq. (1.23), the values of the basic variables x_B must be known. Except for a very few cases, $x_B(x_N)$ cannot be determined in closed form; however, it can be computed for any x_N by an iterative method that solves a system of m nonlinear equations with the same number of unknowns as equations. A procedure for solving the reduced problem starting from the initial feasible solution $x^{k=0}$ is

Step 0 Start with initial feasible solution $x^{k=0}$ and set $x_N^k = x^{k=0}$

Step 1 Use x_N^k to determine the corresponding values of x_B by an iterative method for solving m nonlinear equations $\mathbf{g}(x_B(x_N^k), x_N^k) = \mathbf{0}$.

Step 2 Determine the search direction d^k for the nonbasic variables by an unconstrained minimization scheme.

Step 3 Choose a step size for the line search scheme, β^k, by solving the one-dimensional search problem

$$\text{minimize } F(x_N^k + \beta d^k)$$

with x restricted so that $x_N^k + \beta d^k$ satisfies the bounds on x_N. This one-dimensional search requires repeated applications of Step 1 to evaluate F for the different β values.

Step 4 Test the current point $x^k = (x_B^k, x_N^k)$ for optimality; if not optimal, set $k = k + 1$ and return to Step 1.

Penalty Function Methods. Because unconstrained problems are much easier to solve, the idea of penalty function methods is to transform constrained NLP problems into a sequence of unconstrained NLP problems by adding one or more functions of the constraints to the objective function and to delete the constraints. Using a penalty function, a constrained NLP problem is transformed to an unconstrained problem as

$$\left.\begin{array}{r}\text{minimize } f(x)\\ \text{subject to } \mathbf{g}(x)\end{array}\right\} \Rightarrow \text{minimize } L[f(x), \mathbf{g}(x)]$$

where $L[f(x), \mathbf{g}(x)]$ is a *penalty function.* Various forms of penalty functions have been proposed which can be found elsewhere (McCormick, 1983; Gill, Murray, and Wright, 1981). The penalty function is minimized by stages for a series of values of parameters associated with the penalty function. For many of the penalty functions, the Hessian of the penalty function becomes increasingly *ill-conditioned* (i.e., the function value is extremely sensitive to a small change in the parameter value) as the solution approaches the optimum. This section briefly describes a penalty function method called the *augmented Lagrangian method.*

The augmented Lagrangian method adds a quadratic penalty term to the Lagrangian function [Eq. (1.18)], to obtain

$$L_A(x, \lambda, \psi) = f(x) + \sum_{i=1}^{m} \lambda_i \mathbf{g}_i(x) + \frac{\psi}{2} \sum_{i=1}^{m} \mathbf{g}_i^2(x) = \mathbf{f}(x) + \lambda^T \mathbf{g}(x) + \frac{\psi}{2} \mathbf{g}(x)^T \mathbf{g}(x) \tag{1.25}$$

where ψ is a positive penalty parameter. Some desirable properties of Eq. (1.25) are discussed by Gill, Murray, and Wright (1981).

For ideal circumstances, x^* can be computed by a single unconstrained minimization of the differentiable function [Eq. (1.25)]. However, in general, λ^* is not available until the solution has been determined. An augmented Lagrangian method, therefore, must include a procedure for estimating the Lagrange multipliers. Gill, Murray, and Wright (1981) present the following algorithm:

Step 0 Select initial estimates of the Lagrange multipliers $\lambda^{k=0}$, the penalty parameter ψ, and an initial point $x^{k=0}$. Set $k = k + 1$ and set the maximum number of iterations as J.

Step 1 Check to see if x^k satisfies optimality conditions or if $k > J$. If so, terminate the algorithm.

Step 2 Minimize the augmented Lagrangian function, $L_A(x, \lambda, \psi)$, in Eq. (1.25). Procedures to consider unboundedness must be considered. The best solution is denoted as x^{k+1}.

Step 3 Update the multiplier estimate by computing λ^{k+1}.

Step 4 Increase the penalty parameter ψ if the constraint violations at x^{k+1} have not decreased sufficiently from those at x^k.

Step 5 Set $k = k + 1$ and return to Step 1.

1.4 DYNAMIC PROGRAMMING (DP)

1.4.1 Dynamic Programming Concept

Dynamic programming (*DP*) transforms a sequential or multistage decision problem that may contain many interrelated decision variables into a series of single-stage problems, each containing only one or a few variables. In other words, the DP technique decomposes an N-decision problem into a sequence of N separate, but interrelated, single-decision subproblems. Decomposition is very useful in solving large complex problems by breaking a problem down into a series of smaller subproblems and then combining the solutions of the smaller problems to obtain the solution of the whole model through composition. Using decomposition, a problem is solved more efficiently, resulting in significant computational savings. As a rule of thumb, computations increase exponentially with the number of variables, but only linearly with the number of subproblems. Books that deal with dynamic programming are Cooper and Cooper (1981), Denardo (1982), and Drefus and Law (1977).

Exhaustive enumeration possesses three main shortcomings: (1) it would become impractical if the number of alternative combinations is large, (2) the optimal course of action cannot be verified, even if it is obtained in the early computation, until all the combinations are examined, and (3) infeasible combinations (or solutions) cannot be eliminated in advance. DP can overcome the shortcomings of an exhaustive enumeration procedure using the following concepts:

1. The problem is decomposed into subproblems, and the optimal alternative is selected for each subproblem sequentially so that it is never necessary to enumerate all combinations of the problem in advance.
2. Because optimization is applied to each subproblem, nonoptimal combinations are automatically eliminated.
3. The subproblems should be linked together in a special way so that it is never possible to optimize over infeasible combinations.

Referring to Fig. 1.4, the basic elements in the DP formulation are introduced as follows.

1. *Stages* (n) are the points of the problem where decisions are to be made. If a decision-making problem can be decomposed into N subproblems, there will be N stages in the DP formulation.
2. *Decision variables* (d_n) are courses of action to be taken for each stage. The number of decision variables d_n in each stage is not necessarily equal to one.
3. *State variables* (S_n) are variables describing the state of a system at any stage n. A state variable can be discrete or continuous, finite, or infinite. Referring to Fig. 1.4, at any stage n there are input states S_n and output states S_{n+1}. The state variables of the system in a DP model have the function

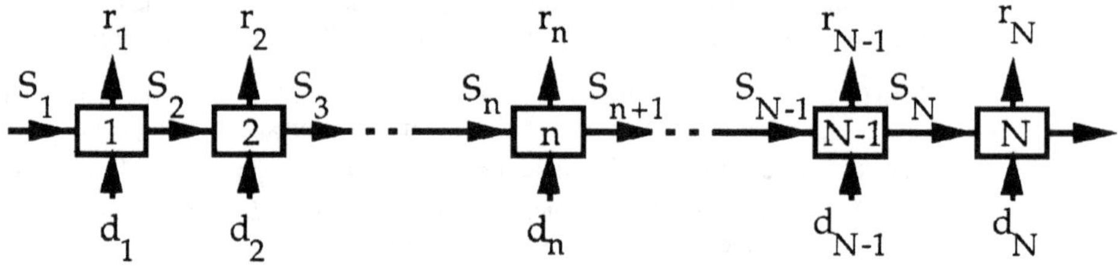

FIGURE 1.4 Sequential representation of serial dynamic programming problems.

of linking succeeding stages so that, when each stage is optimized separately, the resulting decision is automatically feasible for the entire problem. Furthermore, it allows one to make optimal decisions for the remaining stages without having to check the effect of future decisions for decisions previously made.

4. *Stage return* (r_n) is a scalar measuring the effectiveness of the decision in each stage. It is a function of the input state, the output state, and the decision variables of a particular stage. That is, $r_n = r(S_n, S_{n+1}, d_n)$.
5. *Stage transformation* (t_n) is a single-valued transformation that expresses the relationships among the input state, the output state, and the decision. In general, through the stage transformation, the output state at any stage n can be expressed as the function of the input state and the decision as

$$S_{n+1} = t_n(S_n, d_n) \tag{1.26}$$

The basic features that characterize all DP problems are as follows:

1. The problem is divided into stages, with decision variables at each stage.
2. Each stage has a number of states associated with it.
3. The effect of the decision at each stage is to produce return, according to the stage return function, and to transform the current state variable into the state variable for the next stage, through the state transformation function.
4. Given the current state, an optimal policy for the remaining stages is independent of the policy adopted in previous stages. This is called *Bellman's principle of optimality,* which serves as the backbone of the DP algorithm.
5. The solution begins by finding the optimal decision for each possible state in the last stage (called the *backward recursive*) or in the first stage (called the *forward recursive*). A forward algorithm computationally advances from the first to the last stage, whereas a backward algorithm advances from the last stage to the first.
6. A recursive relationship that identifies the optimal policy for each state at any stage n can be developed, given the optimal policy for each state at the next stage, $n + 1$. This backward recursive equation, referring to Fig. 1.4, can be written as

$$f^*_n(S_n) = \underset{d_n}{\text{opt}} \left\{ r_n(S_n, d_n) \, o \, f^*_{n+1}(S_{n+1}) \right\}$$

$$= \text{opt}\{r_n(S_n, d_n) \, o \, f^*_{n+1}[t_n(S_n, d_n)]\} \tag{1.27}$$

where o represents an algebraic operator that can be +, −, ×, or ÷, whichever is appropriate to the problem. The recursive equation for a forward algorithm is stated as

$$f^*_n(S_n) = \text{opt}\{r_n(S_n, d_n) \, o \, f^*_{n=1}(S_{n-1})\} \tag{1.28}$$

An increase in the number of discretizations and/or state variables would increase the number of evaluations of the recursive formula and required core memory per stage. This problem of rapid growth of computer time and core memory requirement associated with multiple state variable DP problems is referred to as the *curse of dimensionality.* From the problem solving viewpoint, the problem of increased computer time is of much less concern than that of the increased computer storage requirement. An increase in required computer core memory might result in exceeding the available storage capacity of a particular computer facility in use, and the problem cannot be solved. On the other hand, an increase in computer time requires one to be more patient for the final result. Therefore, the rapid growth in computer memory requirements associated with multiple state variable problems can make the difference between solvable and unsolvable problems.

1.4.2 Discrete Differential Dynamic Programming

Discrete differential dynamic programming (DDDP) is an iterative DP procedure that is specially designed to overcome some of the difficulties of the DP approach previously mentioned. The DDDP uses the same recursive equation as DP to search among the discrete states in the state-stage domain. Instead of searching over the entire state-stage domain for the optimum, as is the case for DP, the DDDP examines only a portion of the state-stage domain saving computer time and memory (Chow, et al., 1975). This optimization procedure is solved through iterations of trial states and decisions to search for the optimum for a system subject to the constraints that the trial states and decisions should be within the respective *admissible domain,* i.e., feasible in the state and decision spaces.

In the DDDP, the first step is to assume a trial sequence of admissible decisions called the *trial policy,* and the state vectors of each stage are computed accordingly. This sequence of states within the admissible state domain for different stages is called the *trial trajectory.* An alternative to the above procedure is first to assume a trial trajectory and then use it to compute the trial policy. Several states, located in the neighborhood of a trial trajectory, can be introduced to form a band called a *corridor* around the trial trajectory (see Fig. 1.5). It is a common practice to discretize the state space into uniform increments, called the *state increment,* where the total number of discretizations, referred to as *grid points* or *lattice points,* for each state variable is the same. The decisions, therefore, have to be made with respect to the method of discretizing the state variables. The interval of the decision variable is dependent on the corresponding interval of the state variable.

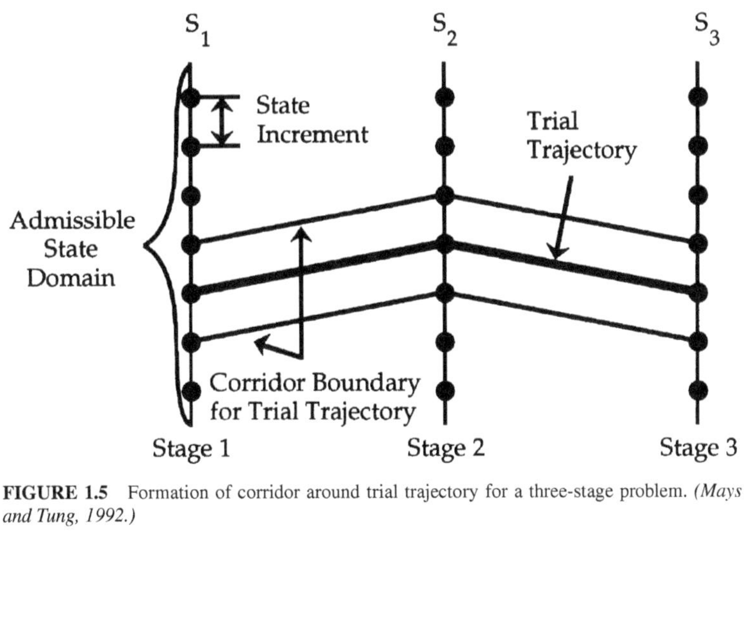

FIGURE 1.5 Formation of corridor around trial trajectory for a three-stage problem. *(Mays and Tung, 1992.)*

The traditional DP algorithm is applied in a DDDP framework to the states within the corridor using the recursive relationship for an improved trajectory to define a new policy, or set of decisions, within the introduced corridor. The improved trial trajectory is then adopted as the new trajectory to form a new corridor. This process of corridor formation, optimization with respect to the states within the corridor and trace-back to obtain an improved trajectory for the entire system defines an *iteration.*

For a corridor size, the iteration is repeated until the relative change of the optimal objective function value between two successive iterations is sufficiently small,

$$|f_k - f_{k-1}|/f_{k-1} \leq \epsilon \tag{1.29}$$

where ϵ is the specified tolerance level. At this point in the optimization procedure, the size of the state increment can be reduced to set up a smaller corridor formed around the improved trajectory for the last iteration completed in which the states or lattice points are closer together. The iterations continue reducing the size of the state increment through the system accordingly until a specified minimum corridor size is reached. Figure 1.6 is a flowchart showing the DDDP algorithm.

Although the use of DDDP partially circumvents the problem associated with the curse of dimensionality of the regular DP, it introduces other considerations involved with choosing the initial trajectory, changing the lattice point or state spacing, and converging to a local optimum instead of the global optimum. The major factors affecting the performance of the DDDP include:

1. The number of lattice points;
2. The initial state increment along with the number of lattice points that determine corridor width;
3. The initial trial trajectory used to establish the location of the corridors within the state space for the first iteration; and
4. The reduction rate of the state increment size that determines the corridor width for various iterations.

These factors are somewhat dependent upon one another for expedient and efficient use of the DDDP for water resources problems. The choice of the initial trial trajectory and the initial corridor width are interdependent. For example, if a small corridor width is chosen in conjunction with an initial trial trajectory that is far from the optimal region, more iterations are required to move the trajectory into the optimal or near-optimal region. Under these circumstances, it is possible for DDDP to converge to a solution that is far from the optimal. The number of lattice points and the initial state increment together determine the initial corridor width. Using a combination of a small number of lattice points and a small initial state increment could also result in local optimal solutions.

The effect of choosing a poor initial trajectory can be reduced if a large number of lattice points and/or large initial state increments are used. In essence, the better the initial trajectory, the smaller the number of lattice points and initial state increments, or simply, the smaller the initial corridor width can be used. Very small state increments with more lattice points can also be used to establish a small corridor width. This can result in improved convergence; however, increasing the number of lattice points increases the computation time and storage requirements. Computation time can be improved by increasing the reduction rate of the state increment at each iteration; however, too large a reduction rate of the state increment may miss the optimal region, thus not providing the optimal solution. Choosing a large initial state increment and a large reduction rate of the state increment size may be advantageous. However, when the initial state increment is too large, resulting in large corridor widths, unnecessary computations are performed in regions of the state space far from the optimal.

1.4.3 Differential Dynamic Programming

This section presents the differential dynamic programming (DDP) method for discrete-time optimal control problems. The term DDP used by Jacobson and Mayne (1970) broadly refers to stagewise NLP

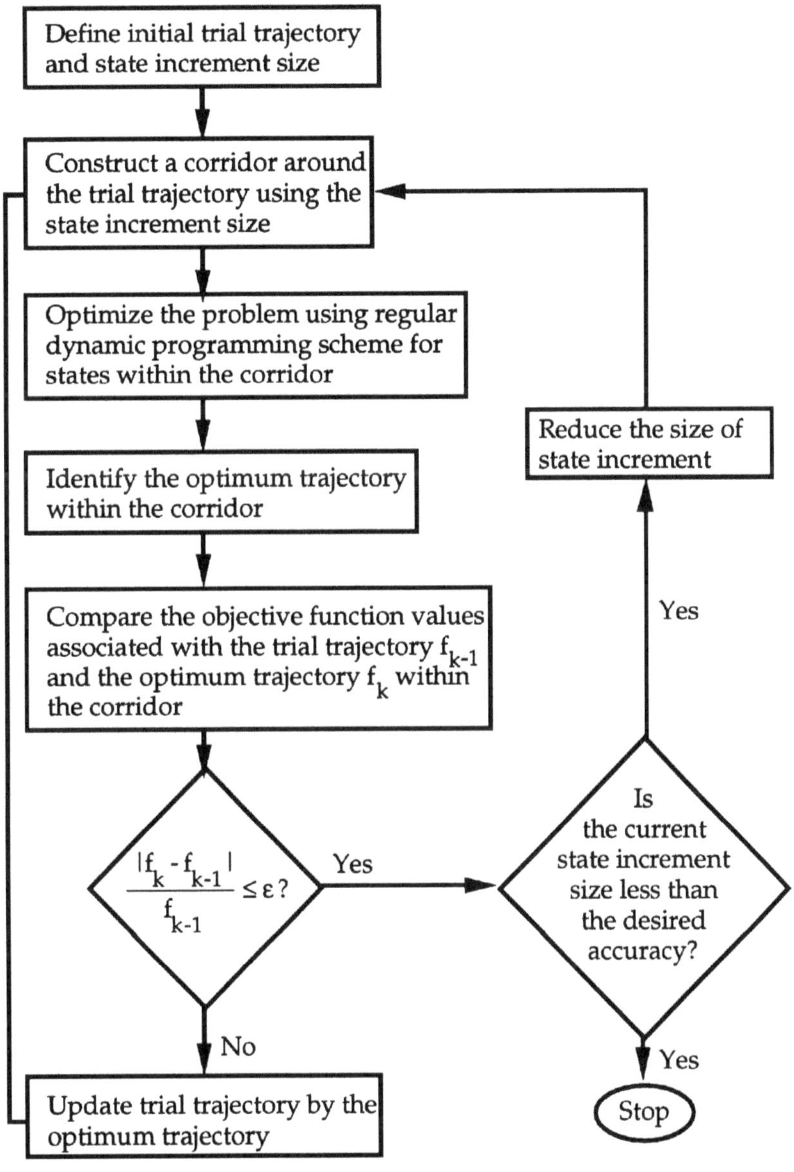

FIGURE 1.6 General algorithm for discrete differential dynamic programming. *(Mays and Tung, 1992.)*

procedures. Earlier representative works to develop DDP procedures for unconstrained discrete-time control problems include Bellman and Dreyfus (1962), Dyer and McReynolds (1970), Jacobson and Mayne (1970), and Yakowitz (1989). Ohno (1978) and Murray and Yakowitz (1981) contributed to DDP techniques for constrained optimal control problems.

The basic *optimal control problem* is stated as follows

$$\min_{u} Z = \sum_{t=1}^{T} f_t(x_t, u_t; t) \tag{1.30}$$

subject to

$$x_{t+1} = \mathbf{g}_t(x_t, u_t; t) \qquad t = 1, \ldots, T \tag{1.31}$$

The objective of DDP is to minimize a quadratic approximation instead of solving the actual control problem using the algorithm approach defined in Fig. 1.7. Yakowitz and Rutherford (1984) pointed out the following properties of DDP for unconstrained problems:

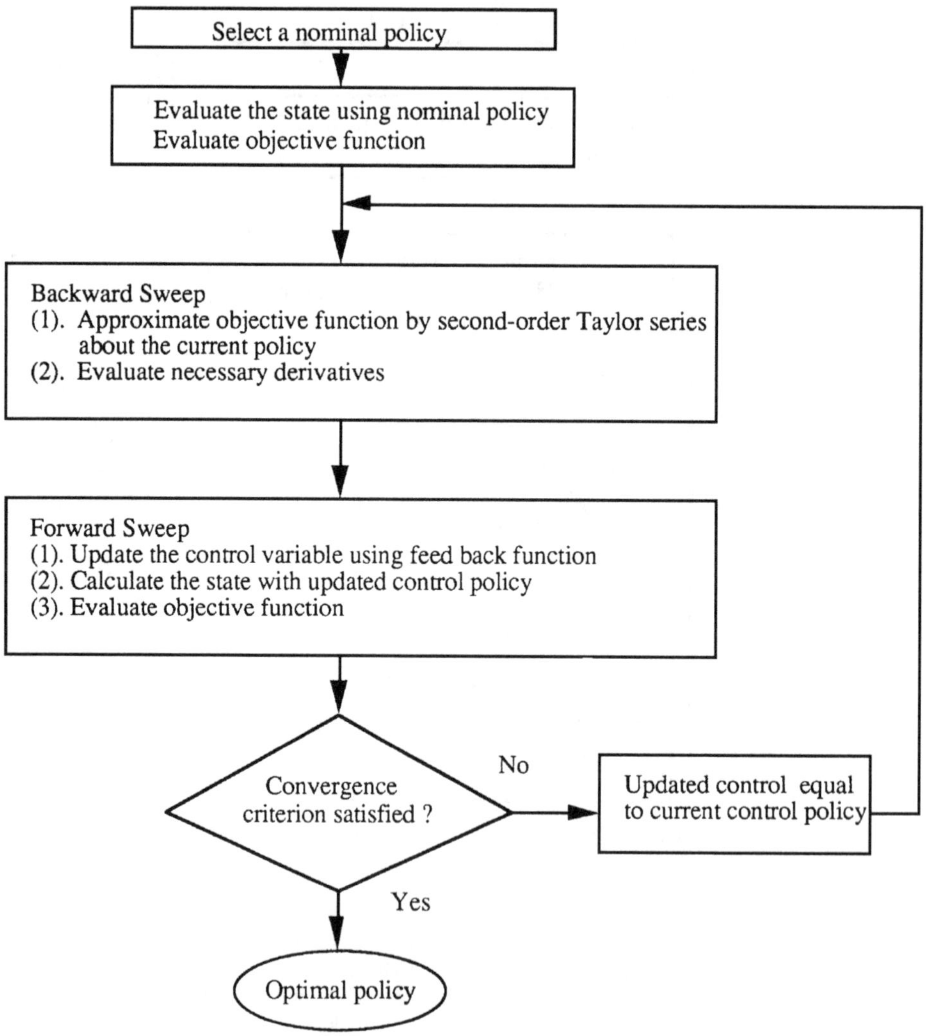

FIGURE 1.7 Flowchart of the unconstrained DDP method.

1. DDP overcomes the curse of dimensionality in that the computational requirements grow as N^3T and memory requirements as NnT, where n and N are, respectively, the state and control variable dimensions, and T is the number of decision times.
2. Under lenient conditions, DDP is globally convergent.
3. No discretization of state or control spaces is required.
4. The convergent rule of the DDP algorithm is quadratic for control problems in which the Hessian matrix of the objective function is convex in a neighborhood of the solution.

Algorithm Definition. Define the current or known control policy as $\bar{u}_t$ for $t = 1, \ldots, T$ and the current or known state trajectory $\bar{x}_t$ for $t = 1, \ldots, T + 1$ with the initial state x_1. For any function $W(x, u)$ defined by the control and state variables, let $QP(W(x, u))$ denote the linear and quadratic path of the Taylor's series expansion of $W(\ )$ about $(\bar{u}, \bar{x})$. The quadratic path for time T where the DDP backward recursion begins is

$$L(x, u; T) = QP((f(x, u; T)) = \frac{1}{2}\delta x^T\left(\frac{\partial^2 f}{\partial x^2}\right)\delta x + \delta x^T\left(\frac{\partial^2 f}{\partial x \partial u}\right)\delta u + \frac{1}{2}\delta u^T\left(\frac{\partial^2 f}{\partial u^2}\right)\delta u + \left(\frac{\partial f}{\partial u}\right)^T \delta u + \left(\frac{\partial f}{\partial x}\right)^T \delta x \tag{1.32}$$

where $\delta x = (x - \bar{x}_T)$ and $\delta u = (u - \bar{u}_T)$ are state and input perturbations, $(\partial f/\partial u)$ and $(\partial f/\partial x)$ are the gradients and $(\partial^2 f/\partial x^2)$, $(\partial^2 f/\partial x \partial u)$, and $(\partial^2 f/\partial u^2)$ are the Hessian of $f(x, u; T)$ which are evaluated at $\bar{x}_T$ and $\bar{u}_T$. Equation (1.32) can be presented in a more compact form as

$$L(x, u; T) = \delta x^T A_T \delta x + \delta x^T \mathbf{B}_T \delta u + \delta u^T \mathbf{C}_T \delta u + \mathbf{D}_T \delta u + \mathbf{E}_T \delta x \tag{1.33}$$

The idea of DDP is to minimize the quadratic approximation instead of the actual control problem value function, thereby obtaining an amenable function that is at the expense of involving truncation error. A necessary condition that a control u^* minimizes $L(x, u; T)$ is

$$\nabla_u L(x, u; T) = 2\mathbf{C}_T \delta u + \mathbf{B}_T \delta x + \mathbf{D}_T = \mathbf{0} \tag{1.34}$$

Then, the optimal control u^* can be found from Eq. (1.34), assuming that $\mathbf{C}_T$ is nonsingular, as

$$u^* = \bar{u}_T + \alpha_T + \beta_T \delta x \tag{1.35}$$

where $$\alpha_T = \left(-\frac{1}{2}\right)\mathbf{C}_T^{-1}\mathbf{D}_T \text{ and } \beta_T = \left(-\frac{1}{2}\right)\mathbf{C}_T^{-1}\mathbf{B}_T$$

The optimal value function is

$$F(x; T) = \min_u Z = \min f(x_T, u_T; T) \tag{1.36}$$

which is approximated by the quadratic as

$$\begin{aligned} V(x; T) &= L((x, u^*; T)) \\ &= L(x, \bar{u}_T + \alpha_T + \beta_T \delta x; T) \end{aligned} \tag{1.37}$$

By substituting Eq. (1.35) into Eq. (1.33), the quadratic $V(x; T)$ can be expressed as

$$V(x; T) = \delta x^T \mathbf{P}_T \delta x + \mathbf{Q}_T \delta x \tag{1.38}$$

where $$\mathbf{P}_T = \mathbf{A}_T - \left(\frac{1}{4}\right)\mathbf{B}_T^T \mathbf{C}_T^{-1} \mathbf{B}_T \tag{1.39}$$

$$\mathbf{Q}_T^T = -\left(\frac{1}{2}\right)\mathbf{D}_T^T\mathbf{C}_T^{-1}\mathbf{B}_T + \mathbf{E}_T^T \tag{1.40}$$

The DDP backward recursion procedure is performed for $t = T, T - 1, \ldots, 1$ using the quadratic

$$L(x, u; t) = QP\{f(x, u; t) + V[\mathbf{g}(x, u; t); t + 1]\} \tag{1.41}$$

where $V(x; t + 1)$ is the quadratic approximate optimal return function defined as

$$V(x; t + 1) = (\delta x)^T\mathbf{P}_{t+1}\delta x + \mathbf{Q}_{t+1}\delta x \tag{1.42}$$

Similar to Eq. (1.33) the quadratic can be expressed as

$$L(x, u; t) = \delta x^T\mathbf{A}_t\delta x + \delta u^T\mathbf{B}_t\delta x + \delta u^T\mathbf{C}_t\delta u + \mathbf{D}_t^T\delta u + \mathbf{E}_t^T\delta x \tag{1.43}$$

The coefficients $\mathbf{A}_t$, $\mathbf{B}_t$, $\mathbf{C}_t$, $\mathbf{D}_t^T$, and $\mathbf{E}_t^T$ are

$$\mathbf{A}_t = \frac{1}{2}\left(\frac{\partial^2 f}{\partial x^2}\right)_t + \left(\frac{\partial \mathbf{g}}{\partial x}\right)_t^T \mathbf{P}_{t+1}\left(\frac{\partial \mathbf{g}}{\partial x}\right)_t + \frac{1}{2}\sum_{i=1}^{n}(\mathbf{Q}_{t+1})_i\left(\frac{\partial^2 \mathbf{g}}{\partial x_i^2}\right) \tag{1.44}$$

$$\mathbf{B}_t = \left(\frac{\partial^2 f}{\partial x \partial u}\right)_t + 2\left(\frac{\partial \mathbf{g}}{\partial x}\right)_t^T \mathbf{P}_{t+1}\left(\frac{\partial \mathbf{g}}{\partial x}\right)_t + \frac{1}{2}\sum_{i=1}^{n}(\mathbf{Q}_{t+1})_i\left(\frac{\partial^2 \mathbf{g}}{\partial x_i \partial u_i}\right) \tag{1.45}$$

$$\mathbf{C}_t = \frac{1}{2}\left(\frac{\partial^2 f}{\partial u^2}\right)_t + \left(\frac{\partial \mathbf{g}}{\partial x}\right)_t^T \mathbf{P}_{t+1}\left(\frac{\partial \mathbf{g}}{\partial x}\right)_t + \frac{1}{2}\sum_{i=1}^{n}(\mathbf{Q}_{t+1})_i\left(\frac{\partial^2 \mathbf{g}}{\partial u_i^2}\right) \tag{1.46}$$

$$\mathbf{D}_t^T = \left(\frac{\partial f}{\partial u}\right)_t + \mathbf{Q}_{t+1}^T\left(\frac{\partial \mathbf{g}}{\partial x}\right)_t \tag{1.47}$$

$$\mathbf{E}_t^T = \left(\frac{\partial f}{\partial u}\right)_t + \mathbf{Q}_{t+1}^T\left(\frac{\partial \mathbf{g}}{\partial x}\right)_t \tag{1.48}$$

The first-order derivatives of $f(x, u; t)$ in the above equations are components of the gradient of $f(x, u; t)$; the second-order derivatives are the components of the Hessian of $f(x, u; t)$; the first-order derivatives of $\mathbf{g}(x, u; t)$ are components of the Jacobian of $\mathbf{g}(x, u; t)$; and the second-order derivatives of $\mathbf{g}(x, u; t)_i$ for $1 \le i \le n$ are the blocks of the Hessian matrices of the coordinates of $\mathbf{g}(x, u; t)$. All derivatives are evaluated about the current states and controls.

The first-order necessary condition for the optimality is

$$\nabla_u L(x, u; t) = \mathbf{0} \tag{1.49}$$

so that the minimizing strategy for the quadratic L(x, u; t) is

$$\delta u(x; t) = \alpha_t + \beta_t(x - \bar{x}_t) \tag{1.50}$$

where

$$\alpha_t = -\frac{1}{2}\mathbf{C}_t^{-1}\mathbf{D}_t \tag{1.51}$$

and

$$\beta_t = -\frac{1}{2}\mathbf{C}_t^{-1}\mathbf{B}_t \tag{1.52}$$

The approximating polynomial for the optimal return function is

$$V(x; t) = L(x, u(x, t); t)$$
$$= (\delta x)^T \mathbf{P}_t (x - \bar{x}_t) + \mathbf{Q}_t^T \delta x \tag{1.53}$$

where
$$\mathbf{P}_t = \mathbf{A}_t - \frac{1}{4}\mathbf{B}_t^T \mathbf{C}_t^{-1} \mathbf{B}_t \tag{1.54}$$

and
$$\mathbf{Q}_t = -\frac{1}{2}\mathbf{D}_t \mathbf{C}_t^{-1} \mathbf{B}_t + \mathbf{E}_t^T \tag{1.55}$$

These equations are necessary for the DDP backward recursion. α_t and β_t for $1 \le t \le T$ must be stored for use in the forward sweep. The forward sweep determines the successor DDP policy by successively selecting controls according to the rule $u\,(x^*;\, t)$ and then calculating the successor state at each time so that $u_1^* = u(x_t^*;\, 1)$ and $x_2^* = \mathbf{g}(x_1^*, u_1^*;\, 1)$. Then the following is for $t = 2, \ldots, T$,

$$u_t^* = u(x_t^*;\, t) + \bar{u}_t \tag{1.56}$$

and
$$x_{t+1}^* = \mathbf{g}(x_t^*, u_t^*;\, t) \tag{1.57}$$

For the next DDP iteration the DDP successor control is the current control sequence $\bar{u}$.

1.5 OPTIMAL CONTROL

1.5.1 Concepts

Linear quadratic (*LQ*) problems find an optimal control law that minimizes a *quadratic performance index* subject to a *dynamic linear system.* Because the quadratic performance index is a function of states and controls at all times, the LQ problem is to find an optimal control law such that the states and controls are as close as possible to the desired states and controls, respectively. There are several methods of solving the LQ problem. One way is to use the DP technique to derive an analytical optimal control law which is a linear function of the current state. Books on optimal control theory include: Anderson and Moore (1990); Athans and Falb (1966); Auslander et al. (1974); Brogan (1974); Dyer and McReynolds (1970); Jacobson (1980); Luus (1967); and Maybeck (1979, 1982).

If the desired states are zero, the LQ problem is a *linear quadratic regulator* (*LQR*) problem which regulates the states to zero. If the desired states are not zero, the LQ problem is a *linear quadratic tracker* (*LQT*) problem which forces the system to track the desired states at all times. For the LQR problem (or LQT) problem, if the dynamic linear system equation is deterministic, the LQR is called *LQRD* (or LQTD for LQT); if the dynamic linear system equation is stochastic, the LQR is called *LQRS* (or LQTS for LQT). In essence, the LQR problem is a special case of the LQT problem.

The combination of LQ problems with a Kalman-Bucy filter (Maybeck, 1982) leads to the *linear quadratic-Gaussian* (*LQG*). The *Kalman-Bucy filter* gives an optimum state estimate by minimizing the mean square error about the state. Because the optimal control law for LQ problems is a linear function of the current state, if the current state is directly known without uncertainty, the optimal control law can directly apply. If the current state is not directly known, the current state can be estimated by the Kalman-Bucy filter. The optimal control law applies by using the estimate of the state by the Kalman-Bucy filter.

Consider a continuous dynamic system defined by the following *state-space model* as

$$\dot{x} = f(x(t), u(t);\, t) \tag{1.58}$$

in which $x(t)$ and $u(t)$ are the vectors of state and control of the system, respectively. The problem is, at a given initial state, to use $u(t)$ to guide the system such that the following *performance index is* minimized,

$$J = F(x(t_f), t_f) + \int_{t_o}^{tf} L(x(t), u(t), t)\, dt \qquad (1.59)$$

in which F and L are real and scalar-valued functions; t_o and t_f are the initial and terminal times, respectively. The minimal cost from time t_o to the terminal time t_f is defined as

$$g(x(t), t_f - t) = \min_{u(t)} \left\{ F(x(t_f), t) + \int_{t}^{tf} L(x(t), u(t), t)\, dt \right\} \qquad (1.60)$$

which is sometimes called the *optimal cost-to-go.*

The well-known Hamilton-Jacoby-Bellman (or *Hamilton-Jacobi equation*) is

$$\frac{\partial g}{\partial t_r} = \min_{u(t)} \left\{ L(x(t), u(t), t) + [\nabla_x g]^T \dot{x} \right\} \qquad (1.61)$$

with the boundary condition

$$g(x(t), t_r)|_{t_r = 0} = F(x(t_f), t_f) \qquad (1.62)$$

The *optimal control* is the one that minimizes the term in the pair of brackets on the right-hand side of the Hamilton-Jacobi-Bellman equation. If there is no constraint on $u(t)$, the necessary condition for optimality can be derived by taking derivatives,

$$\frac{\partial L}{\partial u} + \frac{\partial f}{\partial u} \nabla_x g = \mathbf{0} \qquad (1.63)$$

Consider a linear dynamic system

$$\dot{x}(t) = \mathbf{A}(t)x(t) + \mathbf{B}(t)u(t) \qquad (1.64)$$

with loss function

$$L = [x(t) - x_D(t)]^T \mathbf{P}(t)[x(t) - x_D(t)] + u(t)^T \mathbf{R}(t)u(t)$$

and

$$F = [x(t_f) - x_D(t_f)]^T \mathbf{H}(t)[x(t_f) - x_D(t_f)] \qquad (1.65)$$

in which x_D denotes the desired value of the state variable and $\mathbf{P}(t)$, $\mathbf{R}(t)$, and $\mathbf{H}(t)$ are weighting matrices.

The derivatives $\partial f/\partial u$ and $\partial \mathbf{L}/\partial u$, respectively, are

$$\frac{\partial f}{\partial u} = \mathbf{B}^T(t) \qquad (1.66)$$

$$\frac{\partial \mathbf{L}}{\partial u} = 2\mathbf{R}(t)u(t) \qquad (1.67)$$

The optimal control policy for $u(t)$ is

$$u^*(t) = -\frac{1}{2}\mathbf{R}^{-1}\mathbf{B}^T \nabla_x g(x(t), t_r) \qquad (1.68)$$

In order to obtain $g(x(t), t_f - t)$ for this linear system with quadratic loss function, the Hamilton-Jacobi-Bellman differential equation for $g(x(t), t_f - t)$ is solved after substitution of the optimal control policy. In order to solve this equation, the classic method is to assume a quadratic function form for $g(x(t), t_f - t)$ which has some unknown coefficients. Then substitute it into the Hamilton-Jacobi-Bellman equation to obtain the so-called Riccati-type equation, which is then solved for the unknown coefficients in the quadratic function (Brogan, 1974).

1.5.2 Linear Quadratic Regulator (LQR)

Linear Quadratic Regulator LQR with Deterministic System Equation (LQRD). The purpose of LQRD is to guide a linear system during a fixed period such that the states of the system are as close as possible to the zero states with the least control energy. It is to find an optimal law that minimizes a quadratic performance index such as

$$\mathbf{J}(x(0)) = x(N)^T\mathbf{P}(N)x(N) + \sum_{k=0}^{N-1} [x(k)^T\mathbf{P}(k)x(k) + u^T(k)\mathbf{R}(k)u(k)] \tag{1.69}$$

subject to a dynamic and time-variant linear system

$$x(k+1) = \mathbf{A}(k)x(k) + \mathbf{B}(k)u(k) \tag{1.70}$$

$$x(0) = x_o \tag{1.71}$$

in which T is the transpose of a matrix; the vector $x(k)$ of $(m \times 1)$ is the state at time k; the vector $u(k)$ of $(n \times 1)$ is the control law applied at time k; the $(m \times m)$ weighting matrix $\mathbf{P}(k)$ is symmetric positive semidefinite; the weighting matrix $\mathbf{R}(k)$ of $(n \times n)$ is symmetric positive definite; the matrices $\mathbf{A}(k)$ of $(m \times m)$ and $\mathbf{B}(k)$ of $(m \times n)$ describe the system; k is the time sequence from 0 to N; and x_o is a given initial state. Equation (1.70) can be called a *system transition equation* or transition equation.

An analytic solution of the optimal control law can be derived (Jacobson et al., 1980) by using dynamic programming

$$u^*(k) = \mathbf{L}(k)x(k) \tag{1.72}$$

in which

$$\mathbf{L}(k) = -(\mathbf{R}(k) + \mathbf{B}^T(k)\mathbf{K}_c(k+1)\mathbf{B}(k))^{-1}\mathbf{B}^T(k)\mathbf{K}_c(k+1)\mathbf{A}(k) \tag{1.73}$$

in which the matrix $\mathbf{K}_c(k)$ satisfies the recursive Riccati equation,

$$\mathbf{K}_c(k) = \mathbf{A}^T(k)\mathbf{K}_c(k+1)[\mathbf{I} - \mathbf{B}(k)(\mathbf{B}^T(k)\mathbf{K}_c(k+1)\mathbf{B}(k) + \mathbf{R}(k))\text{-}\mathbf{B}^T(k)\mathbf{K}_c(k+1)]\mathbf{A}(k) + \mathbf{P}(k) \tag{1.74}$$

with the terminal condition

$$\mathbf{K}_c(N) = \mathbf{Q}(N) \tag{1.75}$$

The minimized performance index is given by

$$J^*(x(0)) = x^T(0)\mathbf{K}_c(0)x(0) \tag{1.76}$$

The corresponding minimized cost-to-go performance index is

$$J^*(x(k)) = x^T(k)\mathbf{K}_c(k)x(k) \tag{1.77}$$

The result shows that the optimal control law is a linear function of the current state. If the current state is known without uncertainty, the optimal control law can directly apply. This fact accounts for the great popularity of the LQ problem. It can be seen that the transition equation is deterministic.

1.5.3 Linear Quadratic Gaussian (LQG)

When the state can only be directly observed with uncertainty and the system transition equation uncertainty is also considered, the LQG problem can be formulated and solved on the basis of the *separation theorem* which is an important concept in modern linear control theory. Essentially, LQG can be divided into two parts, the *estimator* part and the *actuator* part (Bertsekas, 1987). The estimator part is an optimal solution of the problem of estimating the state $x(k)$ assuming no control takes place, and the actuator part is an optimal solution of the control problem assuming perfect state information prevails. The estimator can be achieved by using the Kalman-Bucy filtering which estimates the optimal state by minimizing the mean squared error of the estimated state. It can be shown that this estimate is also the expected value of the state condition on the current and previous measurements.

1.6 INTERFACING OPTIMIZERS WITH PROCESS SIMULATORS

1.6.1 Problem Formulation

A general process optimization problem in water resources may be formulated in an NLP framework in terms of state or dependent variables y and control or independent variables x:

$$\text{Minimize } f(y, x) \tag{1.78}$$

subject to process (simulation) equations

$$\mathbf{g}(y, x) = \mathbf{0} \tag{1.79}$$

and additional constraints for design, operation, budget, etc. on the dependent variable y and independent variables x:

$$\underline{w} \leq w(y, x) \leq \overline{w} \tag{1.80}$$

The process simulation equations for hydrosystems applications basically consist of the governing physical equations that simulate a physical process such as conservation of mass, energy, and momentum. These equations are typically large in number, sparse, and nonlinear expressions of the governing partial differential equations. Conceptually, the simplest of the real-world problems cannot be solved in this manner as a result of their size. The existing NLP codes cannot solve such large sparse problems. An alternative approach is to use the appropriate process simulator to solve the constraint equation [Eq. (1.79)] each time the objective and constraints are evaluated in the optimizer. The major advantage of such an approach is the reduced size of the nonlinear optimization problem, so that only a small subset of the constraint equations is treated by the optimizer. The basic idea is that the optimizer only handles the following reduced problem:

$$\text{Minimize } f(x(y), y) \tag{1.81}$$

subject to

$$\underline{w} \leq w(y, x(y)) \leq \overline{w} \tag{1.82}$$

as opposed to the much larger problem defined by Eqs. (1.78) to (1.80).

State (or dependent) variables and the control (or independent) variables are implicitly related through the simulator. In essence, the simulator equations are used to express the states in terms of the controls yielding a much smaller nonlinear optimization problem. The reduced gradient $[\partial F/\partial y]$ where $F(y) = f(x(y), y)$ is required by the optimization routines. The reduced problems are solved by combining augmented Lagrangian and reduced gradient procedures. Techniques for computing gradients of the reduced problem functions were developed by Lasdon and Waren (1978):

Step 1 Use the simulator to solve process Eq. (1.79)

Step 2 Solve for π, the Lagrange multiplier vector, using the linear difference equations

$$\pi^T = \left[\frac{\partial \mathbf{g}}{\partial x}\right]^{-1}\left[\frac{\partial f}{\partial x}\right]^T \tag{1.83}$$

Step 3 Evaluate the reduced gradient

$$\left[\frac{\partial F}{\partial y}\right]^T = \left[\frac{\partial f}{\partial y}\right]^T - \pi^T\left[\frac{\partial \mathbf{g}}{\partial y}\right] \tag{1.84}$$

In Eqs. (1.83) and (1.84), all elements of $\partial f/\partial x$, $\partial \mathbf{g}/\partial x$, $\partial f/\partial y$, and $\partial \mathbf{g}/\partial y$ are evaluated at some solution $(\bar{x}, \bar{y})$ and Eq. (1.84) yields the reduced gradient, $\partial F/\partial y$, at $\bar{y}$. The Lagrange multiplier π has its usual shadow price interpretation.

Solution of the simulator for the state variables, given the control variables from the optimizer, does not guarantee that the state variables satisfy the bound constraint in the optimizer. As a result, the bounds on the state variables are combined with the objective function into a penaltylike function called an augmented Lagrangian (Fletcher, 1975). This approach forces the state bounds to be satisfied and also reduces the number of constraints seen by the optimizer. The *augmented Lagrangian* is:

$$\text{minimize } L(y, \mu, \sigma) = F(y) + \frac{1}{2}\sum_i \sigma_i \min\left[0, c_i - \frac{\mu_i}{\sigma_i}\right]^2 + \frac{1}{2}\sum_i \frac{\mu_i}{\sigma_i} \tag{1.85}$$

where i is the index for each bound constraint; σ_i and μ_i are the penalty weights and Lagrange multipliers for the ith bound; and c_i is the violation of the bounds either above the upper or below the minimum defined as $c_i = \min(y_i - \underline{y}_i, \bar{y}_i - y_i)$.

1.7 MULTIOBJECTIVE PROGRAMMING

1.7.1 Multiobjective Concepts

Multiobjective programming deals with problems involving several objectives that are noncommensurate and conflicting with each other. Among the objective functions involved, there is no single one whose importance is overwhelmingly dominant over all others. Under this circumstance, the ideological theme of optimality in the single-objective context is no longer appropriate. The solution to a multiobjective problem is a best compromisable solution, according to the decision-maker's preference, among the objectives and the noninferior solutions to the problem.

A *noninferior solution* to a multiobjective programming problem is a feasible solution to which there is no other feasible solution that will yield an improvement in one objective without causing degradation to at least one other objective (Cohon, 1978). The collection of such inferior solutions allows the assessment of tradeoff among conflicting objectives. Such tradeoff is often measured by the negative of the slope of the noninferior set, called the *marginal rate of transformation* (*MRT*), representing the *feasible tradeoff* between objectives.

To obtain the solution to a multiobjective programming problem, the preference of the decision-maker among the conflicting objectives must be known. Information concerning the decision-maker's preference is commonly called the *utility function*, which is a function of the objective function values. Geometrically, the utility function can be depicted as a series of *indifference curves.*

The utility of a decision-maker will be the same for combinations of solutions that fall on the same indifference curve. The negative slope of an indifference curve is the *marginal rate of substitution* (*MRS*) indicating the decision-maker's *desired tradeoff* among different objectives. The best compromised solution to a multiobjective programming problem is a unique set of alternatives that possess the property of maximizing the decision-maker's utility and are elements of the noninferior solution set. Referring to Fig. 1.8, the best compromised solution occurs at the point where the indifference curve and noninferior solution set are tangent. In other words, at the best compromised solution, MRT equals to MRS.

Mathematically, multiobjective programming problems can be expressed in terms of *vector optimization* as

maximize $$f(x) = [f_1(x), f_2(x), \ldots, f_K(x)] \tag{1.86a}$$

subject to

$$\mathbf{g}(x) = \mathbf{0} \tag{1.86b}$$

in which $f(x)$ is a vector of K objective functions, $\mathbf{g}(x) = \mathbf{0}$ are vectors of constraints, and x is a vector of decision variables.

1.7.2 Types of Methods

There are many methods proposed to derive the solution to a multiobjective programming problem. Basically, they can be classified into two categories (Cohon, 1978): *generating techniques* and *techniques incorporating knowledge of decision-maker's preference.* The purpose of generating techniques is to provide decision-makers with information about the inferior solution set or feasible tradeoff to the multiobjective problem. System analysts play the role of information providers and do not actively engage in decision-making. On the other hand, techniques in the other category explicitly incorporate the decision-maker's preference to search for the best compromised solution. Detailed descriptions of various techniques for solving multiobjective problems can be found elsewhere (Cohon, 1978; Goicoechea, et al., 1982; Chankong and Haimes, 1983; and Steuer, 1986). In the following sections, some frequently used techniques in multiobjective water resource management are briefly discussed.

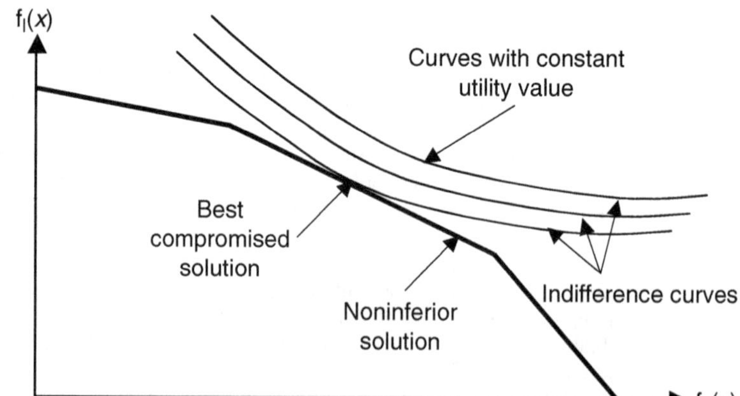

FIGURE 1.8 Noninferior solution and indifference curves.

1.7.3 Noninferior Solution Generating Techniques

Two commonly used generating techniques are the *weighting method* and the *constraint method.* The weighting method casts the multiobjective programming problem, Eqs. (1.86a–b), into

$$\text{maximize} \sum_{k=1}^{K} w_k f_k(x) \tag{1.87a}$$

subject to

$$\mathbf{g}(x) = \mathbf{0} \tag{1.87b}$$

where w_k is the weight on the kth objective. The weighted method is based on the fact that the trade-off between two different objectives $f_{k'}(x)$ and $f_{k''}(x)$ is represented by the respective weights as $-w_{k'}/w_{k''}$. The optimal solution to Eqs. (1.87a–b) is noninferior if it is unique under strictly positive-valued weights. Through systematic assignment of positive-valued weights to each objective function and solving Eqs. (1.87a–b) repeatedly, the noninferior solution set can be derived.

The constraint method arbitrarily selects one objective and places all other objective functions into the constraint set as

$$\text{maximize } f_h(x) \tag{1.88a}$$

subject to

$$f_k(x) \geq \underline{f}_k \qquad \text{for all} \qquad k \neq h \tag{1.88b}$$

$$\mathbf{g}(x) = \mathbf{0} \tag{1.88c}$$

in which $f_h(x)$ is the selected objective and $\underline{f}_k$ is the specified lower bound to be achieved by the kth objective. Note that the specified lower bound $\underline{f}_k$ should be that the optimal solution to Eqs. (1.88a–c) exists. The noninferior solution set to the multiobjective programming problem, Eqs. (1.86a–b) can be obtained by systematically varying $\underline{f}_k$ and solving Eqs. (1.88a–c) accordingly.

Realize that both the weighted method and constraint method derive the inferior solution set to the multiobjective programming problem by repeatedly solving a single objective problem. This computational burden can be greatly reduced if appropriate parametric programming algorithms can be applied. For more detailed descriptions on the theory and algorithm of the two methods and other generating techniques, readers are referred to Cohon (1978) and Chankong and Haimes (1983).

1.7.4 Goal Programming

The *goal programming* procedure for a multiobjective programming problem is to identify when the goal is underachieved, overachieved, or both. For each objective, two types of goals can be used in a goal programming formulation, depending on the nature of the goal: *two-sided goal* or *one-sided goal.* Referring to Fig. 1.9, a two-sided goal considers a penalty when the goal is not met exactly by the problem solution, whereas a one-sided goal incurs a penalty only when the specified goal is underachieved or overachieved, but not both, by the solution. For example, for the objective of profit maximization, the goal type is one-sided because the penalty would not occur if the solution resulted in a higher benefit than the specified goal; for the objective of maintaining the groundwater table at a fixed level, the goal type is a two-sided one because undesirable consequences would occur if such a level were not maintained.

In the goal programming formulation for a multiobjective problem, Eq. (1.86) is transformed into

$$\text{minimize} \sum_{k=1}^{K} (w_k^+ d_k^+ + w_k^- d_k^-) \tag{1.89a}$$

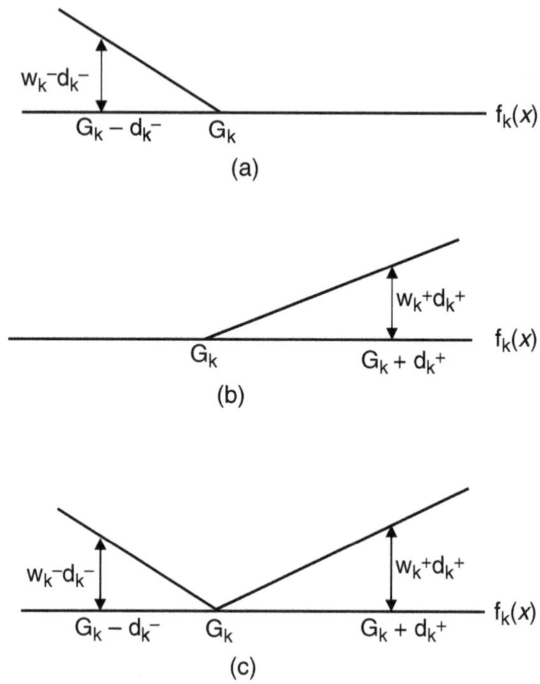

FIGURE 1.9 Types of goals: (*a*) one-sided; (*b*) one-sided; (*c*) two-sided.

subject to

$$f_k(x) - d_k^+ + d_k^- = G_k \qquad \text{for} \qquad k = 1, 2, \ldots, K \tag{1.89b}$$

$$\mathbf{g}(x) = \mathbf{0} \tag{1.89c}$$

in which d_k^+ and d_k^- are the additional nonnegative decision variables representing, respectively, the deviations due to overachieving and underachieving the specified goal G_k for the kth objective, $f_k(x)$; w_k^+ and w_k^- are weighting factors associated with two types of deviation, respectively; x and $\mathbf{g}(x)$ are the vectors of the original decision variables and constraints, respectively. The two nonnegative decision variables d_k^+ and d_k^- satisfy the condition that both will not be simultaneously greater than zero. The weighting factors w_k^+ and w_k^- can be used to indicate the relative importance of different types of deviations for the various objective functions considered in the problem. The solution technique for solving a goal programming problem depends on the nature of the original objective functions $f_k(x)$, and the constraints $\mathbf{g}(x)$.

The objective function of the goal programming formulation given above assumes that goals for all objectives are two-sided. Modifications should be made according to the goal type associated with each objective. For an objective function with a one-sided goal for which only underachievement is to be penalized, the term $w_k^+ d_k^+$ associated with the objective should be removed. Similarly, the term $w_k^- d_k^-$ is removed for the objective to which only overachievement is undesirable. It should be noted that the two deviation variables d_k^+ and d_k^- represent the status of the objective function with respect to its goal G_k. Therefore, regardless of the goal type, they must remain in the constraint equations corresponding to the objectives.

1.7.5 Utility Function Methods

Utility function techniques are based on the *multiattribute utility theory* (*MAUT*) which was designed to evaluate the *utility function* representing the preference of a decision-maker in terms of multiple attributes. In the context of the MAUT framework, the objectives of a multiobjective programming problem are attributes and decision variables are alternatives of choice. Therefore, the utility function of a decision-maker for a multiobjective programming problem can be viewed as a superobjective function to which the problem strives to maximize. Using the MAUT, the multiobjective programming model, Eqs. (1.86a–b), can be expressed as

$$\text{maximize } u[f_1(x), f_2(x), \ldots, f_K(x)] \tag{1.90a}$$

subject to

$$\mathbf{g}(x) = \mathbf{0} \tag{1.90b}$$

in which $u[\quad]$ is the utility function depending on the values of objective functions. The decision-maker prefers a feasible solution x_1 to x_2 if $u[f_1(x_1), f_2(x_1), \ldots, f_K(x_1)] > u[f_1(x_2), f_2(x_2), \ldots, f_K(x_2)]$.

The crux of solving a multiobjective programming problem using MAUT is the development of a utility, $u[\quad]$, which adequately represents the preference behavior of the decision-maker for the multiobjective problem under consideration. For practicality reasons, the MAUT avoids direct assessment of $u[f_1, f_2, \ldots, f_K]$ because deriving a utility function for a single attribute, $u_k(f_k)$, is difficult enough. Many multiattribute utility functions used in practice are of the following general form

$$u[f_1, f_2, \ldots, f_K] = h[u_1(f_1), u_2(f_2), \ldots, u_K(f_K)] \tag{1.91}$$

which indicates that the utility function can be decomposed in terms of uniattribute utility functions. Decomposition reduces the dimensionality of the problem, and hence greatly simplifies the task of assessing the utility function. Zeleny (1982) lists several commonly used forms for utility decomposition.

Two fundamental concepts are essential in the development of utility functions: *preferential independence* and *utility independence*. Zeleny (1982) offers the following definitions for the two. A pair of attributes f_a and f_b is preferentially independent of attribute f_c if the tradeoff between f_a and f_b is not affected by the level of f_c. Attribute f_a is utility independent of attribute f_b when conditional preferences for lotteries on f_a, given f_b, do not depend on the level of f_b. After information about the decision-maker's preference is extracted, it is important to verify the compliance of the two independence conditions. Zeleny provides an excellent elaboration on the subject of multiattribute utility measurement.

Inasmuch as the value of the utility function represents the degree of the decision-maker's preference, then the decision-maker would be indifferent to different combinations of objective function values that yield the same utility value. Therefore, geometric representation of constant utility in the objective function space is the indifference curve or surface. On an indifference curve, the change in utility value is zero; that is,

$$du = \sum_{k=1}^{K} \left(\frac{\partial u}{\partial f_k}\right) df_k \tag{1.92}$$

From Eq. (1.92), the tradeoff between the objectives f_k and f_1 can be computed as

$$\text{MRS}_{k1} = -\frac{\partial f_1}{\partial f_k} = -\frac{\partial u/\partial f_k}{\partial u/\partial f_1} \quad \text{for} \quad k \neq 1 \tag{1.93}$$

where MRS_{k1} is the marginal rate of substituting f_k for f_1.

1.7.6 Surrogate Worth Tradeoff Technique

The surrogate worth tradeoff (SWT) method is described by Haimes and Hall (1974), Haimes et al. (1975); and Haimes (1977). Because of the difficulty in defining the utility function for real-life problems, the method utilizes the approximated utility based on the values of objective functions and the MRS.

Using the SWT method, the multiobjective programming problem in Eq. (1.86) is written as

$$\text{Maximize } f_1(x) \tag{1.94a}$$

subject to

$$f_k(x) \geq f_k^o \qquad \text{for all} \qquad k \neq 1 \tag{1.94b}$$

$$\mathbf{g}(x) = \mathbf{0} \tag{1.94c}$$

in which f_k^o are prespecified values for the various objective functions. Because the Lagrange multiplier associated with a constraint indicates the rate of change of the objective function with respect to the constraint's RHS (Sec. 1.3.2), therefore, the Lagrange multiplier λ_k corresponding to the constraint equation [Eq. (1.94b)] represents the tradeoff between the objective functions $f_k(x)$ and $f_1(x)$. Of interest in a multiobjective problem are those negative-valued λ's. Haimes and Hall (1974) show that the values of Lagrange multipliers λ_k for constraint equation (1.94b) is the MRT between $f_k(x)$ and $f_1(x)$, MRT_{k1}, indicating the amount of $f_1(x)$ that can be gained by sacrificing one unit of $f_k(x)$. In general, the MRT_{k1} is a nondecreasing function of f_k^o.

When a tradeoff between $f_k(x)$ and $f_1(x)$ exists then $\lambda_k < 0$. To avoid the difficult task of assessing the true utility function, the SWT method requires the decision-maker to use an ordinal scale to represent a preference according to the relative difference between the feasible tradeoff MRT_{k1} and the decision-maker's desirable tradeoff MRS_{k1} to sacrifice $f_k(x)$ for gaining $f_1(x)$. The value assigned to the MRT_{k1} is called the *surrogate worth* SW_{k1} which, in turn, depends on f_k^o. In the SWT method, the *surrogate worth function* $\text{SW}_{k1}(f_k^o)$ is applied to a monotonic decreasing function of f_k^o. The assignment of surrogate worth value can be made using

$$\text{SW}_{k1} = \begin{cases} > \min(0, 10) & \text{if} \quad \text{MRS}_{k1} > \text{MRT}_{k1} \\ > \max(-10, 0) & \text{if} \quad \text{MRS}_{k1} < \text{MRT}_{k1} \end{cases} \tag{1.95}$$

with +10 indicating complete favor for substituting $f_k(x)$ for $f_1(x)$ and −10 indicating complete disfavor. The best compromised solution for $f_k(x)$ can be obtained when $\text{SW}_{k1} = 0$. This can be achieved by solving Eq. (1.94a–c) for $\lambda_k = \text{MRT}_{k1}$ under different f_k^o's in Eq. (1.94b) and requesting the corresponding surrogate worth from the decision-maker (see Fig. 1.10).

The process can be repeated by considering different objective functions from which the best compromised objective function values, f_k^* for all $k \neq 1$, can be obtained. Then, the solution to the entire multiobjective problem can be obtained by solving Eqs. (1.94a–c) with f_k^o in Eq. (1.94b) replaced by f_k^*.

1.8 GLOBAL OPTIMIZATION TECHNIQUES

To optimize a complex hydrosystem involving large a number of interrelated decision variables, it is generally difficult to be certain that the true optimal solution can be obtained within a reasonable amount of time. In dealing with such "hard" problems, one could opt to obtain the optimal solution with the anticipation to consume very large amounts of computational time (such as DP, IP, NLP, etc.) or to reach a quick but good solution that is suboptimal through some approximate or heuristic algorithms. Simulated annealing (SA) algorithms and genetic algorithms (GA) are two types of high quality *general algorithms* that are independent of the nature of the problem. Both SA and GA, by

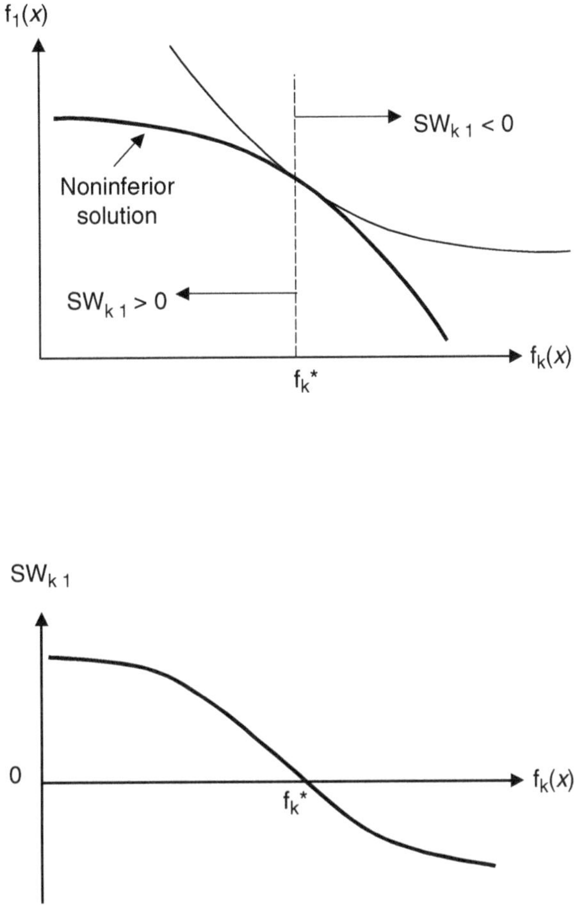

FIGURE 1.10 Identification of best compromised solution by surrogate worth tradeoff method.

nature, are randomization algorithms with local search based on stepwise improvement of the objective function through exploring the neighborhood domain of the current solution. NLP using line search in essence is local search. Quality of the local optimum could be strongly dependent on the initial solution and, in practice, it is difficult to assess the time needed to reach the global optimum. To avoid being trapped in a local optimum, local search algorithms can (1) try a large number of initial solutions; (2) introduce a more complex neighborhood structure to search for a larger part of the solution space; and (3) accept a limited transition to explore the solution space in which the solution is inferior (Aarts and Korst, 1989). SA and GA are gaining more and more applications for water resource systems optimization (see Tables 1.3 and 1.4). In the following descriptions of the two algorithms, the optimization problem is assumed to be of minimization type.

1.8.1 Simulated Annealing (SA) Algorithms

SA algorithms solve optimization problems by using an analogy to the physical annealing process of decreasing temperature to lower energy in a solid to a minimum level. The annealing process involves

TABLE 1.3 List of Sample Applications of Simulated Annealing to Water Resource Problems

Application type	References
Irrigation and agricultural management	Kuo et al. (2001, 2003).
Capacity expansion	Sanchez and Andreu (2001).
Groundwater development and management	Rao et al. (2004); Cunha (1999); Wang and Zheng (1998).
Groundwater remediation	Skaggs et al. (2001a, b).
Monitoring and sampling	Dixon et al. (1999), Nunes et al. (2004); Lee and Ellis (1996).
Water distribution network	Cunha and Sousa (1999).

TABLE 1.4 List of Sample Applications of Genetic Algorithm to Water Resource Problems

Application type	References
Watershed management	Nicklow and Muleta (2001); Harrell and Ranjithan (2003); Yeh and Labadie (1997).
Pipe system calibration	Kapelan et al. (2003); Lingireddy and Ormsbee (2002); Vitkovsky et al. (2000).
Water distribution system design	Morley et al. (2001); Wu and Simpson (2001); Reis et al. (1997); Savic and Walters (1997).
Water network rehabilitation	Dandy and Engelhardt (2001); Halhal et al. (1997).
Groundwater monitoring	Reed and Minsker (2004).
Groundwater system identification	Tsai et al. (2003); Aral et al. (2001); Heidari and Ranjithan (1998).
Groundwater remediation	Maskey et al. (2002); Zheng and Wang (2002); Guan and Aral (1999).

two steps: (1) increase the temperature of a heat bath to a maximum value at which the solid melts; and (2) carefully decrease the temperature of the heat bath to allow particles to arrange themselves in a more structured lattice to achieve a minimum energy level. If the temperature of the heat bath decreases too rapidly (called quenching), the solid could be frozen into a meta-stable state in which the solid would be brittle. The connection between optimization and the annealing process was first noted by Pincus (1970) and formalized as an optimization technique by Kirkpatrick et al. (1983).

SA algorithms employ random search which not only accepts solutions that improve the current solution, but also accepts solutions that are inferior with a specified probability according to the Metropolis criterion

$$\Pr(\text{accept } x_j^{(k-1)}) = \exp\left[-\frac{(f(x_j^{(k-1)}) - f(x_*^{(k)}))}{T}\right] \tag{1.96}$$

where (k) is the optimal solution in the kth iteration; $(k - 1)$ is the jth trial solution of the $(k + 1)$th iteration; $f()$ is the objective function value analogous to the energy level of the solid; and T is the control parameter analogous to the temperature of the heat bath. Aarts and Korst (1989), through the framework of Markov chains, prove the asymptotic convergence of SA algorithms under the condition of the existence of unique stationary probability.

The implementation of SA algorithms is remarkably easy as shown in Fig. 1.11 which involves the following components: (1) generation of possible solutions to explore solution space; (2) evalu-

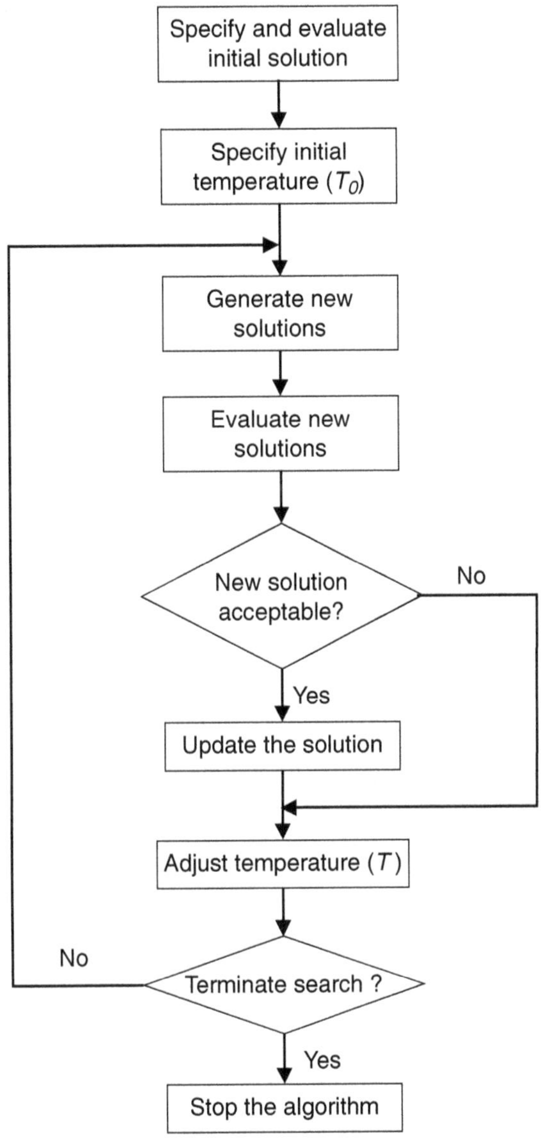

FIGURE 1.11 Algorithm of simulated annealing.

ation of the objective function; and (3) definition of the annealing schedule specified by the initial control parameter (T_0), decrement of control parameter (ΔT), and convergence criterion.

Mechanisms for generating trial solutions in each interation involve introducing random changes with the intent to cover the solution domain to be searched. One can generate trial solutions using

$$x_j^{(k-1)} = x_*^{(k)} + \boldsymbol{C}u, \quad \text{for } j = 1, 2, \ldots, \text{J} \tag{1.97}$$

in which $\boldsymbol{C}$ is the matrix controlling the range of the domain to be searched, $\boldsymbol{u}$ is the vector of the random number for a specified range that is consistent with $\boldsymbol{C}$, and J is the total number of trial solutions in the $(k+1)$th iteration. As the domain of the search changes with the iteration, elements in $\boldsymbol{C}$ or $\boldsymbol{u}$ could vary with the value of control parameter T. Ways to specify $\boldsymbol{C}$ and $\boldsymbol{u}$ can be found elsewhere (Vanderbilt and Louie, 1984; Parks, 1990).

SA algorithms only require evaluating the objective function value at each generated trial solutions; computational efficiency of the entire process could become an important issue because implementation of the algorithms anticipates a large number of function evaluations. Many optimization problems in hydrosystem engineering involve objective functions whose values are dependent on the physical constraints defined by complex flow simulations. In such cases, it is worthy of the efforts to search and implement more computationally efficient procedures. To handle constraints in a SA algorithm, the simplest way is to reject the trial solution if it leads to constraint violation. However, the approach would not work properly if (1) the constraints are of equality type and (2) the feasible space is disjoint. In either case constraints can be transformed to an unconstrained problem by introducing a penalty into the objective function due to constraint violation, such as Eq. (1.25), as

$$f_A(x) = f(x) + \frac{1}{T} w^T v \tag{1.98}$$

where $\boldsymbol{w}$ is the weighting vector for each constraint and $\boldsymbol{v}$ is the vector containing the magnitude of constraint violation. In general, SA algorithms would gradually decrease the value of control parameter T as the iteration proceeds. The inverse dependence of the control parameter of the penalty term would force the search more and more heavily toward the feasible region as the search progresses.

In the annealing schedule, three parameters are to be determined: (1) initial value of control parameter (T_0), (2) decrement of control parameter (ΔT), and (3) convergence criteria. In the implementation of the SA algorithm, T_0 is generally set to be high enough to allow virtually all trial solutions to be accepted. This process is analogous to having the temperature in the heat bath high enough to "melt" the solid. It is equivalent to the acceptance probability for T_0 close to one. It is clear that the value of T_0 will depend on the scale of the objective function and, hence, is problem-specific. For practical problems, the value of T_0 can be estimated by conducting an initial search in the solution domain in which all solutions are accounted for that lead to deterioration (or increase) in the objective function. Based on the average increment in objective function value for the inferior solutions, the initial control parameter T_0, according to Eq. (1.96), can be estimated as

$$T_0 = -\frac{\overline{\Delta f}}{\ln(P_0)} \tag{1.99}$$

in which P_0 is the target acceptance probability to start an SA algorithm. Kirkpatrick et al. (1983) suggested $P_0 = 0.8$. Aarts and Korst (1989) proposed an alternative estimation equation that takes into account both inferior and improved solutions.

From the annealing process viewpoint, one can restore the system to an equilibrium state by using a large decrement in the control parameter with a large number of trial solutions or a smaller decrement with a smaller number of trial solutions. A simple way to specify the decrement of the control parameter is by using $T^{(k+1)} = \alpha T^{(k)}$ (called the exponential cooling scheme) with α generally in the range of 0.8 ~0.99. Alternatively, the linear cooling scheme $T^{(k+1)} = T^{(k)} - \Delta T$ was found to be comparable to the exponential cooling scheme (Randelman and Grest, 1986). A review of decrements in the control parameter can be found in Laarhoven and Aarts (1987).

To terminate the execution of a SA algorithm, one can specify the final value of the control parameter to be used or the total number of trial solutions to be generated. Alternatively, one can halt the algorithm if lack of improvement in the objective function value is identified, which can be (1) no improvement can be found in all trial solutions at one control parameter and (2) the acceptance ratio falls below a specified value P_f.

1.8.2 Genetic Algorithms (GAs)

The GAs are a class of computer-based search procedures that emulate the biological evaluation processes of natural selection and genetics. Since its introduction by Goldberg in 1983, this innovative search procedure has been applied for optimization in a wide variety of areas (Goldberg, 1989). In the area of water resources engineering, some sample applications are listed in Table 1.13. The GAs have been demonstrated to be robust search procedures that outperform the conventional optimization procedures, in particular for problems with high-dimensionality, discontinuities, and noises.

Using GA for optimization, analogies are made between the properties of an optimization problem and the biological process of gene selection and reproduction. The solution space for an optimization problem can be treated as the environment in which potential solutions can be considered as genes. The degree of fitness of each gene can be measured by the value of the objective function of an optimization problem. In each iteration of a GA search, several genes representing solutions are generated from the population. These genes compete for their survival based on their fitness, the one that is more fit is more likely to survive and influence the next generation. Through this competition, the population evolves to contain high-performance individuals. In a GA, iteration is represented by a generation. In a GA, decision variables are analogous to biological genes and are represented by coded structures either in the form of real number or binary codes. A point in the solution space is represented by a collection of genes and the coded genes are juxtaposed to form an individual or chromosome.

Like any optimal seeking algorithm, a GA requires the specification of an initial population (the first generation) of potential solutions. This can be done by random generation of a population or by specifying an initial solution. In the initial population, the fitness of an individual is evaluated with respect to the objective function. Those individuals with a high level of fitness will have a higher chance of being selected to produce offspring than those individuals with a lower level of fitness. The selection can be conducted in many different ways, such as stochastic universal sampling (Baker, 1987), which behaves as does a roulette wheel with the slot size apportioned to each individual's relative fitness in the population. The principle of the selection is to prefer better solutions to worse ones. This very feature of the selection procedure in a GA is similar to the Metropolis acceptance criterion of the SA algorithm. The implementation of such a selection procedure will prevent the solutions being trapped at the local optimum.

Upon the completion of selection of individuals in a generation, individuals selected will mate to produce the next generation's population. During the mating, biological processes of gene crossover and mutation could take place. Again, fitness of individuals in the population of the new generation will be evaluated based on which selection and mating processes will be repeated. A schematic diagram of the GA is shown in Fig. 1.12. Through this repeated fitness-selection-reproduction cycle, the population generated by the GA will evolve toward an improved solution.

Two major mechanisms are involved for those selected individuals to generate offspring in the next generation, i.e., crossover and mutation. Crossover allows information exchange between two chosen individuals forming the parents. In GA, crossover is done by randomly selecting positions in the chromosomes of two individuals to swap their coded genes to produce the offspring. The rate of crossover has been suggested as 0.6 to 1 (Goldberg, 1989). Mutation then is a process by which new genetic material can be introduced into the population. It is applied on a bit-by-bit basis with the mutation rate specified by the user.

1.9 SELECTED APPLICATIONS IN WATER RESOURCES

Inasmuch as the existing literature is too extensive to be exhaustively listed, this section only provides some sample references applying different optimization techniques to a variety of water resource management, planning, and design problems. The applications are categorized according to the optimization technique used and the type of problem addressed. Tables 1.5 to 1.13, respectively,

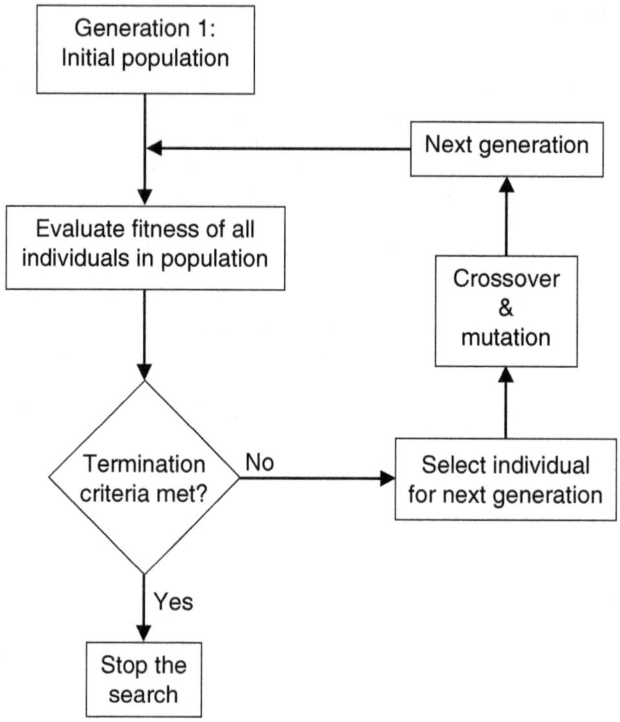

FIGURE 1.12 Schematic diagram of genetic algorithms.

list applications of linear programming, nonlinear programming, dynamic programming, differential dynamic programming, multiobjective programming, linear quadratic control techniques, and applications of interfacing optimizers with simulators to various water resource problems, including reservoir operation and management, irrigation and agricultural management, conjunctive use management, water distribution systems, floodplain management, urban drainage systems, water quality management, groundwater management, and hydrologic and hydraulic modelings.

TABLE 1.5 List of Sample Applications of Linear Programming to Water Resource Problems

Application type	References
Conjunctive use	Paudyal and Das Gupta (1990); Tyagi (1988); Kaushal et al. (1985); Chaturvedi and Chaube (1985); Louie et al. (1982).
Water distribution systems	Kessler et al. (1990); Ormsbee and Kessler (1990); Swamee and Shamir (1990); Kessler and Shamir (1989); Fujiwara and Dey (1988); Fujiwara et al. (1987); Morgan and Goulter (1985); Bhave (1983a,b); Khanjani and Busch (1983); Quindry et al. (1981); Shamir (1979).
Floodplain management	Weisz and Day (1974).
Groundwater management	Piper et al. (1989); Chau (1988); Lindner et al. (1988); Gorelick (1988); Atwood and Gorelick (1985); Schafer (1984); Tyagi and Narayana (1984); Willis and Liu (1984); Helweg and Arora (1983); Heidari (1982); Aguado et al. (1974).
Urban drainage	Neugebauer and Schilling (1991); Elimam et al. (1989); Yeh et al. (1980).
Agriculture/irrigation	Buller et al. (1991); Chaudhry and Young (1989); Batie et al. (1989); Kaushal et al. (1985); Tyagi and Narayana (1984); Howitt (1983); Moore et al. (1974).
Hydrological modeling	Loaiciga and Church (1990); Lall and Olds (1987); Wang and Yu (1986); Bardsley (1984); Han and Rao (1980).
Reservoir operation and management	Reznicek and Simonovic (1992); Reznicek et al. (1991); Tao and Lennox (1991); Lognathan and Bhattacharya (1990); Diaz and Fontane (1989); Tollow (1989); Kuczera (1989); Bhaskar and Whitlatch (1987); Lele (1987); Martin (1987); Moy et al. (1986); Pogge et al. (1985); Yeh (1985).
Water quality management	Lall and Lin (1991); Walker et al. (1990); Ponnambalam et al. (1990); Hayes (1989); Eheart and Park (1989); Bose et al. (1988); Fontaine and Lesht (1987); McBean (1987); Valocchi and Eheart (1987); Jenq et al. (1984); Olufeagba and Flake (1981).
Estuary management	Martin (1987).

TABLE 1.6 List of Sample Applications of Integer Programming to Water Resource Problems

Application type	References
Hydropower	Trezos (1991)
Drought management	Martin (1991)

TABLE 1.7 List of Sample Applications of Network Optimization to Water Resource Problems

Application type	References
Reservoir operation	Sabet and Creel (1991); Afshar et al. (1991)

TABLE 1.8 List of Sample Applications of Nonlinear Programming to Water Resource Problems

Application type	References
Conjunctive use	Kiashyap and Chandra (1982); Rydzewski and Rashid (1981); Matsukawa et al. (1992).
Water distribution systems	Ostfelt and Shamir (1993 a, b); Brion and Mays (1991); Lansey and Basnet (1991); Duan et al. (1990); Lansey and Mays (1989); Lansey et al. (1989); Chiplunkar et al. (1986); Rowell and Barnes (1982); Whitlatch and Asplund (1981).
Groundwater management	Liu (1988); Schafer (1984); Larson et al. (1977).
Urban drainage	Musselman and Talavage (1979); Lott (1976).
Reservoir operation and management	Sabet and Creel (1991); Cantiller (1986); Rosenthal et al. (1983); Ford et al. (1981).
Water quality management	Nix and Heaney (1988); Gorelick et al. (1984); Loucks (1978); Spofford (1976); McNamara (1976); Bundgaard-Nielsen and Hwang (1976).
Estuary management	Tung et al. (1990).
Water treatment plant	Wu and Chu (1991).
Hydropower	Tejada-Guibert et al. (1990).

TABLE 1.9 List of Sample Applications of Dynamic Programming to Water Resource Problems

Application type	References
Conjunctive use	Onta et al. (1991); Becklynck et al. (1983).
Water distribution systems	Lansey and Awumah (1994); Martin (1990); Zessler and Shamir (1989); Flynn and Marino (1989); Ormsbee et al. (1989).
Floodplain management	Hopkins et al. (1982); Morin et al. (1981); Hopkins et al. (1976).
Groundwater management	Cleveland and Yeh (1991); Jones et al. (1987).
Urban drainage	Kuo et al. (1991); Li and Matthew (1990); Han and Rao (1987); Nouh (1987); Han et al. (1980); Wenzel et al. (1979); Froise and Burges (1978); Mays et al. (1976); Mays and Bedient (1982).
Water supply systems	Martin (1987); Buras (1985).
Agriculture/irrigation	Tsakiris and Kiountouzis (1984); Jensen et al. (1980); Rhenals and Bras (1981)
Hydrological/hydraulic modeling	Zavesky and Goodman (1988).
Reservoir operation and management	Braga Jr. et al. (1991); Piccardi and Soncini-Sessa (1991); Vedula and Mohan (1990); Kuo et al. (1990); Druce (1990); Kelman et al. (1990); Alaouze (1989); Liango (1988); Whitlatch et al. (1988); Foufoula-Georgiou and Kitanidis (1988); Yeh (1976); Maidment and Chow (1981).
Water quality management	Joshi and Modak (1989, 1987); Stedinger and Bell-Graf (1979); Shih and Meier (1972).
Environmental	Liebman and Lynn (1966).

TABLE 1.10 List of Sample Applications of Differential Dynamic Programming to Water Resources Problems

Application type	References
Groundwater management	Chang et al. (1992); Culver and Shoemaker (1992); Jones et al. (1987); Liao and Shoemaker (1991); Makinde-Odusola and Marino (1989); Whiffen and Shoemaker (1993).
Reservoir operation and management	Murray and Yakowitz (1979), Carriaga and Mays (1995a, 1995b).
Estuary management	Li and Mays (1995).

TABLE 1.11 List of Sample Applications of Multiobjective Programming to Water Resources Problems

Application type	References
Groundwater management	Duckstein et al. (1994); Chang et al. (1992); Culver and Shoemaker (1992); Jones et al. (1987); Liao and Shoemaker (1991); Makinde-Odusola and Marino (1989); Whiffen and Shoemaker (1993).
Reservoir operation and management	Crawley and Dandy (1993); Loganatham and Bhattacharya (1990); Can and Houck (1984).
Estuary management	Mao and Mays (1994).

TABLE 1.12 List of Sample Applications of Linear Quadratic Control to Water Resources Problems

Application type	References
Reservoir operation and management	Hooper et al. (1991); Georgakakos and Marks (1987); Wasimi and Kitanidis (1983).
Estuary management	Zhao and Mays (1995a, b).

TABLE 1.13 List of Sample Applications of Interfacing Optimizers with Simulators

Application type	References
Water distribution system	Brion and Mays (1991); Lansey and Mays (1989); Cullinane et al. (1992).
Groundwater management	Gorelick et al. (1984); Wanakule et al. (1986)
Reservoir operation and management	Unver and Mays (1990).
Estuary management	Bao and Mays (1994a, b).

REFERENCES

Aarts, E., and J. Korst (1989). *Simulated Annealing and Boltzmann Machines*, Wiley, New York.

Afshar, A., M. A. Marino, and A. Abrishamchi (1991). "Reservoir Planning for Irrigation District," *Journal of Water Resources Planning and Management,* ASCE, 117(1): 74–85.

Aguado, F., I. Remson, M. F. Pikul, and W. A. Thomas (1974). "Optimum Pumping for Aquifer Dewatering," *Journal of Hydraulic Division,* ASCE, 100: 860–877.

Alaouze, C. M. (1989) "Reservoir Releases to Uses with Different Reliability Requirements," *Water Resources Bulletin,* 25(6): 1163–1168.

Anderson, B. and J. B. Moore (1990). *Optimal Control of Linear Quadratic Methods,* Prentice-Hall, Englewood Cliffs, N.J.

Aral, M. M., J. B. Guan, M. L. Maslia (2001). "Identification of Contaminant Source Location and Release History in Aquifers," *Journal of Hydrologic Engineering*, ASCE, 6 (3): 225–234.

Athans, M. and P. L. Falb (1986). *Optimal Control,* McGraw-Hill, New York.

Atwood, D. F. and S. M. Gorelick (1985). "Optimal Hydraulic Containment of Contaminated Ground Water," *Proceedings of the Fifth National Symposium and Exposition on Aquifer Restoration and Ground Water Monitoring,* The Fawcett Center, Columbus, Ohio, May 21–24, pp. 328–344.

Auslander, D. M., Y. Takahashi, and M. J. Rabins (1974). *Introduction to Systems and Control,* McGraw-Hill, New York.

Baker, J. E. (1987). "Reducing Bias and Inefficiency in the Selection Algorithm," *Genetic Algrithms and Their Applications: Proceedings of the Second International Conference on Genetic Algorithms,* Lawrence Erlbaum, Hillsdale, NJ, pp.14–21.

Bao, Y. and L. W. Mays (1994a). "Optimization of Freshwater Inflows to the Lavaca-Tres Palacios Estuary," *Journal of Water Resources Planning and Management,* ASCE, 120(2): 218–236, March/April.

Bao, Y. and L. W. Mays (1994b). "New Methodology for Optimization of Freshwater Inflows to Estuaries," *Journal of Water Resources Planning and Management,* ASCE, 120(2): 199–217, March/April.

Bardsley, W. E. (1984). "Conservative Estimation of Groundwater Volumes: Application of Linear Programming to Tritium Data," *Journal of Hydrology,* 67 (1–4): 183–193.

Batie, S. S., R. A. Kramer, and W. E. Cox (1989). "Economic and Legal Analysis of Strategies for Managing Agricultural Pollution of Groundwater," Virginia Polytechnic Institute and State University, Blacksburg, Va., Dept. of Agricultural Economics, Report, October.

Becklynck, J., M. Besbes, P. Combes, P. Hubert, and G. Marsily (1983). "Managing the Integration of Groundwater and Surface Water Resources: A Case Study of Supplying Potable Water to the Lille Metropolitan Area, Ground Water in Water Resources Planning," *Proceedings of a Symposium*, Koblenz, West Germany, Aug. 28–Sept. 3, IAHS Publication no. 142, vol. II. : 685–695.

Bellman, R. and S. Dreyfus (1962). *Applied Dynamic Programming,* Princeton University Press, Princeton, N.J.

Bertsekas, D. P. (1987). *Dynamic Programming: Deterministic and Stochastic Models,* Prentice-Hall, Inc., Englewood Cliffs, N.J.

Bhaskar, N. R. and E. E. Whitlatch (1987). "Comparison of Reservoir Linear Operation Rules Using Linear and Dynamic Programming," *Water Resources Bulletin,* 23(6): 1027–1036.

Bhave, P. R. (1983a). "Optimization of Gravity-Fed Water Distribution Systems: Theory," *Journal of Environmental Engineering,* ASCE, 109 (1): 189–205.

Bhave, P. R. (1983b). "Optimization of Gravity-Fed Water Distribution Systems: Application," *Journal of Environmental Engineering,* ASCE, 109(2): 383–395.

Bose, S. K., P. Ray, and B. K. Dutta (1988). "Water Quality Management of the Hooghly Estuary—A Linear Programming Model," *Water Science and Technology,* 20(6/7): 235–242.

Bounds, D. G. (1987). "New Optimization Methods from Physics and Biology," *Nature*, 329: 215–218.

Braga Jr., B. P. F., W. W.-G. Yeh, L. Becker, and M. T. L. Barrow (1991). "Stochastic Optimization of Multiple-Reservoir-System Operation," *Journal of Water Resources Planning and Management,* ASCE, 117(4): 471–481.

Brion, L. M. and L. W. Mays (1991). "Methodology for Optimal Operation of Pumping Stations in Water Distribution Systems," *Journal of Hydraulic Engineering,* ASCE, 117(11): 1551–1571.

Brogan, W. L. (1974). *Modern Control Theory,* Quantum Publishers, New York.

Buller, O., H. L. Manges, L. R. Stone, and J. R. Williams (1991). "Modeled Crop Water Use and Soil Water Drainage," *Agricultural Water Management,* 19(2): 117–134.

Bundgaard-Nielsen, M. and C. L. Hwang (1976). "A Review on Decision Models in Economics of Regional Water Quality Management," *Water Resources Bulletin,* 12(3): 461–479.

Buras, N. (1985). "Application of Mathematical Programming in Planning Surface Water Storage," *Water Resources Bulletin,* AWRA, 21(6): 1013–1020.

Buys, J. D. (1972). "Dual Algorithms for Constrained Optimization Problems," Ph.D. thesis, University of Leiden, Netherlands.

Can, E. K. and M. H. Houck (1984). "Real-Time Reservoir Operations by Goal Programming," *Journal of Water Resources Planning and Management,* ASCE, 110(3): 297–309.

Cantiller, R. R. (1986). "Nonlinear Approach to the Determination of Reservoir Operating Rules," Ph.D. dissertation, Purdue University, Dept. of Civil Engineering, Lafayette, Ind.

Carriaga, C. C. and L. W. Mays (1995a). "Optimal Control Approach for Sedimentation Control in Alluvial Rivers," *Journal of Water Resources Planning and Management,* ASCE, 121(6): 408–417, Nov./Dec.

Carriaga, C. C. and L. W. Mays (1995b). "Optimization Modeling for Sedimentation in Alluvial Rivers," *Journal of Water Resources Planning and Management,* ASCE, 121(3): 251–259, Nov./Dec.

Chang, L.-C. (1990). "The Application of Constrained Optimal Control Algorithms to Groundwater Remediation," Ph.D. dissertation, Cornell University, Ithaca, N.Y.

Chang, L.-C., C. A. Shoemaker, and P. L.-F. Liu (1992). "Application of a Constrained Optimal Control Algorithm to Groundwater Remediation," *Water Resources Research,* 28(12): 3157–3173.

Chankong, V. and Y. Y. Haimes (1983). *Multiobjective Decisionmaking: A Theory and Methodology,* Elsevier Science, New York.

Chaturvedi, M. C. and U. C. Chaube (1985). "Irrigation System Study in International Basin," *Journal of Water Resources Planning and Management,* ASCE, 111(2): 137–148.

Chau, T. S. (1988). "Analysis of Sustained Ground-Water Withdrawals by the Combined Simulation-Optimization Approach," *Groundwater,* 26(4): 454–463.

Chaudhry, M. A. and R. A. Young (1989). "Valuing Irrigation Water in Punjab Province, Pakistan: A Linear Programming Approach," *Water Resources Bulletin,* AWRA, 25(5): 1055–1061.

Chiplunkar, A. V., S. L. Mehndiratta, and P. Khanna (1986). "Looped Water Distribution System Optimization for Single Loading," *Journal of Environmental Engineering,* ASCE, 112(2): 264–279.

Cleveland, T. G. and W. W-G. Yeh (1991). "Optimal Configuration and Scheduling of Ground-Water Tracer Test," *Journal of Water Resources Planning and Management,* ASCE, 117(1): 37–51.

Cohon, J. L. (1978). *Multiobjective Programming and Planning,* Academic, New York.

Cooper, L. L. and M. W. Cooper (1981). *Introduction to Dynamic Programming,* Pergamon, Elmsford, N.Y.

Crawley, P. D. and G. C. Dandy (1993). "Optimal Operation of Multiple-Reservoir System," *Journal of Water Resources Planning and Management,* ASCE, 119(1): 1–17.

Cullinane, M. J., K. Lansey, and L. W. Mays (1992). "Optimization—Availability Based Design of Water Distribution Networks," *Journal of Hydraulic Engineering,* ASCE, 118(3): 420–441, Mar.

Culver, T. B. (1991). "Dynamic Optimal Control of Groundwater Remediation with Management Periods: Linearized and Quasi-Newton Approaches," Ph.D. dissertation, Cornell University, Ithaca, N.Y.

Culver, T. and C. Shoemaker (1992). "Dynamic Optimal Control for Groundwater Remediation with Flexible Management Periods," *Water Resources Research,* 28(3): 629–641.

Culver, T. B. and C. A. Shoemaker (1992). "Dynamic Optimal Control for Groundwater Remediation with Flexible Management Periods," *Water Resources Research,* AGU, 28(3): 629–641.

Cunha, M. D. (1999). "On Solving Aquifer Management Problems with Simulated Annealing Algorithms," *Water Resources Management,* 13 (3): 153–169.

Cunha, M. D. and J. Sousa (1999). "Water Distribution Network Design Optimization: Simulated Annealing Approach," *Journal of Water Resources Planning and Management,* ASCE, 125 (4): 215–221.

Dandy, G. C. and M. Engelhardt (2001). "Optimal scheduling of water pipe replacement using genetic algorithms," *Journal of Water Resources Planning and Management,* ASCE, 127 (4): 214–223.

Denardo, E. V. (1982). *Dynamic Programming Theory and Applications,* Prentice-Hall, Englewood Cliffs, N.J.

Dennis, J. E. and R. B. Schnable (1983). *Numerical Methods for Unconstrained Optimization,* Prentice-Hall, Englewood Cliffs, N.J.

Diaz, G. E. and D. G. Fontane (1989). "Hydropower Optimization via Sequential Quadratic Programming," *Journal of Water Resources Planning and Management,* ASCE, 115(6): 715–734.

Dixon, W., G. K. Smyth, and B. Chiswell (1999). "Optimized Selection of River Sampling Sites," *Water Research,* 33 (4): 971–978.

Dreyfus, S. and A. Law (1977). *The Art and Theory of Dynamic Programming,* Academic, New York.

Druce, D. J. (1990). "Incorporating Daily Flood Control Objectives into a Monthly Stochastic Dynamic Programing Model for a Hydroelectric Complex," *Water Resources Research,* AGU, 26(1): 5–11.

Duan, N., L. W. Mays, and K. E. Lansey (1990). "Optimal Reliability-Based Design of Pumping and Distribution Systems," *Journal of Hydraulic Engineering,* ASCE, 116(2): 249–268.

Duckstein, L., W. Treichel, and S. El Magnouni (1994). "Ranking Ground-Water Management Alternative by Multicriterion Analysis," *Journal of Water Resources Planning and Management,* ASCE, 120(4): 546–565.

Dyer, P. and S. McReynolds (1970). *The Computational Theory of Optimal Control,* Academic, New York.

Edgar, T. F. and D. M. Himmelblau (1988). *Optimization of Chemical Processes,* McGraw-Hill, New York.

Eheart, J. W. and H. Park (1989). "Effects of Temperature Variation on Critical Stream Dissolved Oxygen," *Water Resources Research,* 25(2): 145–151.

Elimam, A. A., C. Charalambous, and F. H. Ghobrial (1989). "Optimum Design of Large Sewer Networks," *Journal of Environmental Engineering,* ASCE, 15(6): 1171–1190.

Fletcher, R. (1980). *Practical Methods of Optimization, Vol. 1, Unconstrained Optimization,* John Wiley, New York.

Flynn, L. E., and M. A. Marino (1989). "Aqueduct and Reservoir Capacities for Distribution Systems," *Journal of Water Resources Planning and Management,* ASCE, 115(5): 547–565.

Fontaine, T. D. and B. M. Lesht (1987). "Improving the Effectiveness of Environmental Management Decisions with Optimization and Uncertainty Analysis Techniques," in *Systems Analysis in Water Quality Management.* Pergamon, New York, pp. 31–41.

Ford, D. T., R. Garland, and C. Sullivan (1981). "Operation Policy Analysis: Sam Rayburn Reservoir," *Journal of the Water Resources Planning and Management Division,* ASCE, 107(WR2): 339–350.

Foufoula-Georgiou, E. and P. K. Kitanidis (1988). "Gradient Dynamic Programming for Stochastic Optimal Control of Multidimensional Water Resources Systems," *Water Resources Research,* 24(8): 1345–1359.

Froise, S. and S. J. Burges (1978). "Least-Cost Design of Urban-Drainage Networks," *Journal of Water Resources Planning and Management,* ASCE, 104(WR1): 75–92.

Fujiwara, O. and D. Dey (1988). "Method for Optimal Design of Branched Networks on Flat Terrain," *Journal of Environmental Engineering,* 114(6): 1464–1475.

Fujiwara, O., B. Jenchaimahakoon, and N. C. P. Edirisinghe (1987). "Modified Linear Programming Gradient Method for Optimal Design of Looped Water Distribution Networks," *Water Resources Research,* 23(6): 977–982.

Garfinkel, R. S. and G. L. Nemhauser (1972). *Integer Programming,* John Wiley, New York.

Gass, S. I. (1985). *Linear Programming: Methods and Application,* McGraw-Hill, New York.

Georgakakos, A. P. and D. H. Marks, "A New Method for the Real-Time Operation of Reservoir Systems," *Water Resources Research,* 23(7): 1376–1390, 1987.

Gill, P. E., W. Murray, and M. H. Wright (1981). *Practical Optimization,* Academic, London and New York.

Goldberg, D. E. (1989). *Genetic Algorithms in Search, Optimization and Machine Learning*, Addison-Wesley, Reading, Mass.

Gorelick, S. M. (1988). *Review of Groundwater Management Models, Efficiency in Irrigation: The Conjunctive Use of Surface and Groundwater Resources,* Winrock International, Arlington, Va., pp. 103–121.

Gorelick, S. M., C. I. Voss, P. E. Gill, W. Murray, M. A. Saunders, and M. M. Wright (1984). "Aquifer Reclamation Design: The Use of Containment Transport Simulation Combined with Nonlinear Programming," *Water Resources Research,* AGU, 20(4): 415–427.

Guan, J. and M. M. Aral (1999). "Optimal Remediation with Well Locations and Pumping Rates Selected as Continuous Decision Variables," *Journal of Hydrology*, 221 (1–2): 20–42.

Haimes, Y. Y. (1977). *Hierarchical Analysis of Water Resource Systems: Modeling and Optimization of Large-Scale Systems*, McGraw-Hill, New York.

Haimes, Y. Y. and W. A. Hall. (1974). "Multiobjectives in water resources systems analysis: The surrogate worth trade-off method." Water Resources Research, 10(4).

Haimes, Y.Y., W. A. Hall, and H. T. Freeman (1975). *Multiobjective Optimization in Water Resources Systems*, Elsevier, Amsterdam.

Halhal, D., G. A. Walters, D. Ouazar, and D. A. Savic (1997). "Water Network Rehabilitation with Structured Messy Genetic Algorithm," *Journal of Water Resources Planning and Management*, ASCE, 123 (3): 137–146.

Hall, W. and W. Butcher (1968). "Optimal Timing of Irrigation," *Journal of Irrigation and Drainage Division,* ASCE, 94(2): 267–275.

Han, J. and A. R. Rao (1980). "Optimal Parameter Estimation and Investigation of Objective Functions of Urban Runoff Models," Purdue University Water Resources Research Center, Lafayette, Ind., Technical Report no. 135.

Han, J. and A. R. Rao (1987). "Optimal Design of Storm Drain Systems," *Computational Hydrology '87,* Lighthouse Publications, Mission Viejo, Cal., pp. C4–C8.

Han, J., A. R. Rao, and M. H. Houck (1980). "Least Cost Design of Urban Drainage Systems," Purdue University Water Resources Research Center, Lafayette, Ind., Technical Report no. 138.

Harrell, L. J. and S. R. Ranjithan (2003). "Detention Pond Design and Land Use Planning for Watershed Management," *Journal of Water Resources Planning and Management*, ASCE, 129 (2): 98–106.

Hayes, D. F. (1989). "Systems Approach to Water Quality Management During Drought Periods in the Cumberland River Basin," *Proceedings of the Nineteenth Mississippi Water Resources Conference,* Water Resources Research Institute, Mississippi State University, Mississippi State, Miss.

Heidari, M. (1982). "Application of Linear System Theory and Linear Programming to Groundwater Management in Kansas," *Water Resource Bulletin,* 18: 1003–1012.

Heidari, M. and S. R. Ranjithan (1998). "A hybrid optimization approach to the estimation of distributed parameters in two-dimensional confined aquifers," *Journal of American Water Resources Association*, 34 (4): 909–920.

Helweg, O. and S. Arora (1983). "Controlling Ground Water Degradation in the Tulare Lake Basin," *Water Resources Bulletin,* 19(5): 837–844.

Hillier, F. S. and G. J. Lieberman (1990). *Introduction to Operation Research,* 5th ed., McGraw-Hill, New York.

Himmelblau, D. M. (1972). *Applied Nonlinear Programming,* McGraw-Hill, New York.

Hooper, E. R., A. P. Georgakakos, and D. P. Lettenmaier (1991). "Optimal Stochastic Operation of Salt River Project, Arizona," *Journal of Water Resources Planning and Management,* ASCE, 117(5): 566–587.

Hopkins, L. D., E. D. J. Brill, J. C. Liebman, and H. G. J. Wenzel (1976). "Flood Plain Management Through Allocation of Land Uses—A Dynamic Programming Model," Illinois Water Resources Center, Urbana, Ill., *Research Report* no. 117.

Hopkins, L. D., E. D. J. Brill, and B. Wong (1982). "Generating Alternative Solutions for Dynamic Programming Models of Water Resources Problems," *Water Resources Research,* 18(4): 782–790.

Howitt, R. E. (1983). "Calibration of Large-Scale Economic Models of the Agricultural Sector and Reliable Estimation of Regional Derived Demand for Water," University of California, Davis, California Water Resources Center Report.

Jacobson, D. H., D. H. Martin, M. Pachter, and T. Geveci (1980). "Extensions of Linear-Quadratic Control Theory," in A. V. Balakrishnan, and M. Thoma, eds., *Lecture Notes in Control and Information Sciences,* Springer-Verlag, New York.

Jacobson, D. and D. Mayne (1970). *Differential Dynamic Programming,* Elsevier, New York.

Jenq, T-R., M. L. Granstrom, S-F. Hsueh, and C. G. Uchrin (1984). "Phosphorus Management LP Model Case Study," *Water Resources Bulletin,* 20(4): 511–520.

Jensen, P. and J. W. Barnes (1980). *Network Flow Programming,* Wiley, New York.

Jensen, P. A., H. W. Chu, and D. D. Cochard, (1980). "Network Flow Optimization for Water Resources Planning with Uncertainties in Supply and Demand," University of Texas, Center for Research in Water Resources, Technical Report, CRWR-172, .

Jones, L., R. Willis, and W. W. G. Yeh (1987). "Optimal Control of Nonlinear Groundwater Hydraulics Using Differential Dynamic Programming," *Water Resources Research,* 23(11): 2097–2106.

Joshi, V. and P. Modak (1987). "Reliability Based Discrete Differential Dynamic Programming Model for River Water Quality Management under Financial Constraint," *Systems Analysis in Water Quality Management,* Pergamon, New York, pp. 15–22.

Joshi, V. and P. Modak (1989). "Heuristic Algorithms for Waste Load Allocation in a River Basin," *Water Science and Technology,* 21(8/9): 1057–1064.

Kapelan, Z. S., D. A. Savic and G. A. Walters (2003). "A Hybrid Inverse Transient Model for Leakage Detection and Roughness Calibration in Pipe Networks," *Journal of Hydraulic Research*, 41 (5): 481–492.

Karmarkar, N. (1984). "A New Polynomial-Time Algorithm for Linear Programming," *Combinatorica,* 4(4): 373–395, .

Kaushal, M. P., S. D. Khepar, and S. N. Panda (1985). "Saline Groundwater Management and Optimal Cropping Pattern," *Water International,* 10(2): 86–91.

Kelman, J., J. R. Stedinger, L. A. Cooper, E. Hsu, and S-Q. Yuan (1990). "Sampling Stochastic Dynamic Programming Applied to Reservoir Operation," *Water Resources Research,* 26(3): 447–454.

Kessler, V., L. Ormsbee, and U. Shamir (1990). "Methodology for Least-Cost Design of Invulnerable Water Distribution Networks," *Civil Engineering Systems,* 7(1): 20–28.

Kessler, A. and U. Shamir (1989). "Analysis of the Linear Programming Gradient Method for Optimal Design of Water Supply Networks," *Water Resources Research,* 25(7): 1469–1480.

Khanjani, M. J. and J. R. Busch (1983). "Optimal Irrigation Distribution Systems with Internal Storage," *Transactions of the American Society of Agricultural Engineers,* 26(3): 743–747.

Khatchian, L. (1979). "A Polynomial Algorithm in Linear Programming," *Soviet Mathematics,* Doklady 20, pp. 191–194.

Kiashyap, D. and S. Chandra (1982). "Distributed Conjunctive Use Model for Optimal Cropping Pattern, Optimal Allocation of Water Resources," IAHS Publication No. 135, *Proceedings* of a Symposium held at the First Scientific General Assembly of the IAHS, Exeter, England, July 19–30, pp. 377–384.

Kirkpatrick, S., C. D. Gerlatt, Jr., and M. P. Vecchi (1983). "Optimization by Simulated Annealing," *Science*, 220: 671–680.

Kuczera, G. (1989). "Fast Multireservoir Multiperiod Linear Programing Models," *Water Resources Research,* 25(2): 169–176.

Kuo, J. T., N. S. Hsu, W. S. Chu, S. Wan, and Y. J. Lin (1990). "Real-Time Operation of Tanshui River Reservoirs," *Journal of Water Resources Planning and Management,* ASCE, 116(3): 349–361.

Kuo, J-T., B. C. Yen, and G-P. Hwang (1991). "Optimal Design for Storm Sewer System with Pumping Stations," *Journal of Water Resources Planning and Management,* ASCE, 117(1): 11–27.

Kuo, S. F., C. W. Liu, and S. K. Chen (2003). "Comparative Study of Optimization Techniques for Irrigation Project Planning," *Journal of American Water Resources Association,* 39 (1): 59–73.

Kuo, S.F., C. W. Liu and G. P. Merkley (2001). "Application of the Simulated Annealing Method to Agricultural Water Resource Management," *Journal of Agricultural Engineering Research,* 80 (1): 109-124.

Laarhoven, P. J. M. van and E. H. L. Aarts (1987). *Simulated Annealing: Theory and Applications*, Reidel, Dordrecht, Holland.

Lall, U. and Y. C. Lin (1991). "Groundwater Management Model for Salt Lake County, Utah with Some Water Rights and Water Quality Considerations," *Journal of Hydrology,* 123(3–4): 367–393.

Lall, U. and J. Olds (1987). "Parameter Estimation Model for Ungaged Streamflows," *Journal of Hydrology,* 92(3–4): 245–262.

Land, A. and A. Doig (1960). "An Automatic Method of Solving Discrete Programming Problems," *Econometrica,* 28(3).

Lansey, K. E. and K. Awumah (1994). "Optimal Pump Operations Considering Pump Switches," *Journal of Water Resources Planning and Management,* ASCE, 120(1): 17–35.

Lansey, K. E. and C. Basnet (1991). "Parameter Estimation for Water Distribution Networks," *Journal of Water Resources Planning and Management,* ASCE, 117(1): 126–144.

Lansey, K. E., N. Duan, L. W. Mays, and Y-K. Tung (1989). "Water Distribution System Design under Uncertainties," *Journal of Water Resources Planning and Management,* ASCE, 115(5): 630–645.

Lansey, K. E. and L. W. Mays (1989). "Optimization Model for Water Distribution System Design," *Journal of Hydraulic Engineering,* ASCE, 115(10): 1401–1418.

Lapidus, L. and R. Luus (1967). *Optimal Control of Engineering Processes,* Blaisdell Publishing Company, Waltham, Mass.

Larson, S. P., T. Maddock III, and S. S. Papadopulos (1977). "Optimization Techniques Applied to Ground-Water Development," *Proceedings* of Symposium on the Optimal Development and Management of Groundwater, Birmingham, England, July, International Association of Hydrogeology, pp. E57–E66.

Lasdon, L. S., R. L. Fox, and M. W. Ratner (1974). "Nonlinear Optimization Using the Generalized Reduced Gradient Method," *Revue Francaise d'Automatique, Informatique et Recherche Operationnelle,* V-3, November, pp. 73–104.

Lasdon, L. S., and A. D. Waren (1978). "Generalized Reduced Gradient Software for Linearly and Nonlinearly Constrained Problems," in *Design and Implementation of Optimization Software,* H. J. Greenberg (ed.), Sijthoff and Noordhoff, Rockville, Md., pp. 363–397.

Lee, Y. M. and J. H. Ellis (1996). "Comparison of Algorithms for Nonlinear Integer Optimization: Application to Monitoring Network Design," *Journal of Environmental Engineering*, ASCE, 122 (6): 524–531.

Lele, S. M. (1987). "Improved Algorithms for Reservoir Capacity Calculation Incorporating Storage-Dependent Losses and Reliability Norm," *Water Resources Research,* 23(10): 1819–1823.

Li, G. and R. G. S. Matthew (1990). "New Approach for Optimization of Urban Drainage Systems," *Journal of Environmental Engineering,* 116(5): 927–944.

Li, G. and L. W. Mays (1995). "Differential Dynamic Programming for Estuarine Management," *Journal of Water Resources Planning and Management,* ASCE, 121(6): 455–463, November/December.

Liango, Q. (1988). "Reliability Constrained Markov Decision Programming and Its Practical Application to the Optimization of Multipurpose Reservoir Regulation, Computational Methods in Water Resources," in *Numerical Methods for Transport and Hydrologic Processes,* vol. 2, Elsevier, New York, pp. 417–422.

Liao, L. Z. and C. A. Shoemaker (1991). "Convergence in Unconstrained Discrete-Time Differential Dynamic Programming," *IEEE Trans. Automatic Control,* 36(6): 692–706.

Liebman, J. S., L. S. Lasdon, L. Schrage, and A. Waren (1986). *Modeling and Optimization with GINO,* Scientific Press, Palo Alto, Cal.

Liebman, J. and W. Lynn (1966). "The Optimal Allocation of Stream Dissolved Oxygen," *Water Resources Research,* 2(3): 581–591.

Lindner, W., K. Lindner, and G. Karadi (1988). "Optimal Groundwater Management in Two-Aquifer Systems," *Water Resources Bulletin,* 24(1): 27–33.

Lingireddy, S. and L. E. Ormsbee (2002). "Hydraulic Network Calibration Using Genetic Optimization," *Civil Engineering and Environmental Systems*, 19 (1): 13–39.

Liu, C. C. K. (1988). "Solute Transport Modeling in Heterogeneous Soils: Conjunctive Application of Physically Based and System Approaches," *Journal of Contaminant Hydrology,* 3(1): 97–111.

Loaiciga, H. A. and R. L. Church (1990). "Linear Programs for Nonlinear Hydrologic Estimation," *Water Resources Bulletin,* 26(4): 645–656.

Loganathan, G. V. and D. Bhattacharya (1990). "Goal-Programming Techniques for Optimal Reservoir Operations," *Journal of Water Resources Planning and Management,* ASCE, 116(6): 820–838.

Lott, D. L. (1976). "Optimization Model for the Design of Urban Flood-Control Systems," Texas University at Austin, Center for Research in Water Resources, Technical Report, CRWR-141.

Loucks, D. P. (1978). "Planning Comprehensive Water Quality Protection Systems," *American-Soviet Symposium on Use of Mathematical Models to Optimize Water Quality Management* (Kharkov and Rostov-on-Don, USSR; December 9–16, 1975), Environmental Research Laboratory, U.S. Environmental Protection Agency, Gulf Breeze, Fla., Report no. EPA-600/9-78-024, 1–33.

Louie, P. W. F., W. W.-G. Yeh, and N.-S. Hsu (1982). "An Approach to Solving a Basin-Wide Water Resources Management Planning Problem with Multiple Objectives," California Water Resources Center, Contribution no. 184.

Luenberger, D. G. (1984). *Introduction to Linear and Nonlinear Programming,* Addison-Wesley, Reading, Mass.

Maidment, D. R. and V. T. Chow (1981). "Stochastic State Variable Dynamic Programming for Reservoir System Analysis," *Water Resources Research,* 17(6): 1578–1584.

Makinde-Odusola, B. A. and M. A. Marino (1989). "Optimal Control of Groundwater by the Feedback," *Water Resources Research,* 25(6): 1341–1352.

Martin, Q. W. (1987a). "Estimating Freshwater Inflow Needs for Texas Estuaries by Mathematical Programming," *Water Resources Research,* 23(2): 230–238.

Martin, Q. W. (1987b). "Hierarchical Algorithm for Water Supply Expansion," *Journal of Water Resources Planning and Management,* 113(5): 677–695.

Martin, Q. W. (1987c). "Optimal Daily Operation of Surface-Water Systems," *Journal of Water Resources Planning and Management,* 113(4): 453–470.

Martin, Q. W. (1990). "Linear Water-Supply Pipeline Capacity Expansion Model," *Journal of Hydraulic Engineering,* 116(5): 675–690.

Martin, Q. W. (1991). "Drought Management Plan for Lower Colorado River in Texas," *Journal of Water Resources Planning and Management,* ASCE, 117(6): 645–661.

Maskey, S., A. Jonoski, and D. P Solomatine (2002). "Groundwater Remediation Strategy Using Global Optimization Algorithms," *Journal of Water Resources Planning and Management*, ASCE, 128 (6): 431–440.

Mao, N. and L. W. Mays (1994). "Goal Programming Models for Determining Freshwater Inflows to Bays and Estuaries," *Journal of Water Resource Planning and Management,* ASCE, 120(3): 316–329, May/June.

Matsukawa, J., B. A. Finney, and R. Willis (1992). "Conjunctive Use Planning in Mad River Basin, California," *Journal of Water Resources Planning and Management,* ASCE, 118(2): 115–132.

Maybeck, P. S. (1979). *Stochastic Models, Estimation, and Control,* vol. 1, Academic Press, New York.

Maybeck, P. S. (1982). *Stochastic Models, Estimation, and Control,* vol. 3, Academic Press, New York.

Mays, L. W. and P. B. Bedient (1982). "Model for Optimal Size and Location of Detention," *Journal of the Water Resources Planning and Management Division,* 108(WR3): 270–285.

Mays, L. W. and Y. K. Tung (1992). *Hydrosystems Engineering and Management,* McGraw-Hill, New York.

Mays, L. W., H. G. Wenzel, and J. C. Liebman (1976). "Model for Layout and Design of Sewer Systems," *Journal of the Water Resources Planning and Management Division,* ASCE, 102(WR2): 385–405.

McBean, E. A. (1987). "Developing Management Positions for Acidic Emission Reduction Negotiations," *Systems Analysis in Water Quality Management.* Pergamon, New York, pp. 51–59.

McCormick, G. P. (1983). *Nonlinear Programming: Theory, Algorithms, and Applications,* John Wiley, New York.

McNamara, J. R. (1976). "An Optimization Model for Regional Water Quality Management," *Water Resources Research,* 12(2): 125–134.

Metropolis, N., A. W. Rosenbluth, M. N. Rosenbluth, A. H. Teller, and E. Teller (1958). "Equations of State Calculations by Fast Computing Machines," *Journal of Chemical Physics* 21: 1087–1092.

Moore, C. V., H. H. Snyder, and P. Sun (1974). "Effects of Colorado River Water Quality and Supply on Irrigated Agriculture," *Water Resources Research,* 10(2): 137–144.

Morgan, D. R. and I. C. Goulter (1985). "Optimal Urban Water Distribution Design," *Water Resources Research,* 21(5): 642–652.

Morin, T. L., W. L. J. Meier, and K. S. Nagaraj (1981). "Optimal Mix of Adjustments to Floods," Purdue University Water Resources Research Center, Lafayette, Ind., Technical Report no. 139.

Morley, M.S., R. M. Atkinson, A. Savicm, and G. A. Walters (2001). "GAnet: Genetic Algorithm Platform for Pipe Network Optimisation," *Advances in Engineering Software*, 32 (6): 467–475.

Moy, W. S., J. L. Cohon, and C. S. ReVelle (1986). "Programming Model for Analysis of the Reliability, Resilience, and Vulnerability of a Water Supply Reservoir," *Water Resources Research,* 22(4): 489–498.

Murray, D. M. and S. J. Yakowitz (1979). "Constrained Differential Dynamic Programming and its Application to Multireservoir Control," *Water Resource Research,* 15(5): 1017–1027.

Murray, M. and S. J. Yakowitz (1981). "The Application of Optimal Control Methodology to Nonlinear Programming Problems," *Math. Programming,* vol. 21, pp. 331–347.

Murtaugh, B. A. and M. A. Saunders (1978). "Large-Scale Linearly Constrained Optimization," *Mathematical Programming,* 14: 41–72.

Musselman, K. J., and J. J. Talavage (1979). "Interactive Multiple Objective Optimization," Purdue University Water Resources Research Center, Lafayette, Ind., Technical Report no. 121.

Nemhauser, G. L. and L. A. Wolsey (1988). *Integer and Combinatorial Optimization,* Wiley, New York.

Neugebauer, K., W. Schilling, and J. Weiss (1991). "Network Algorithm for the Optimum Operation of Urban Drainage Systems," *Water Science and Technology,* 24(6): 209–216.

Nicklow, J. W. and M. K. Muleta (2001). "Watershed Management Technique to Control Sediment Yield in Agriculturally Dominated Areas," *Water International*, 26 (3): 435–443.

Nix, S. J. and J. P. Heaney (1988). "Optimization of Storm Water Storage-Release Strategies," *Water Resources Research,* 24(11): 1831–1838.

Nouh, M. (1987). "Storm Sewer Design Sensitivity Analysis Using ILSD-2 Model," *Journal of Water Resources Planning and Management,* 113(1): 151–158.

Nunes, L. M., E. Paralta, M. C. Cunha, and L. Ribeiro (2004). "Groundwater Nitrate Monitoring Network Optimization with Missing Data," *Water Resources Research*, 40 (2): 24.

Ohno, K. (1978). "A New Approach of Differential Dynamic Programming for Discrete Time Systems," *IEEE Trans. Automat. Control,* AC-23: 37–47.

Olufeagba, B. J. and R. H. Flake (1981). "Modelling and Control of Dissolved Oxygen in an Estuary," *Ecological Modelling,* 14(1): 79–94.

Onta, P. R., A. Das Gupta, and R. Harboe (1991). "Multistep Planning Model for Conjunctive Use of Surface-Water and Ground-Water Resources," *Journal of Water Resources Planning and Management,* 117(6): 662–678.

Ormsbee, L. and A. Kessler (1990). "Optimal Upgrading of Hydraulic-Network Reliability," *Journal of Water Resources Planning and Management,* 116(6): 784–802.

Ormsbee, L. E., T. M. Walski, D. V. Chase, and W. W. Sharp (1989). "Methodology for Improving Pump Operation Efficiency," *Journal of Water Resources Planning and Management,* 115(2): 148–164.

Ostfelt, A. and U. Shamir (1993a). "Optimal Operation of Multiquality Networks. II: Steady Condition," *Journal of Water Resources Planning and Management,* ASCE, 119(6): 645–662.

Ostfelt, A. and U. Shamir (1993b). "Optimal Operation of Multiquality Networks. II: Unsteady Condition," *Journal of Water Resources Planning and Management,* ASCE, 119(6): 663–684.

Parks, G. T. (1990). "An Intelligent Stochastic Optimization Routine for Nuclear Fuel Cycle Design," *Nuclear Technology*, 89, 233–246.

Paudyal, G. N. and A. Das Gupta (1990). "Irrigation Planning by Multilevel Optimization," *Journal of Irrigation and Drainage Engineering,* 116(2): 273–291.

Piccardi, C. and R. Soncini-Sessa (1991). "Stochastic Dynamic Programming for Reservoir Optimal Control: Dense Discretization and Inflow Correlation Assumption Made Possible by Parallel Computing," *Water Resources Research,* 27(5): 729–741.

Pincus, M. (1970). "A Monte Carlo Method for the Approximate Solution of Certain Types of Constrained Optimization Problems," *Operations Research,* 18: 1225–1228.

Piper, S., W. Y. Huang, and M. Ribaudo (1989). "Farm Income and Ground Water Quality Implications from Reducing Surface Water Sediment Deliveries," *Water Resources Bulletin,* 25(6): 1217–1230.

Pogge, E. C., Y. S. Yu, and G. T. Wang (1985). "Study Of Multireservoir Operation With Minimum Desirable Flow Constraints," Kansas Water Resources Research Institute, Manhattan, Kan., Contribution no. 251.

Ponnambalam, K., E. A. McBean, and T. E. Unny (1990). "Impacts of Meteorological Variations on Acid Rain Abatement Decisions," *Journal of Environmental Engineering,* 116(6): 1063–1075.

Quindry, G. E., E. D. Brill, and J. C. Liebman (1981). "Optimization of Looped Water Distribution Systems," *Journal of the Environmental Engineering Division,* 107(EE4): 665–679.

Randelman, R. E. and G. S. Grest (1986). "N-City Traveling Salesman Problem—Optimization by Simulated Annealings," *Journal of Statistical Physics*, 45: 885–890.

Rao, S.V.N., V. Sreenivasulu, S. M. Bhallamudi, B. S Thandaveswara, and K. P. Sudheer (2004)."Planning Groundwater Development in Coastal Aquifers," *Hydrological Sciences Journal*, 49 (1): 155–170.

Reed, P. M. and B. S. Minsker (2004). "Striking the Balance: Long-Term Groundwater Monitoring Design for Conflicting Objectives," *Journal of Water Resources Planning and Management*, ASCE, 130 (2): 140–149.

Reis, L. F. R., R. M. Porto, and F. H. Chaudhry (1997). "Optimal Location of Control Valves in Pipe Networks by Genetic Algorithm, *Journal of Water Resources Planning and Management*, ASCE, 123 (6): 317–326.

Reznicek, K. K. and S. P. Simonovic (1992). Issues in Hydropower Modeling Using GEMSLP Algorithm," *Journal of Water Resources Planning and Management,* ASCE, 118(1): 54–70.

Reznicek, K. K., S. P. Simonovic, and C. R. Bector (1991). "Optimization of Short-Term Operation of a Single Multipurpose Reservoir—A Goal Programming Approach," *Canadian Journal of Civil Engineering,* 18(3): 397–406.

Rhenals, A. E. and R. L. Bras (1981). "The Irrigation Scheduling Problem and Evapotranspiration Uncertainty," *Water Resources Research,* 17(5): 1328–1338.

Rockafellar, R. T. (1973a). "A Dual Approach to Solving Nonlinear Programming Problems by Unconstrained Optimization," *Mathematical Programming,* 5: 354–373.

Rockafellar, R. T. (1973b). "The Multiplier Method of Hestenes and Powell Applied to Convex Programming," SIAM, *Journal of Control and Optimization,* 12: 268–285.

Rockafellar, R. T. (1974). "Augmented Lagrangian Multiplier Functions and Duality in Nonconvex Programming," SIAM, *Journal of Applied Mathematics,* 12: 555–562.

Rosenthal, R. E. (1977). "Scheduling Reservoir Releases for Maximum Hydropower Benefit by Nonlinear Programming on a Network," University of Tennessee, College of Business Administration, Knoxville, Tenn., Working Paper no. 43.

Rosenthal, R. E., M. Hanscom, and R. S. Dembo (1983). "Nonlinear Programming Methods in Reservoir System Management," Tennessee University, Knoxville, Tenn., Research Report no. 93.

Rowell, W. F., and J. W. Barnes (1982). "Obtaining Layout of Water Distribution Systems," *Journal of the Hydraulics Division,* ASCE, 108(HY1): 137–148.

Rubin, Y. (1987). *Hierarchical Method for the Design of Water Allocation and Water Distribution Networks Based on Graph-Theory, Irrigation and Water Allocation,* International Association of Hydrological Sciences Press, Institute of Hydrology, Wallingford, Oxfordshire, UK, IAHS Publication no. 169, pp. 207–220.

Rydzewski, J. R. and HA-H. Rashid (1981). "Optimization of Water Resources for Irrigation in East Jordan," *Water Resources Bulletin,* 17(3): 367–372.

Sabet, H. and C. L. Creel (1991). "Network Flow Modeling of Oroville Complex," ASCE, *Journal of Water Resources Planning and Management,* 117(3): 301–320.

Salkin, H. M. (1975). *Integer Programming,* Addison-Wesley, Reading, Mass.

Sanchez, E. and J. Andreu (2001). "Optimization of a Water Resources System Expansion Using the Genetic Algorithm and Simulated Annealing Methods," *Ingenieria Hidraulica en Mexico*, 16 (2): 17–26.

Savic, D. A. and G. A. Walters (1997). "Genetic Algorithms for Least-Cost Design of Water Distribution Networks," *Journal of Water Resources Planning and Management*, ASCE,123 (2): 67–77.

Schafer, J. M. (1984). "Determining Optimum Pumping Rates for Creation of Hydraulic Barriers to Ground Water Pollutant Migration," *Proceedings of the Fourth National Symposium and Exposition on Aquifer Restoration and Ground Water Monitoring*, May 23–25, The Fawcett Center, Columbus, Ohio, pp. 50–60.

Shamir, U. (1979). "Optimization in Water Distribution Systems Engineering," *Mathematical Programming,*" 11: 65–75.

Shih, C. S. and W. L. J. Meier (1972). "Integrated Management of Quantity and Quality of Urban Water Resources," *Water Resources Bulletin,* 8(5): 1006–1017.

Skaggs, R. L., L. W. Mays, and L. W. Vail (2001a). "Simulated Annealing with Memory and Directional Search for Ground Water Remediation Design," *Journal of American Water Resources Association*, 37 (4): 853–866.

Skaggs, R. L., L. W. Mays, and L. W. Vail (2001b). "Application of Enhanced Annealing to Ground Water Remediation Design," *Journal of American Water Resources Association*, 37 (4): 867–875.

Spofford, W. O. J. (1976). "Integrated Residuals Management: A Regional Environmental Quality Management Model," Resources for the Future, Washington, D.C., Reprint 130.

Stedinger, J. R. and J. M. Bell-Graf (1979). "Systems Analysis," Literature Review, *Journal of the Water Pollution Control Federation,* 51(6): 1092–1098.

Swamee, P. K. and A. K. Sharma (1990). "Reorganization of Water-Distribution System," *Journal of Environmental Engineering,* 116(3): 588–600.

Taha, H. (1975). *Integer Programming: Theory, Applications, and Computations,* Academic, New York.

Taha, A. T. (1987). *Operations Research: An Introduction,* Macmillan, New York.

Tao, T. and W. C. Lennox (1991). "Reservoir Operations by Successive Linear Programming," *Journal of Water Resources Planning and Management,* 117(2): 274–280.

Tejada-Guibert, A., J. R. Stedinger, and K. Staschus (1990). "Optimization of Value of CVP's Hydropower Production," *Journal of Water Resources Planning and Management,* ASCE, 116(1): 52–70.

Tennessee Valley Authority, Water Resource Management Methods Staff (1977). "Testing of Different Optimization Methods on a Reservoir Subsystem," Knoxville, Tenn., Report B-16.

Texas Water Development Board (1970). "System Simulation for Management of a Total Water Resource," Report 118.

Tollow, A. J. (1989). "Operation of Water Supply Reservoirs by 'Control Bands' Derived by Simulation," *Hydrological Sciences Journal,* 34(4): 449–463.

Trezos, T. (1991). "Integer Programming Application for Planning of Hydropower Production," *Journal of Water Resources Planning and Management,* ASCE, 117(3): 340–351.

Tsai, F. T. C., N. Z. Sun, and W. W. G. Yeh (2003)."Global-Local Optimization for Parameter Structure Identification in Three-Dimensional Groundwater Modelling," *Water Resources Research,* 39 (2): 28.

Tsakiris, G. and E. Kiountouzis (1984). "Optimal Intraseasonal Irrigation Water Distribution," *Advances in Water Resources,* 7(2): 89–92.

Tung, Y. K., Y. Bao, L. Mays, and G. Ward (1990). "Optimization of Freshwater Inflow to Estuaries," *Journal of Water Resources Planning and Management,* 116(A) July/Aug.

Tyagi, N. K. (1988). "Managing Salinity through Conjunctive Use of Water Resources," *Ecological Modelling,* 40(1): 11–24.

Tyagi, N. K. and V. V. D. Narayana (1984). "Water Use Planning for Alkali Soils under Reclamation," *Journal of Irrigation and Drainage Engineering,* 110(2): 192–207.

Unver, O. I. and L. W. Mays (1990). "Model for Real-Time Optimal Flood Control Operation of a Reservoir System," *Water Resources Management,* Kluwer Academic, vol. 4, pp. 21–46.

Valocchi A. J. and J. W. Eheart (1987). "Incorporating Parameter Uncertainty into Groundwater Quality Management Models," in *Systems Analysis in Water Quality Management,* Pergamon, New York.

Vanderbilt, D., and S. G. Louie (1984). "A Monte Carlo Simulated Annealing Approach to Optimization over Continuous Variables," *Journal of Computer Physics*, 56:259– 271.

Vedula, S. and S. Mohan (1990). "Real-time Multipurpose Reservoir Operation: A Case Study," *Hydrological Sciences Journal,* 35(4): 447–462.

Vitkovsky, J. P., A. R. Simpson, and M. F. Lambert (2000). "Leak Detection and Calibration Using Transients and Genetic Algorithms, *Journal of Water Resources Planning and Management*, ASCE, 126 (4): 262–265.

Walker, D. J., D. T. Noble, and R. S. Magleby (1990). "Effective Cost-Share Rates and the Distribution of Social Costs in the Rock Creek, Idaho, Rural Clean Water Project," *Journal of Soil and Water Conservation,* 45(4): 477–479.

Wanakule, N., L. W. Mays, and L. S. Lasdon (1986). "Optimal Mangement of Large-Scale Aquifers: Methodology and Application," *Water Resources Research,* 22(4): 447–466.

Wang, G.-T. and Y.-S. Yu (1986). "Estimation of Parameters of the Discrete, Linear, Input-Output Model," *Journal of Hydrology,* 85(1/2): 15–30.

Wang, M. and C. Zheng (1998). "Ground Water Management Optimization Using Genetic Algorithms and Simulated Annealing: Formulation and Comparison," *Journal of American Water Resources Association*, 34 (3): 519–530.

Wasimi, S. A. and P. K. Kitanidis (1983). "Real-Time Forecasting and Daily Operation of a Multi-reservoir System During Floods by Linear Quadratic Gaussian Control," *Water Resources Research,* 19(6): 1511–1522.

Weisz, R. N. and J. C. Day (1974). "A Methodology for Planning Land Use and Engineering Alternatives for Flood Plain Management: The Flood Plain Management System Model," U.S. Army Engineer Institute for Water Resources, Fort Belvoir, Va., IWR Paper 74-p2.

Wenzel, H. G. J., B. C. Yen, and W. H. Tang (1979). "Advanced Methodology for Storm Sewer Design—Phase II," Water Resources Center, University of Illinois, Research Report no. 140.

Whiffen, G. J. and C. A. Shoemaker (1993). "Nonlinear Weighted Feedback Control of Groundwater Remediation Under Uncertainty," AGU, *Water Resources Research,* 29(9): 3277–3289.

Whitlatch, E. E. and P. L. Asplund (1981). "Capital Cost of Rural Water Distribution Systems," *Water Resources Bulletin,* 17(2): 310–314.

Whitlatch, E. E., N. R. Bhaskar, and M. J. Martin (1988). "Operating Policies for Reservoir Pumped-Water Interties," *Journal of Water Resources Planning and Management,* 114(6): 687–703.

Willis, R. and P. Liu (1984). "Optimization Model for Ground-Water Planning," *Journal of Water Resources Planning and Management,* 110(3): 333–347.

Winston, W. L. (1987). *Operations Research: Applications and Algorithms,* PWS, Boston.

Wu, M-Y. and W-S. Chu (1991). "System Analysis of Water Treatment Plant in Taiwan," *Journal of Water Resources Planning and Management,* ASCE, 117(5): 536–548.

Wu, Z. Y. and A. R. Simpson (2001). "Competent Genetic-Evolutionary Optimization of Water Distribution Systems," *Journal of Computing in Civil Engineering*, 15 (2): 89–101.

Wunderlich, W. O. (1991). "Chapter 11: System Planning and Operation," in *Hydropower Engineering Handbook,* J. S. Gulliver, and R. E. A. Arndt, eds., McGraw-Hill, New York.

Yakowitz, S. (1989). "Algorithms and Computational Techniques in Differential Dynamic Programming," *Control and Dynamic Systems,* Academic, New York, 31, pp. 75–91.

Yakowitz, S. and B. Rutherford (1984). "Computational Aspects of Discrete-Time Optimal Control," *Applied Mathematics and Computation,* 15: 29–45.

Yeh, C. H. and J. W. Labadie (1997). "Multiobjective Watershed-Level, Planning of Storm Water Detention Systems," *Journal of Water Resources Planning and Management*, ASCE, 123 (6): 336–343.

Yeh, W. W.-G. (1976). "Optimization of Real Time Daily Operation of a Multiple Reservoir System," Report no. UCLA-ENG-7628.

Yeh, W. W.-G. (1985). "Reservoir Management and Operations Models: A State-of-the-Art Review," *Water Resources Research,* 21(12): 1797–1818.

Yeh, W. W.-G., L. Becker, D. Toy, and A. L. Graves (1980). "Central Arizona Project: Operations Model," *Journal of the Water Resources Planning and Management Division,* ASCE, 106(WR2): 521–540.

Zavesky, R. R. and A. S. Goodman (1988). "Water-Surface Profiles Without Energy Loss Coefficients," *Water Resources Bulletin,* 24(6): 1281–1287.

Zessler, U. and U. Shamir (1989). "Optimal Operation of Water Distribution Systems," *Journal of Water Resources Planning and Management,* 115(6): 735–752.

Zhao B. and L. W. Mays (1995a). "Calibration and Control for Estuary Systems Under Uncertainty," submitted to *Water Resources Management,* Kluwer Academic, Hingham, Mass.

Zhao, B. and L. W. Mays (1995b)."Estuary Management by Stochastic Linear Quadratic Optimal Control," *Journal of Water Resources Planning and Management,* ASCE, 121(5): 382–391, September/October.

Zheng, C. M. and P. P. Wang (2002). "A Field Demonstration of the Simulation Optimization Approach for Remediation System Design," *Ground Water*, 40 (3): 258–265.

CHAPTER 2
UNCERTAINTY AND RELIABILITY ANALYSIS

Yeou-Koung Tung
Department of Civil Engineering
Hong Kong University of Science & Technology
Kowloon, Hong Kong

2.1 INTRODUCTION

2.1.1 Uncertainties in Water Resources Engineering

In the analysis and design of water resources engineering systems, uncertainties arise in various aspects including, but not limited to, hydraulic, hydrologic, structural, environmental, and socioeconomic aspects. Of particular interest to water resources engineers are those related to hydraulic and hydrologic uncertainties. Uncertainties attribute mainly to our lack of a perfect understanding with regard to the phenomena and processes involved, and a perfect knowledge of how to determine parameter values for processes that are fairly well understood. *Uncertainty* could simply be defined as the occurrence of events that are beyond our control (Mays and Tung, 1992). Yen and Ang (1971) classified uncertainties into two types: *objective uncertainties* associated with any random process or deducible from statistical samples and *subjective uncertainties* for which no quantitative factual information is available.

More specifically, in water resources engineering analyses and design, uncertainties can arise from several sources (Yen et al., 1986) such as *natural randomness, model uncertainties, parameter uncertainties, data uncertainties,* and *operational uncertainties.* Natural uncertainty is associated with the inherent randomness of natural processes such as the occurrence of precipitation and flood events. In water resources engineering, models ranging from simple empirical equations to sophisticated computer simulations are used. Model uncertainty reflects the inability of the model or design technique to accurately represent the system's true physical behavior. Consequently, using an imperfect model for prediction could result in error, compromising the performance of the engineering design. Parameter uncertainties result from the inability to accurately quantify the inputs to and parameters of a model. Parameter uncertainty can be caused by changes in operational conditions of hydraulic structures, inherent variability of inputs and parameters in time and space, and lack of sufficient data. Data uncertainties include: (1) measurement errors, (2) inconsistency and nonhomogeneity of data, (3) data handling and transcription errors, and (4) inadequate representativeness of the data sample due to time and space limitations. Operational uncertainties include those associated with construction, manufacture, deterioration, maintenance, and human factors.

In general, uncertainty due to inherent randomness of physical processes cannot be eliminated. Other uncertainties such as those associated with the lack of complete knowledge of the process, models, parameters, and data can be reduced through research, data collection, and careful manufacture.

Several expressions have been used to describe the degree of uncertainty of a parameter, a function, a model, or a system. The most complete and ideal description of uncertainty is the probability density function (defined in Sec. 2.2.3) of the quantity subject to uncertainty. However, in most practical problems such a probability function cannot be derived or found precisely.

Another measure of the uncertainty of a quantity is to express it in terms of a *confidence interval.* Use of confidence intervals has two drawbacks: (1) the uncertain quantities may not be normally distributed as assumed in the conventional procedures to determine the confidence interval, and

(2) no means is available to directly combine confidence intervals of individual contributing random components to produce the confidence interval of the system as a whole.

A useful alternative is to use the *statistical moments* associated with a quantity subject to uncertainty. In particular, the second-order moment (see Sec. 2.2.4) is a measure of the dispersion of a random variable.

The existence of various uncertainties (including inherent randomness of natural processes) is the main contributor to potential failure of hydraulic structures. Knowledge of statistical information describing the uncertainties is essential in reliability analysis. Therefore, uncertainty analysis is a prerequisite for reliability analysis.

The main objective of uncertainty analysis is to identify the statistical properties of a model output as a function of stochastic input parameters. Uncertainty analysis provides a formal and systematic framework to quantify the uncertainty associated with the model output. Furthermore, it offers the designer insight regarding the contribution of each stochastic input parameter to the overall uncertainty of the model output. Such knowledge is essential to identify the "important" parameters to which more attention should be given, in order to have a better assessment of their values, and accordingly, to reduce the overall uncertainty of the output.

2.1.2 Reliability of Water Resources Systems

All hydraulic structures placed in a natural environment are subject to various external stresses. The *resistance* or *strength* of a hydraulic structure is its ability to accomplish the intended mission satisfactorily, without failure, when subject to *load* of demands or external stresses. Failure occurs when the resistance of the system is exceeded by the load.

Reliability analysis can be applied to various types of engineering problems such as design of engineering facilities; evaluation of the safety of existing structures; inspection, repair, and other maintenance operations; routing, distribution and allocation, and other planning and management operations; data sampling and measurement network design; real-time prediction and long-term forecasting; and comparison and assessment of techniques and procedures.

The *reliability* of a hydraulic structure is the probability of safety p_s that the load does not exceed the resistance of the structure,

$$p_s = P\,[L \le R] \tag{2.1}$$

in which P[ ] denotes probability. Conversely, the failure probability p_f can be expressed as

$$p_f = P\,[L > R] = 1 - p_s \tag{2.2}$$

Failure of hydraulic structures can be classified as *structural failure* and *performance failure* (Yen and Ang, 1971; Yen et al., 1986). Structural failure involves damage or change of the structure or facility, hindering its ability to function as desired. On the other hand, performance failure does not necessarily involve structural damage, but the performance limit of the structure is exceeded and undesirable consequences occur. Generally, the two types of failure are related. Some hydraulic structures, such as dams and levees, are designed on the concept of structural failure, whereas other hydraulic structures, such as sewers and water-supply networks, are designed on the basis of performance failure.

A common practice for measuring the reliability of a water resources engineering structure is the *return period* or *recurrence interval.* The return period is defined as the long-term average (or expected) time between two successive failure-causing events. Flood frequency analysis using annual maximum flow series is a typical example of this kind. The main disadvantage of using return period is that reliability is measured only in terms of time of occurrence of loads, without considering their interactions with resistance (Melchers, 1987).

Two other types of reliability measures are frequently used in engineering practice to consider the relative magnitudes of resistance and anticipated load (called the *design load*). One measure is the *safety margin* (SM) defined as the difference between the resistance and the anticipated load; that is,

$$\mathrm{SM} = R - L \tag{2.3}$$

The other measure is the *safety factor* (SF) which is the ratio of resistance to load as

$$\mathrm{SF} = \frac{R}{L} \tag{2.4}$$

Several types of safety factors and their applications to hydraulic engineering design are discussed by Yen (1979).

2.1.3 Overview of Uncertainty and Reliability Analysis Methods

Uncertainty analysis techniques involve different levels of mathematical complexity and data requirements. The appropriate technique to be used depends on the nature of the problem at hand including the availability of information, model complexity, and type and accuracy of the results desired.

Several useful analytical approaches for uncertainty analysis including derived distribution techniques and integral transform techniques are described in Sec. 2.3. Although these analytical approaches are rather restrictive in practical applications, they are, however, powerful tools for deriving complete information about a stochastic process in some situations. Also, Sec. 2.3 describes several approximation techniques that are particularly useful for problems involving complex functions that cannot be dealt with analytically. They are primarily developed to estimate the statistical moments about the underlying random processes.

There are two basic probabilistic approaches to evaluate the reliability of a hydraulic structure or system. The most direct approach is a statistical analysis of data of past failure records for similar structures. The other approach is reliability analysis which considers and combines the contribution of each potentially influencing factor. The former is a lumped system approach requiring no knowledge about the behavior of the facility or structure and its load and resistance interaction. However, in many cases this direct approach is impractical because: (1) the sample size may be too small to be statistically reliable, especially for low-probability/high-consequence events, (2) the sample may not be representative of the structure or of the population, and (3) the physical conditions of a hydraulic structure or system may vary with respect to time; that is, the system is *nonstationary.*

There are two major steps in reliability analysis: (1) identify and analyze the uncertainties of each of the contributing factors, and (2) combine the uncertainties of the stochastic factors to determine the overall reliability of the structure. The second step, in turn, may also proceed in two different ways: (1) directly combining the uncertainties of all the factors, and (2) separately combining the uncertainties of the factors belonging to different components or subsystems to first evaluate the respective subsystem reliability and then combine the reliabilities of the different components or subsystems to yield the overall reliability of the structure.

2.2 REVIEW OF PERTINENT PROBABILITY AND STATISTICAL THEORIES

In probability theory, an *experiment* represents, in general, the process of observing random phenomena. A phenomenon is considered *random* if the outcome of an observation cannot be predicted with absolute precision. The totality of all possible outcomes of an experiment is called the *sample space.* An *event* is any subset of outcomes contained in the sample space. Therefore, an event may be an empty (or null) set, subset of the sample space, or the sample space itself. If two events contain no common elements in the sets, then they are *mutually exclusive* or *disjoint events.* If the occurrence of one event depends on the occurrence of another event, then these are called *conditional events.*

Probability is a numeric measure of the likelihood of the occurrence of an event. In general, the probability of the occurrence of an event can be assessed objectively through observations of the occurrence of the event or subjectively on the basis of experience and judgment.

2.2.1 Rules of Probability Computations

In probability computations, the three basic axioms of probability are: (1) $P(A) \geq 0$, (2) $P(S) = 1$ with S being the sample space, and (3) for A and B being mutually exclusive events, then $P(A \cup B) = P(A) + P(B)$. Relaxing the requirement of mutual exclusiveness in axiom (3), the probability of the union of N events can be written as

$$P\left(\bigcup_{i=1}^{N} A_i\right) = \sum_{i=1}^{N} P(A_i) - \sum_{i<j}\sum P(A_i, A_j) + \sum_{i<j<k}\sum\sum P(A_i, A_j, A_k) - \cdots + (-1)^N P(A_1, A_2, \ldots, A_N) \tag{2.5}$$

in which $P(A_i, A_j) = P(A_i \cap A_j)$. If all events involved are mutually exclusive, all but the first summation on the right-hand side of Eq. (2.5) vanish.

If two events are *statistically independent,* this implies that the occurrence of one event has no influence on the occurrence of the other event. The probability of joint occurrence of N independent events is

$$P\left(\bigcap_{i=1}^{N} A_i\right) = P(A_1)\cdot P(A_2)\cdots P(A_N) = \prod_{i=1}^{N} P(A_i) \tag{2.6}$$

The probability that a conditional event occurs is called the *conditional probability.* The conditional probability $P(A \mid B)$ can be computed as

$$P(A \mid B) = \frac{P(A,B)}{P(B)} \tag{2.7}$$

in which $P(A \mid B)$ is the probability of occurrence of event A given that event B occurred. The probability of joint occurrence of N events (not necessarily independent) can be evaluated as

$$P\left(\bigcap_{i=1}^{N} A_i\right) = P(A_1)\cdot P(A_2 \mid A_1)\cdot P(A_3 \mid A_2, A_1)\cdots P(A_N \mid A_{N-1}, A_{N-2}, \ldots, A_2, A_1) \tag{2.8}$$

Sometimes, the event E occurs along with several attribute events A_i that are mutually exclusive ($A_i \cap A_j = \varnothing$, for $i \neq j$) and *collectively exhaustive* ($A_1 \cup A_2 \cup \cdots \cup A_N) = S$. The probability of occurrence of event E, regardless of the attributes, can be computed as

$$P(E) = \sum_{i=1}^{N} P(E, A_i) = \sum_{i=1}^{N} P(E \mid A_i)P(A_i) \tag{2.9}$$

which is called the *total probability theorem.*

2.2.2 Random Variables and Their Distributions

In analyzing statistical characteristics of water resources system performance, many events of interest can be defined by the related random variables. A *random variable* is a real-valued function that is defined on the sample space. A random variable can be discrete or continuous.

The *cumulative distribution function* (*CDF*) or simply *distribution function* (*DF*) of a random variable X is defined as

$$F(x) = P(X \leq x) \tag{2.10}$$

The CDF $F(x)$ is a nondecreasing function of its argument. Furthermore, $\lim_{x \to -\infty} F(x) = 0$ and $\lim_{x \to +\infty} F(x) = 1$.

For a discrete random variable X, the *probability mass function* (*PMF*) of X is defined as

$$p(x) = P(X = x) \tag{2.11}$$

The PMF of any discrete random variable must satisfy two conditions: (1) $p(x_i) \geq 0$ for all x_i's, and (2) $\Sigma_{\text{all } i}\, p(x_i) = 1$.

For a continuous random variable, the *probability density function* (*PDF*) $f(x)$ is defined as

$$f(x) = \frac{dF(x)}{dx} \tag{2.12}$$

in which $F(x)$ is the CDF of X. Similar to the discrete case, any PDF of a continuous random variable must satisfy two conditions: (1) $f(x) \geq 0$, and (2) $\int f(x)\, dx = 1$.

When problems involve multiple random variables, the *joint distribution* and *conditional distribution* are used. For example, in the design and operation of a flood-control reservoir, one often has to consider the flood peak and flood volume simultaneously. In such cases, one would need to develop a *joint PDF* of flood peak and flood volume. For the purpose of illustration, the discussions are limited to problems involving two continuous random variables. For discrete random variables, one simply replaces the integrals by summations.

The joint PDF of two continuous random variables X and Y is denoted as $f_{X,Y}(x, y)$ and the corresponding joint CDF is

$$F_{X,Y}(x, y) = \int_{-\infty}^{x} \int_{-\infty}^{y} f_{X,Y}(x, y)\, dx\, dy \tag{2.13}$$

Two random variables X and Y are statistically independent if and only if $f_{X,Y}(x,y) = f_X(x) f_Y(y)$ and $F_{X,Y}(x, y) = F_X(x) F_Y(y)$. Consequently, a problem involving multiple independent random variables is, in effect, a univariate problem in which each individual random variable can be treated separately.

2.2.3 Statistical Properties of Random Variables

In probability and statistics, the term *population* describes the complete assemblage of all the values representative of a particular random process. A *sample* is any subset of the population. Further, *parameters* in a statistical model are quantities that are descriptive of a population. *Sample statistics* or simply *statistics* are quantities calculated on the basis of sample observations.

Descriptors that are commonly used to assess statistical properties of a random variable are those showing the central tendency, dispersion, and asymmetry of a distribution. These descriptors are related to the *statistical moments* of the random variable. Without losing generality, the following discussion considers continuous random variables. For discrete random variables, the integral sign is replaced by the summation sign and the PDF by the PMF.

The rth moments of a random variable X about any reference point $X = x_0$ is defined, for the continuous case, as

$$E[(X - x_0)^r] = \int_{-\infty}^{\infty} (x - x_0)^r f(x)\, dx \tag{2.14}$$

where $E[\ \ ]$ is a *statistical expectation operator.*

Two types of statistical moments are commonly used: *moments about the origin* where $x_0 = 0$ and the *central moments* where $x_0 = \mu$ with μ being the mean of the random variable defined in Eq.

(2.17). The rth central moment is denoted as $\mu_r = E[(X - \mu)^r]$ and the rth moment about the origin is denoted as $\mu'_r = E[X^r]$. Through the binomial expansion, the central moments μ_r can be obtained from the moments about the origin μ'_r as

$$\mu_r = \sum_{i=0}^{r} (-1)^i \, {}_rC_i \, \mu^i \, \mu'_{r-i} \tag{2.15}$$

where ${}_rC_i = r!/[i!\,(r-i)!]$. Conversely, the moments about the origin μ'_r can be obtained from the central moments μ_r as

$$\mu'_r = \sum_{i=0}^{r} {}_rC_i \, \mu^i \, \mu_{r-i} \tag{2.16}$$

The central tendency is commonly measured by the *expectation* which is defined as

$$E[X] = \mu = \int_{-\infty}^{\infty} x f(x)\, dx \tag{2.17}$$

This expectation is also known as the *mean* of a random variable. Two operational properties of the expectation are useful:

1. The expectation of the sum of several random variables equals the sum of the individual expectations,

$$E\left(\sum_{i=1}^{N} a_i X_i\right) = \sum_{i=1}^{N} a_i \mu_i \tag{2.18}$$

2. If $X_1, X_2, \ldots, X_N$ are statistically independent random variables, then

$$E\left(\prod_{i=1}^{N} X_i\right) = \prod_{i=1}^{N} \mu_i \tag{2.19}$$

Two other types of measures of central tendency are sometimes used in practice. They are the *median* and *mode*. The median of a random variable is the value that splits the area under a PDF into two equal halves. The mode is defined as the value of the random variable at which the PDF is maximum.

The variability of a random variable is measured by the *variance*, which is defined as

$$\text{Var}[X] = \sigma^2 = E[(X - \mu)^2] = \int_{-\infty}^{\infty} (x - \mu)^2 f(x)\, dx \tag{2.20}$$

The variance is the second-order central moment. The positive square root of variance is called the *standard deviation*, which is often used as the measure of the degree of uncertainty associated with a random variable. To compare the degree of uncertainty of two random variables of different units, a dimensionless measure $\Omega = \sigma/\mu$, called the *coefficient of variation*, is useful.

Two important operational properties of the variance are:

1. $\text{Var}[X] = E[X^2] - E^2[X]$ (2.21)
2. When $X_1, X_2, \ldots, X_N$ are independent random variables,

$$\text{Var}\left(\sum_{i=1}^{N} a_i X_i\right) = \sum_{i=1}^{N} a_i^2 \sigma_i^2 \tag{2.22}$$

where a_i is a constant and σ_i is the standard deviation of the random variable X_i.

The asymmetry of the PDF of a random variable is measured by the *skew coefficient* γ, defined as

$$\gamma = \frac{\mu_3}{\mu_2^{1.5}} = \frac{E[(X-\mu)^3]}{\sigma^3} \tag{2.23}$$

The skew coefficient is dimensionless and is related to the third central moment. The sign of the skew coefficient indicates the extent of symmetry of the probability distribution about its mean. If $\gamma = 0$, the distribution is symmetric about its mean; if $\gamma > 0$, the distribution has a long tail to the right; if $\gamma < 0$, the distribution has a long tail to the left. In practice, sample statistical moments higher than three are rarely used because their accuracies decrease rapidly when estimated from the small samples.

When a problem involves two dependent random variables, the degree of linear dependence between the two can be measured by the *correlation coefficient* $\rho(X, Y)$ which is defined as

$$\rho(X,Y) = \frac{\text{cov}(X,Y)}{\sigma_X \sigma_Y} \tag{2.24}$$

where cov(X, Y) is the covariance between random variables X and Y defined as

$$\text{cov}(X,Y) = E[(X-\mu_X)(Y-\mu_Y)] = E(XY) - \mu_X\mu_Y \tag{2.25}$$

It can be easily shown that $\text{cov}(Z_1, Z_2) = \rho(Z_1, Z_2)$ with Z_1 and Z_2 being the standardized random variables. Figure 2.1 graphically illustrates several cases of correlation coefficient. If the two random

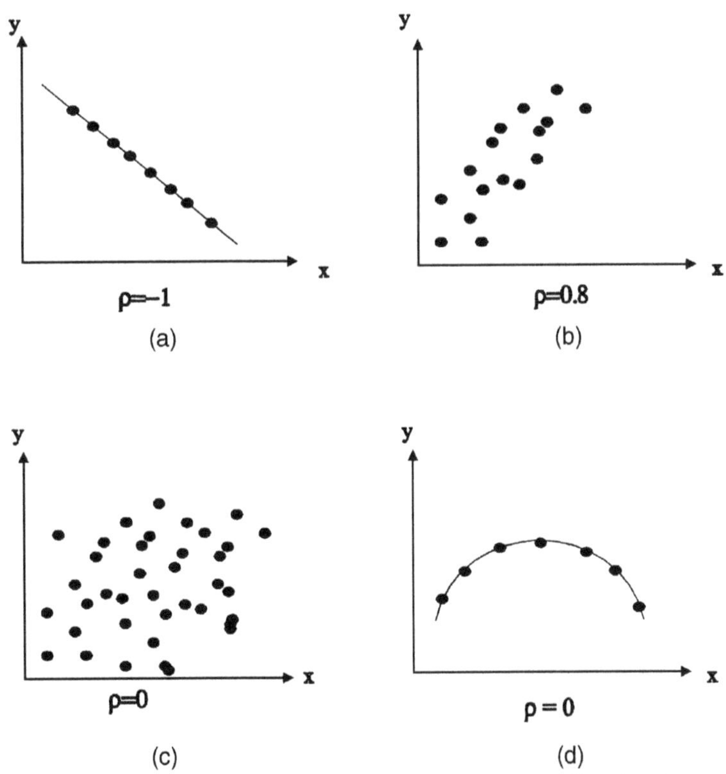

FIGURE 2.1 Different cases of correlation between two random variables: (a) perfectly linearly correlated in opposite direction; (b) strongly linearly corrected in positive direction; (c) uncorrelated in linear fashion; (d) perfectly correlated in nonlinear fashion, but uncorrelated linearly.

variables X and Y are independent, then $\rho(X,Y) = \text{cov}(X,Y) = 0$. However, the reverse is not necessarily true because $\rho(X,Y)$ does not reflect nonlinear dependence between X and Y. If the random variables involved are not independent, Eq. (2.22) can be generalized as

$$\text{Var}\left(\sum_{i=1}^{N} a_i X_i\right) = \sum_{i=1}^{N} a_i^2\sigma_i^2 + 2\sum_{i=1}^{N-1}\sum_{j=i+1}^{N} a_i a_j \,\text{cov}[X_i, X_j] \tag{2.26}$$

2.2.4 Univariate Probability Distributions

This section summarizes two discrete distributions and several commonly used univariate continuous distributions.

Binomial Distribution. The *binomial distribution* is a discrete distribution applicable to a random process with only two possible outcomes. In water resources systems, the state of components or subsystems can be classified as either functioning or failed. Consider a system involving a total of n independent components with each component having two possible states, functioning or failed. If the probability of each component being failed is p, then the probability of having x number of failed components in the system can be computed as

$$p(x) = {}_nC_x p^x q^{n-x} \qquad x = 0, 1, 2, \ldots, n \tag{2.27}$$

where $q = 1 - p$, the probability of being in the functioning state for each component. A random variable X having a binomial distribution with parameters n and p has the expectation $E(X) = np$ and variance $\text{Var}(X) = npq$.

Poisson Distribution. When the number of system components $n \to \infty$ and $p \to 0$ while $np = \nu =$ constant, the binomial distribution becomes a *Poisson distribution* with the PMF defined as

$$p(x) = e^{-\nu}\nu^x/x! \qquad x = 0, 1, 2, \ldots \tag{2.28}$$

where the parameter $\nu > 0$ represents the mean of a discrete random variable X having a Poisson distribution. For a Poisson random variable, the mean and the variance are identical to ν.

The Poisson distribution has been applied widely in modeling the number of occurrences of a random event within a specified time or space interval. Equation (2.28) can be modified as

$$p(x) = e^{-\lambda t}(\lambda t)^x/x! \qquad x = 0, 1, 2, \ldots \tag{2.29}$$

in which the parameter λ is the average rate of occurrence of the random event in an interval $(0, t)$.

Normal Distribution. The *normal distribution* is also called the *Gaussian distribution.* A normal random variable having the mean μ and variance σ^2 has the PDF

$$f(x \mid \mu, \sigma^2) = \frac{1}{\sqrt{2\pi}\,\sigma}\exp\left[-\frac{1}{2}\left(\frac{x-\mu}{\sigma}\right)^2\right] \qquad \text{for} \qquad -\infty < x < \infty \tag{2.30}$$

The normal distribution is bell-shaped and symmetric with respect to $x = \mu$. Therefore, the skew coefficient of a normal random variable is zero. A random variable which is the sum of independent normal random variables is also a normal random variable with mean and variance that can be computed by Eqs. (2.18) and (2.22), respectively.

The normal distribution sometimes provides a viable alternative to approximate the probability of a random variable that is not normally distributed. An important theorem relating to the sum of independent random variables is the *central limit theorem,* which can be loosely stated as: The distribution of the sum of a number of independent random variables, regardless of their individual distribution, can be approximated by a normal distribution as long as none of the variables has a dominant effect on the sum. The larger the number of random variables involved in the summation is, the better the approximation.

Probability computations for normal random variables are made by first transforming to a *standardized variable* Z as

$$Z = (X - \mu)/\sigma \tag{2.31}$$

in which Z has a mean of zero and a variance of one. The PDF of Z, called the *standard normal distribution,* is

$$\phi(z) = \frac{1}{\sqrt{2\pi}} e^{-z^2/2} \qquad -\infty < z < \infty \tag{2.32}$$

Computations of probability for a normal random variable can be made as

$$P(X \le x) = P[(X - \mu)/\sigma \le (x - \mu)/\sigma] = P[Z \le z] = \Phi(z) \tag{2.33}$$

where $\Phi(z)$ is the CDF of the standard normal random variable Z. A table of standard normal probability is shown in Table 2.1. Several algebraic formulas are available for fairly accurate approximation of the normal probability (Abramowitz and Stegun, 1972; Stedinger et al., 1993).

TABLE 2.1 Cumulative Standard Normal Probability, $\Phi(z) = P(Z \le z)$.

z	0.00	0.01	0.02	0.03	0.04	0.05	0.06	0.07	0.08	0.09
−3.4	0.0003	0.0003	0.0003	0.0003	0.0003	0.0003	0.0003	0.0003	0.0003	0.0002
−3.3	0.0005	0.0005	0.0005	0.0004	0.0004	0.0004	0.0004	0.0004	0.0004	0.0003
−3.2	0.0007	0.0007	0.0006	0.0006	0.0006	0.0006	0.0006	0.0005	0.0005	0.0005
−3.1	0.0010	0.0009	0.0009	0.0009	0.0008	0.0008	0.0008	0.0008	0.0007	0.0007
−3.0	0.0013	0.0013	0.0013	0.0012	0.0012	0.0011	0.0011	0.0011	0.0010	0.0010
−2.9	0.0019	0.0018	0.0017	0.0017	0.0016	0.0016	0.0015	0.0015	0.0014	0.0014
−2.8	0.0026	0.0025	0.0024	0.0023	0.0023	0.0022	0.0021	0.0021	0.0020	0.0019
−2.7	0.0035	0.0034	0.0033	0.0032	0.0031	0.0030	0.0029	0.0028	0.0027	0.0026
−2.6	0.0047	0.0045	0.0044	0.0043	0.0041	0.0040	0.0039	0.0038	0.0037	0.0036
−2.5	0.0062	0.0060	0.0059	0.0057	0.0055	0.0054	0.0052	0.0051	0.0049	0.0048
−2.4	0.0082	0.0080	0.0078	0.0075	0.0073	0.0071	0.0069	0.0068	0.0066	0.0064
−2.3	0.0107	0.0104	0.0102	0.0099	0.0096	0.0094	0.0091	0.0089	0.0087	0.0084
−2.2	0.0139	0.0136	0.0132	0.0129	0.0125	0.0122	0.0119	0.0116	0.0113	0.0110
−2.1	0.0179	0.0174	0.0170	0.0166	0.0162	0.0158	0.0154	0.0150	0.0146	0.0143
−2.0	0.0228	0.0222	0.0217	0.0212	0.0207	0.0202	0.0197	0.0192	0.0188	0.0183
−1.9	0.0287	0.0281	0.0274	0.0268	0.0262	0.0256	0.0250	0.0244	0.0239	0.0233
−1.8	0.0359	0.0352	0.0344	0.0336	0.0329	0.0322	0.0314	0.0307	0.0301	0.0294
−1.7	0.0446	0.0436	0.0427	0.0418	0.0409	0.0401	0.0392	0.0384	0.0375	0.0367
−1.6	0.0548	0.0537	0.0526	0.0516	0.0505	0.0495	0.0485	0.0475	0.0465	0.0455
−1.5	0.0668	0.0655	0.0643	0.0630	0.0618	0.0606	0.0594	0.0582	0.0571	0.0559
−1.4	0.0808	0.0793	0.0778	0.0764	0.0749	0.0735	0.0722	0.0708	0.0694	0.0681
−1.3	0.0968	0.0951	0.0934	0.0918	0.0901	0.0885	0.0869	0.0853	0.0838	0.0823
−1.2	0.1151	0.1131	0.1112	0.1093	0.1075	0.1056	0.1038	0.1020	0.1003	0.0985
−1.1	0.1357	0.1335	0.1314	0.1292	0.1271	0.1251	0.1230	0.1210	0.1190	0.1170
−1.0	0.1587	0.1562	0.1539	0.1515	0.1492	0.1469	0.1446	0.1423	0.1401	0.1379
−0.9	0.1841	0.1814	0.1788	0.1762	0.1736	0.1711	0.1685	0.1660	0.1635	0.1611
−0.8	0.2119	0.2090	0.2061	0.2033	0.2005	0.1977	0.1949	0.1922	0.1894	0.1867
−0.7	0.2420	0.2389	0.2358	0.2327	0.2296	0.2266	0.2236	0.2206	0.2177	0.2148
−0.6	0.2743	0.2709	0.2676	0.2643	0.2611	0.2578	0.2546	0.2514	0.2483	0.2451
−0.5	0.3085	0.3050	0.3015	0.2981	0.2946	0.2912	0.2877	0.2843	0.2810	0.2776

TABLE 2.1 Cumulative Standard Normal Probability, $\Phi(z) = P(Z \le z)$ (*Continued*)

z	0.00	0.01	0.02	0.03	0.04	0.05	0.06	0.07	0.08	0.09
−0.4	0.3446	0.3409	0.3372	0.3336	0.3300	0.3264	0.3228	0.3192	0.3156	0.3121
−0.3	0.3821	0.3783	0.3745	0.3707	0.3669	0.3632	0.3594	0.3557	0.3520	0.3483
−0.2	0.4207	0.4168	0.4129	0.4090	0.4052	0.4013	0.3974	0.3936	0.3897	0.3859
−0.1	0.4602	0.4562	0.4522	0.4483	0.4443	0.4404	0.4364	0.4325	0.4286	0.4247
−0.0	0.5000	0.4960	0.4920	0.4880	0.4840	0.4801	0.4761	0.4721	0.4681	0.4641
0.0	0.5000	0.5040	0.5080	0.5120	0.5160	0.5199	0.5239	0.5279	0.5319	0.5359
0.1	0.5398	0.5438	0.5478	0.5517	0.5557	0.5596	0.5636	0.5675	0.5714	0.5753
0.2	0.5793	0.5832	0.5871	0.5910	0.5948	0.5987	0.6026	0.6064	0.6103	0.6141
0.3	0.6179	0.6217	0.6255	0.6293	0.6331	0.6368	0.6406	0.6443	0.6480	0.6517
0.4	0.6554	0.6591	0.6628	0.6664	0.6700	0.6736	0.6772	0.6808	0.6844	0.6879
0.5	0.6915	0.6950	0.6985	0.7019	0.7054	0.7088	0.7123	0.7157	0.7190	0.7224
0.6	0.7257	0.7291	0.7324	0.7357	0.7389	0.7422	0.7454	0.7486	0.7517	0.7549
0.7	0.7580	0.7611	0.7642	0.7673	0.7704	0.7734	0.7764	0.7794	0.7823	0.7852
0.8	0.7881	0.7910	0.7939	0.7967	0.7995	0.8023	0.8051	0.8078	0.8106	0.8133
0.9	0.8159	0.8186	0.8212	0.8238	0.8264	0.8289	0.8315	0.8340	0.8365	0.8389
1.0	0.8413	0.8438	0.8461	0.8485	0.8508	0.8531	0.8554	0.8577	0.8599	0.8621
1.1	0.8643	0.8665	0.8686	0.8708	0.8729	0.8749	0.8770	0.8790	0.8810	0.8830
1.2	0.8849	0.8869	0.8888	0.8907	0.8925	0.8944	0.8962	0.8980	0.8997	0.9015
1.3	0.9032	0.9049	0.9066	0.9082	0.9099	0.9115	0.9131	0.9147	0.9162	0.9177
1.4	0.9192	0.9207	0.9222	0.9236	0.9251	0.9265	0.9278	0.9292	0.9306	0.9319
1.5	0.9332	0.9345	0.9357	0.9370	0.9382	0.9394	0.9406	0.9418	0.9429	0.9441
1.6	0.9452	0.9463	0.9474	0.9484	0.9495	0.9505	0.9515	0.9525	0.9535	0.9545
1.7	0.9554	0.9564	0.9573	0.9582	0.9591	0.9599	0.9608	0.9616	0.9625	0.9633
1.8	0.9641	0.9649	0.9656	0.9664	0.9671	0.9678	0.9686	0.9693	0.9699	0.9706
1.9	0.9713	0.9719	0.9726	0.9732	0.9738	0.9744	0.9750	0.9756	0.9761	0.9767
2.0	0.9772	0.9778	0.9783	0.9788	0.9793	0.9798	0.9803	0.9808	0.9812	0.9817
2.1	0.9821	0.9826	0.9830	0.9834	0.9838	0.9842	0.9846	0.9850	0.9854	0.9857
2.2	0.9861	0.9864	0.9868	0.9871	0.9875	0.9878	0.9881	0.9884	0.9887	0.9890
2.3	0.9893	0.9896	0.9898	0.9901	0.9904	0.9906	0.9909	0.9911	0.9913	0.9916
2.4	0.9918	0.9920	0.9922	0.9925	0.9927	0.9929	0.9931	0.9932	0.9934	0.9936
2.5	0.9938	0.9940	0.9941	0.9943	0.9945	0.9946	0.9948	0.9949	0.9951	0.9952
2.6	0.9953	0.9955	0.9956	0.9957	0.9959	0.9960	0.9961	0.9962	0.9963	0.9964
2.7	0.9965	0.9966	0.9967	0.9968	0.9969	0.9970	0.9971	0.9972	0.9973	0.9974
2.8	0.9974	0.9975	0.9976	0.9977	0.9977	0.9978	0.9979	0.9979	0.9980	0.9981
2.9	0.9981	0.9982	0.9982	0.9983	0.9984	0.9984	0.9985	0.9985	0.9986	0.9986
3.0	0.9987	0.9987	0.9987	0.9988	0.9988	0.9989	0.9989	0.9989	0.9990	0.9990
3.1	0.9990	0.9991	0.9991	0.9991	0.9992	0.9992	0.9992	0.9992	0.9993	0.9993
3.2	0.9993	0.9993	0.9994	0.9994	0.9994	0.9994	0.9994	0.9995	0.9995	0.9995
3.3	0.9995	0.9995	0.9995	0.9996	0.9996	0.9996	0.9996	0.9996	0.9996	0.9997
3.4	0.9997	0.9997	0.9997	0.9997	0.9997	0.9997	0.9997	0.9997	0.9997	0.9998

Source: Devore, 1987.

Lognormal Distribution. The *lognormal distribution* is a commonly used continuous distribution when random variables cannot be negative-valued. A random variable X is lognormally distributed if its logarithmic transform $Y = \ln(X)$ has a normal distribution with mean $\mu_{\ln X}$ and variance $\sigma^2_{\ln X}$. The PDF of a lognormal random variable is

$$f(x \mid \mu_{\ln X}, \sigma^2_{\ln X}) = \frac{1}{\sqrt{2\pi}\,\sigma_{\ln X}\,x} \exp\left[-\frac{1}{2}\left(\frac{\ln(x) - \mu_{\ln X}}{\sigma_{\ln X}}\right)^2\right] \qquad x > 0 \tag{2.34}$$

Statistical properties of a lognormal random variable on the original scale can be computed from those of the log-transformed variable. To compute the statistical moments of X from those of ln (X), the following formulas are useful.

$$\mu_X = \exp\left(\mu_{\ln X} + \frac{\sigma^2_{\ln X}}{2}\right) \tag{2.35}$$

$$\sigma^2_X = \mu^2_X\,[e^{\sigma^2_{\ln X}} - 1] \tag{2.36}$$

$$\gamma_X = \Omega^3_X + 3\Omega_X \tag{2.37}$$

Equation (2.37) indicates that lognormal distributions are always positively skewed. Conversely, the statistical moments of ln X can be computed from those of X by

$$\mu_{\ln X} = \frac{1}{2}\ln\left[\frac{\mu^2_X}{1+\Omega^2_X}\right] = \ln(\mu_X) - \frac{1}{2}\sigma^2_{\ln X} \tag{2.38}$$

$$\sigma^2_{\ln X} = \ln(1 + \Omega^2_X) \tag{2.39}$$

Because the sum of normal random variables is normally distributed, the product of independent lognormal random variables is also lognormally distributed. In cases that two lognormal random variables are correlated with a correlation coefficient $\rho_{X,Y}$ in the original space, then the covariance terms in the log-transformed space must be included in determining the variance. From $\rho_{X,Y}$, the correlation coefficient in the log-transformed space can be computed as

$$\rho_{\ln X,\,\ln Y} = \frac{\ln(1 + \rho_{X,Y}\,\Omega_X\,\Omega_Y)}{\sqrt{\ln(1 + \Omega^2_X)\,\ln(1 + \Omega^2_Y)}} \tag{2.40}$$

Gamma Distribution and Its Variations. The *gamma distribution* is another rather versatile continuous distribution for a positive-valued random variable. The two-parameter gamma distribution has the PDF defined as

$$f(x|\alpha, \beta) = \frac{\beta}{\Gamma(\alpha)}\,(\beta x)^{\alpha - 1} e^{-\beta x} \qquad \text{for} \qquad x > 0 \tag{2.41}$$

in which $\alpha > 0$ and $\beta > 0$ are parameters and $\Gamma(\ )$ is a gamma function. When $\alpha = 1$, the gamma distribution reduces to the *exponential distribution* with the PDF defined as

$$f(x|\beta) = \beta\exp(-\beta x) \qquad \text{for} \qquad x > 0 \tag{2.42}$$

The exponential distribution is widely used for describing the life span of various electronic and mechanical components. It plays an important role in reliability mathematics using time-to-failure analysis (see Sec. 2.6).

Two variations of the gamma distribution are frequently used in hydrologic frequency analysis, namely, *Pearson* and *log-Pearson type 3 distributions.* The log-Pearson type 3 distribution is recommended for use by the U.S. Interagency Advisory Committee on Water Data (1982) as the standard distribution for flood frequency analysis. The Pearson type 3 distribution is a three-parameter gamma distribution with the third parameter being the lower bound of the random variable whereas the log-Pearson type 3 distribution is simply a three-parameter log-gamma distribution. Stedinger et al. (1993) provide a good summary of these two distributions.

Extremal Distributions. Water resources engineering reliability analysis often focuses on the statistical characteristics of extreme events. The work on statistics of extremes was pioneered by Fisher and Tippett (1928) and expanded by Gumbel (1958). Three types of asymptotic distributions of extremes have been derived based on the different characteristics of the underlying distribution (Haan, 1977):

Type I—Parent distributions are unbounded in the direction of extremes and all statistical moments exist.

Type II—Parent distributions are unbounded in the direction of extremes but all moments do not exist. Thus type II is of little use in practical engineering applications.

Type III—Parent distributions are bounded in the direction of the desired extreme.

Type I extremal distribution is sometimes referred to as a *Gumbel distribution,* a *Fisher-Tippett distribution,* or a *double exponential distribution.* The CDF and PDF of type I extremal distributions have, respectively, the following forms

$$F(x|x_o, \beta) = \exp\left[-\exp \mp \left(\frac{x - x_o}{\beta}\right)\right] \qquad -\infty < x, x_o < \infty \tag{2.43a}$$

$$f(x|x_o, \beta) = \frac{1}{\beta}\exp\left[\mp\left(\frac{x - x_o}{\beta}\right) - \exp \mp \left(\frac{x - x_o}{\beta}\right)\right] \qquad -\infty < x, x_o < \infty \tag{2.43b}$$

in which $\beta \geq 0$; $-$ applies to the largest extreme whereas $+$ is for the smallest extreme.

Type III extremal distribution is widely used for modeling the smallest extreme. The resulting distribution is also known as the *Weibull distribution* having a PDF defined as

$$f(x|x_o, \alpha,\beta) = \frac{\alpha}{\beta}\left(\frac{x - x_o}{\beta}\right)^{\alpha - 1} \exp -\left(\frac{x - x_o}{\beta}\right)^{\alpha}, x \geq x_o, \alpha, \beta > 0 \tag{2.44}$$

The Weibull distribution is very versatile and can take many shapes. The CDF of the Weibull distribution can be derived as

$$F(x|x_o, \alpha, \beta) = 1 - \exp -\left(\frac{x - x_o}{\beta}\right)^{\alpha} \tag{2.45}$$

Recently, the *generalized extreme value distribution* (*GEV*) has been used in hydrologic frequency analysis. The GEV distribution provides an expression that encompasses all three types of extreme value distributions. Detailed descriptions of the GEV distribution are provided by Stedinger et al. (1993).

2.2.5 Multivariate Probability Distributions

Multivariate probability distributions are extensions of univariate probability distributions which jointly account for more than one random variable. *Bivariate* and *trivariate* distributions are special cases when two and three random variables, respectively, are involved. The fundamental bases of multivariate probability distributions are described in Sec. 2.2.3. There are several ways to construct a multivariate distribution, and detailed descriptions can be found in Johnson and Kotz (1976). In this section, two commonly used multivariate distributions, namely, multivariate normal and multivariate lognormal, are described.

Multivariate Normal Distribution. The *multivariate normal* PDF for N correlated normal random variables is defined as

$$f(\mathbf{x}) = \frac{|\mathbf{C}^{-1}|^{1/2}}{(2\pi)^{N/2}} \exp\left[-\frac{1}{2}(\mathbf{x} - \boldsymbol{\mu})^T\mathbf{C}^{-1}(\mathbf{x} - \boldsymbol{\mu})\right] \qquad -\infty < \mathbf{x} < \infty \tag{2.46}$$

in which $\boldsymbol{\mu} = (\mu_1, \mu_2, \ldots, \mu_N)$, an $N \times 1$ vector of means $\mathbf{C} = E[(\mathbf{X} - \boldsymbol{\mu})(\mathbf{X} - \boldsymbol{\mu})^T]$ is an $N \times N$ covariance matrix and T represents the transpose. In terms of the standard normal random variables $Z_i = (X_i - \mu_i)/\sigma_i$ the *standardized multivariate normal PDF* can be expressed as

$$\phi(\mathbf{z}|\mathbf{R}) = \frac{|\mathbf{R}^{-1}|^{1/2}}{(2\pi)^{N/2}} \exp\left(-\frac{1}{2}\mathbf{z}^T\mathbf{R}^{-1}\mathbf{z}\right) \qquad -\infty < \mathbf{z} < \infty \tag{2.47}$$

in which $\mathbf{R} = E[\mathbf{Z}\mathbf{Z}^T]$ is a $N \times N$ correlation matrix for the multivariate normal variables.

Evaluation of the probability of multivariate normal random variables involves multidimensional integration as

$$\Phi(\mathbf{z}|\mathbf{R}) = P(Z_1 \leq z_1, Z_2 \leq z_2, \ldots, Z_N \leq z_N|\mathbf{R})$$

$$= \int_{-\infty}^{z_1} \int_{-\infty}^{z_2} \cdots \int_{-\infty}^{z_N} \phi(\mathbf{z}|\mathbf{R})\, dz \tag{2.48}$$

Accurate evaluation for $\Phi(\mathbf{z}|\mathbf{R})$ is generally difficult, and is often resolved by approximations. Several practical methods have been proposed to compute the bounds of multivariate normal probability based on univariate or bivariate normal probabilities (Rackwitz, 1978; Ditlevsen, 1979, 1982, 1984; Bennett and Ang, 1983).

Multivariate Lognormal Distribution. The *bivariate lognormal* PDF is

$$f(x_1, x_2) = \frac{1}{2\pi x_1 x_2\, \sigma_1'\, \sigma_2' \sqrt{1-\rho'^2}}\, e^{-Q'/[2(1-\rho'^2)]} \qquad \text{for} \qquad x_1, x_2 > 0 \tag{2.49}$$

in which

$$Q' = \frac{[\ln(x_1) - \mu_1']^2}{\sigma_1'^2} + \frac{[\ln(x_2) - \mu_2']^2}{\sigma_1'^2} - 2\rho' \frac{[\ln(x_1) - \mu_1'][\ln(x_2) - \mu_2']}{\sigma_1'\sigma_2'}$$

where μ' and σ' are the mean and standard deviation of log-transformed random variables, subscripts 1 and 2 indicate the random variables X_1 and X_2, respectively, and ρ' is the correlation coefficient of the two log-transformed random variables. After log-transformation is made, properties of multivariate lognormal random variables follow exactly as the multivariate normal case.

2.3 METHODS FOR UNCERTAINTY ANALYSIS

From the previous discussions of uncertainty in water resources systems modeling and analysis, it is not difficult to find that many quantities of interest are functionally related to several variables that are subject to varying levels of uncertainty. The methods applicable for uncertainty analysis are dictated by the information available on the stochastic variables involved and the functional relationships among the variables. In principle, it would be ideal to derive the exact probability distribution of the model output as a function of those of the stochastic variables. In this section, some analytical and approximate methods for uncertainty analysis are described.

2.3.1 Analytical Methods

Derived Distribution Method. The *derived distribution method* determines the distribution of a random variable W which is related to another random variable X, whose PDF and/or CDF are known, through the functional relationship $W = g(X)$. The CDF of W can be obtained as

$$H_W(w) = P[W \leq w] = F_X[g^{-1}(w)] \tag{2.50}$$

where $g^{-1}(w)$ represents the inverse function of g. The PDF of W, $h_W(w)$, then, can be obtained according to Eq. (2.12) by taking the derivative of $H_W(w)$ with respect to w.

In the case that the functional relation between X and W is either strictly increasing or strictly decreasing, the PDF of W can be derived directly from the PDF of X as

$$h_W(w) = f_X(x)\ |dx/dw| \tag{2.51}$$

in which $|dx/dw|$ is called the *Jacobian.* This derived distribution method is also called the *transformation of variables technique.*

Consider a multivariate case in which N random variables $W_1, W_2, \ldots, W_N$ are related to N other random variables $X_1, X_2, \ldots, X_N$ through a system of N equations as

$$W_i = g_i\,(X_1, X_2, \ldots, X_N) \qquad \text{for} \qquad i = 1, 2, \ldots, N$$

When the functions $g_1(.), g_2(.), \ldots, g_N(.)$ satisfy a monotonic relationship, the joint PDF of random variables W can be directly obtained from the following equation

$$h(w_1, w_2, \ldots, w_N) = f(x_1, x_2, \ldots, x_N)\ |\mathbf{J}| \tag{2.52}$$

where $f(x_1, x_2, \ldots, x_N)$ and $h(w_1, w_2, \ldots, w_N)$ are the joint PDFs of Xs and Ws, respectively; and $|\mathbf{J}|$ is the absolute value of the determinant of the $N \times N$ *Jacobian matrix*

$$|\mathbf{J}| = \begin{bmatrix} \dfrac{\partial x_1}{\partial w_1} & \dfrac{\partial x_1}{\partial w_2} & \cdots & \dfrac{\partial x_1}{\partial w_N} \\ \vdots & \vdots & \ddots & \vdots \\ \dfrac{\partial x_N}{\partial w_1} & \dfrac{\partial x_N}{\partial w_2} & \cdots & \dfrac{\partial x_N}{\partial w_N} \end{bmatrix} \tag{2.53}$$

In theory, the above concept of deriving the PDF of a random variable as a function of the PDFs of other random variables is simple and straightforward. However, the success of implementing such procedures largely depends on the functional relation, the form of the PDFs involved, and the analyst's mathematical skill. It is in common situations in which analytical derivations are virtually impossible.

Integral Transform Techniques. *Integral transformation techniques* originally found their applications in univariate statistical analysis. The well-known integral transforms are *Fourier transform, Laplace transform,* and *exponential transform.* One useful transform technique, but less known to water resources engineering, is the *Mellin transform* (Epstein, 1948; Park, 1987).

The Fourier transform of a PDF leads to the *characteristic function* and the Laplace and exponential transforms yield the *moment generating function.* The Laplace transform is the same as the exponential transform except that the Laplace transform is only applicable to nonnegative random variables. Once the characteristic function or moment generating function is derived, statistical moments of a random variable can be computed. One major advantage of integral transforms is that, if such transforms of a PDF exist, the relationship between the PDF and its integral transform is unique (Kendall et al., 1987).

In dealing with a multivariate problem in which a random variable is a function of several random variables, the convolution property of these integral transforms becomes analytically powerful, especially when the stochastic variables in the model are independent. Fourier and exponential transforms are powerful in treating the sum and difference of random variables, whereas the Mellin transform is attractive to the quotient and product of random variables (Griffin, 1975; Springier, 1979).

The Fourier transform of a PDF, $f(x)$, is defined as

$$\mathscr{F}_x(s) = \int_{-\infty}^{\infty} e^{isx} f(x)\, dx = E\,[e^{isx}] \tag{2.54}$$

where s is the parameter in the transformed space and $i = \sqrt{-1}$. The resulting Fourier transform $\mathscr{F}(s)$ of a PDF is called the characteristic function. The characteristic function of a random variable is unique and always exists for all values of the argument s. Therefore, given a characteristic function of a random variable, its probability density function can be uniquely determined through the inverse transform. The characteristic functions of some commonly used PDFs are shown in Table 2.2. Using the characteristic function, the rth-order moment about the origin of the random variable X can be obtained as

$$E[X^r] = \mu_r' = \frac{1}{i^r}\left[\frac{d^r\,\mathscr{F}_x(s)}{d\,s^r}\right]_{s=0} \tag{2.55}$$

The Fourier transform is particularly useful when random variables are related linearly. In such cases, the convolution property of the Fourier transform can be applied to derive the characteristic function of the resulting random variable. More specifically, consider that $W = X_1 + X_2 + \cdots + X_N$ and all X's are independent random variables with known PDF, $f_i(x_i)$, $i = 1, 2, \ldots, N$. The characteristic function of W then can be obtained as

$$\mathscr{F}_W(s) = \mathscr{F}_1(s)\,\mathscr{F}_2(s)\cdots\mathscr{F}_N(s) \tag{2.56}$$

which is the product of the characteristic function of each individual random variable X. The resulting characteristic function for W can be used in Eq. (2.55) to obtain the statistical moments of any order for the random variable W.

The *Laplace/exponential transforms* of a PDF $f(x)$ are defined, respectively, as

$$\mathscr{L}_x(s) = \int_0^{\infty} e^{sx} f(x)\, dx \tag{2.57}$$

$$\mathscr{E}_x(s) = \int_{-\infty}^{\infty} e^{sx} f(x)\, dx \tag{2.58}$$

The transformed functions given by Eqs. (2.57) and (2.58) of a PDF are called moment generating functions. The moment generating functions of commonly used PDFs are shown in Table 2.2. Similar to the characteristic function, statistical moments of a random variable X can be derived, from its moment generating function, as

$$E[X^r] = \mu_r' = \left[\frac{d^r\,\mathscr{E}_x(s)}{d\,s^r}\right]_{s=0} \tag{2.59}$$

Basically, the moment generating function has properties similar to the characteristic function such as the convolution properties described by Eq. (2.58). However, it possesses two deficiencies: (1) the moment generating function of a random variable may not always exist for all distribution functions and all values of s and (2) the correspondence between a PDF and moment generating function may not necessarily be unique.

The *Mellin transform* of a function $f(x)$, for $x > 0$, is defined as (Griffin, 1975; Springier, 1979)

$$M_X(s) = M[f(x)] = \int_0^{\infty} x^{s-1} f(x)\, dx \qquad x > 0 \tag{2.60}$$

where $M_X(s)$ is the Mellin transform of the function $f(x)$. A one-to-one correspondence exists between $M_X(s)$ and $f(x)$. The Mellin transform provides an alternative way to find the moments of any order of a nonnegative random variable.

TABLE 2.2 Characteristic Functions of Some Commonly Used Distributions

Distribution function	PDF	Range	Characteristic function	Moment generating
Binomial	$p(x) = {}_nC_x p^x q^{n-x}$	$x = 0, 1, 2, \ldots, n$	$(q + pe^{is})^n$	$(q + pe^{s})^n$
Poisson	$p(x) = e^{-\nu}\nu^x/x!$	$x = 0, 1, 2, \ldots$	$\exp\{\nu(e^{is} - 1)\}$	$\exp\{\nu(e^{s} - 1)\}$
Normal	$f(x) = \frac{1}{\sqrt{2\pi}\sigma} \exp\left[-\frac{1}{2}\left(\frac{x-\mu}{\sigma}\right)^2\right]$	$-\infty < x < \infty$	$\exp\{i\mu s - 0.5s^2\sigma^2\}$	$\exp\{\mu s + 0.5\sigma^2 s^2\}$
Gamma	$f(x) = \frac{\beta}{\Gamma(\alpha)} (\beta x)^{\alpha-1} e^{-\beta x}$	$x > 0$	$\left(\frac{\beta}{\beta - is}\right)^\alpha$	$\left(\frac{\beta}{\beta - s}\right)^\alpha$
Extreme-value I	$f(x) = \frac{1}{\beta}\exp\left[\mp\left(\frac{x - x_o}{\beta}\right) - \exp \mp \left(\frac{x - x_o}{\beta}\right)\right]$	$-\infty < x < \infty$	$e^{i\alpha s}\Gamma(1 \mp i\beta s)$	$e^{\alpha s}\Gamma(1 \mp \beta s)$

The Mellin transform of the convolution of the PDFs associated with two independent random variables in a product form is equal to the product of the Mellin transform of two individual PDFs as

$$M_W(s) = M[f(w)] = M[g(x)*h(y)] = M_X(s)\, M_Y(s) \tag{2.61}$$

in which * is the convolution operator. Equation (2.61) can be extended to a general case involving more than two independent random variables.

From this convolution property of the Mellin transform and its relationship to statistical moments, one can immediately see the advantage of the Mellin transform as a tool for obtaining the moments of a random variable that is related to other random variables in a multiplicative fashion. In addition to the convolution property, which is of primary importance, the Mellin transform also has several useful properties as summarized in Table 2.3. Furthermore, the Mellin transforms of some commonly used PDFs are tabulated in Table 2.4. Tung (1990) applied the Mellin transform to uncertainty analysis of hydraulic and hydrologic problems.

TABLE 2.3 Mellin Transform of Products and Quotients of Random Variables (Park, 1987)

Random variable	PDF given	$M_W(s) =$
$W = X$	$f(x)$	$M_X(s)$
$W = X^b$	$f(x)$	$M_X(bs - b + 1)$
$W = 1/X$	$f(x)$	$M_X(2 - s)$
$W = XY$	$f(x), g(y)$	$M_X(s)M_Y(s)$
$W = X/Y$	$f(x), g(y)$	$M_X(s)M_Y(2 - s)$
$W = aX^bY^c$	$f(x), g(y)$	$a^{s-1}M_X(bs - b + 1)M_Y(cs - c + 1)$

a, b, c: constants; *X, Y, W:* random variables.
Source: Park, 1987.

TABLE 2.4 Mellin Transforms for Some Commonly Used Probability Density Functions

Probability	PDF	Range	Mellin transform
Uniform	$f(x) = \dfrac{1}{b-a}$	$a \le x \le b$	$\dfrac{b^s - a^s}{s(b-a)}$
Exponential	$f(x) = \beta \exp(-\beta x)$	$x > 0$	$\beta^{1-s}\,\Gamma(s)$
Gamma	$f(x) = \dfrac{\beta}{\Gamma(\alpha)}(\beta x)^{\alpha-1}e^{-\beta x}$	$x > 0$	$\dfrac{\beta^{1-s}\,\Gamma(\alpha+s-1)}{\Gamma(\alpha)}$
Triangular	$f(x) = \dfrac{2(x-a)}{(c-a)(b-a)}$	$a \le x \le b$	
	$f(x) = \dfrac{2(c-x)}{(c-a)(c-b)}$	$b \le x \le c$	$\dfrac{2}{(c-a)s(s+1)}\left[\dfrac{c(c^s-b^s)}{c-b} - \dfrac{a(b^s-a^s)}{b-a}\right]$
Standard beta	$f(x) = \dfrac{\Gamma(\alpha+\beta)}{\Gamma(\alpha)\Gamma(\beta)}x^{\alpha-1}(1-x)^{\beta-1}$	$0 \le x \le 1$	$\dfrac{\Gamma(\alpha+\beta)\Gamma(\alpha+s-1)}{\Gamma(\alpha)\Gamma(\alpha+\beta+s-1)}$
Standard normal	$f(x) = \dfrac{1}{\sqrt{2\pi}}e^{-x^2/2}$	$-\infty < x < \infty$	$2^{(s-1)/2}\,\Gamma(s/2)/\sqrt{\pi}$, for s = odd 0, for s = even

Although the Mellin transform is useful for uncertainty analysis, it possesses one drawback that should be pointed out: under some combinations of distribution and functional form, the resulting transform may not be analytic for all values of s. This could occur especially when quotients or variables with negative exponents are involved. under such circumstances, other transforms such as the Laplace or Fourier transform could be used to find the moments.

2.3.2 Approximation Methods

Most of the models or design procedures used in water resources engineering analyses are nonlinear and highly complex. This basically limits analytical derivation of the statistical properties of model output using the methods described in Sec. 2.3.1. Engineers frequently have to resort to methods that yield approximations to the statistical properties of uncertain model output. In the following, three methods that are useful for uncertainty analysis are described.

First-Order Variance Estimation Method. The *first-order variance estimation* (*FOVE*) *method* is also called the *variance propagation method* (Berthouex, 1975). It estimates uncertainty in terms of the variance of system output, which is evaluated on the basis of statistical properties of the system's stochastic variables. The method approximates the function involving stochastic variables by the Taylor series expansion.

Consider that a hydraulic or hydrologic design quantity W is related to N stochastic variables $\mathbf{X}$ as

$$W = g(\mathbf{X}) = g(X_1, X_2, \ldots, X_N) \tag{2.62}$$

where $\mathbf{X}$ is an N-dimensional column vector of stochastic variables. The Taylor series expansion of the function $g(\mathbf{X})$ with respect to a selected point $\mathbf{X} = \mathbf{x}_o$ in the parameter space can be expressed as

$$W = g(\mathbf{x}_o) + \sum_{i=1}^{N} \left(\frac{\partial g}{\partial X_i} \right)_{\mathbf{x}_o} (X_i - x_{io}) + \frac{1}{2} \sum_{i=1}^{N} \sum_{j=1}^{N} \left(\frac{\partial^2 g}{\partial X_i \, \partial X_j} \right)_{\mathbf{x}_o} (X_i - x_{io})(X_j - x_{jo}) + \epsilon \tag{2.63}$$

in which ϵ represents the higher-order terms. The partial derivative terms are called the *sensitivity coefficients;* each represents the rate of change of model output W with respect to a unit change of each variable around $\mathbf{x}_o$. Dropping the second- and higher-order terms, the preceding equation can be expressed, in matrix/vector form, as

$$W \approx g(\mathbf{x}_o) + \mathbf{s}_o^T \cdot (\mathbf{X} - \mathbf{x}_o) \tag{2.64}$$

in which $\mathbf{s}_o$ is the vector of sensitivity coefficients evaluated at $\mathbf{X} = \mathbf{x}_o$. The mean and variance of W, by the first-order approximation, can be expressed, respectively, as

$$E[W] \approx g(\mathbf{x}_o) + \mathbf{s}_o^T \cdot (\boldsymbol{\mu} - \mathbf{x}_o) \tag{2.65}$$

and

$$\text{Var}[W] \approx \mathbf{s}_o^T \mathbf{C}(\mathbf{X}) \mathbf{s}_o \tag{2.66}$$

in which $\boldsymbol{\mu}$ and $\mathbf{C}(\mathbf{X})$ are the vector of means and covariance matrix of stochastic variables $\mathbf{X}$, respectively.

The common practice of the first-order variance method is to take the expansion point $\mathbf{x}_o = \boldsymbol{\mu}$ and the mean and variance of W can be estimated as

$$E[W] \approx g(\boldsymbol{\mu}) \tag{2.67}$$

and

$$\text{Var}[W] \approx \mathbf{s}^T \mathbf{C}(\mathbf{X}) \mathbf{s} \tag{2.68}$$

in which $\mathbf{s}$ is an N-dimensional vector of sensitivity coefficients evaluated at $\mathbf{x}_o = \boldsymbol{\mu}$. When all the stochastic variables are independent, the variance of model output W can be approximated as

$$\text{Var}[W] \approx \mathbf{s}^T \mathbf{D}\, \mathbf{s} = \sum_{i=1}^{N} s_i^2 \sigma_i^2 \tag{2.69}$$

in which $\mathbf{D} = \text{diag}(\sigma_1^2, \sigma_2^2, \ldots, \sigma_N^2)$, a diagonal matrix of variances of the stochastic variables. The ratio $s_i^2\, \sigma_i^2/\text{var}[W]$ indicates the proportion of overall uncertainty in the model output contributed by the uncertainty associated with variable X_i.

In general, $E[g(\mathbf{X})] \neq g(E[\mathbf{X}])$ unless $g(\mathbf{X})$ is a linear function of $\mathbf{X}$. Improvement of the accuracy can be made by incorporating higher-order terms in the Taylor series expansion. However, as the higher-order terms are included, the mathematical complication, and the required information, increase rapidly. The method can be expanded to include the second-order term for estimating the mean to account for the presence of correlation between stochastic variables.

The first two moments are used in uncertainty analysis for practical engineering design. The FOVE method does not require knowledge of the PDF of stochastic variables which simplifies the analysis. However, this advantage is also a disadvantage of the method because it is insensitive to the distributions of stochastic variables in the uncertainty analysis.

The computational effort associated with the FOVE method largely depends on how the sensitivity coefficients are calculated. For simple analytical functions the computation of the derivatives is a trivial task. However, for functions that are complex and/or implicit in the form of computer programs, or charts and figures, the task of computing the derivatives could become cumbersome. In such cases point estimation (PE) techniques can be used to circumvent the problems.

Rosenblueth's Probabilistic Point Estimation Method. *Rosenblueth's point estimation (PE) method* was first developed for handling stochastic variables that are symmetric (Rosenblueth, 1975), which was later extended to treat nonsymmetric random variables (Rosenblueth, 1981). Referring to Fig. 2.2, the basic idea of Rosenblueth's PE method is to approximate the original PDF or PMF of the random variable X by assuming that the entire probability mass of X is concentrated at two points x_- and x_+. The four unknowns, that is, the locations of x_- and x_+ and the corresponding probability masses p_- and p_+, are determined in such manner that the first three moments of the original random variable X are preserved. The solutions for x_-, x_+, p_-, and p_+ are

$$x_- = \mu - z_- \sigma \tag{2.70a}$$

$$x_+ = \mu + z_+ \sigma \tag{2.70b}$$

$$p_+ = \frac{z_-}{z_+ + z_-} \tag{2.70c}$$

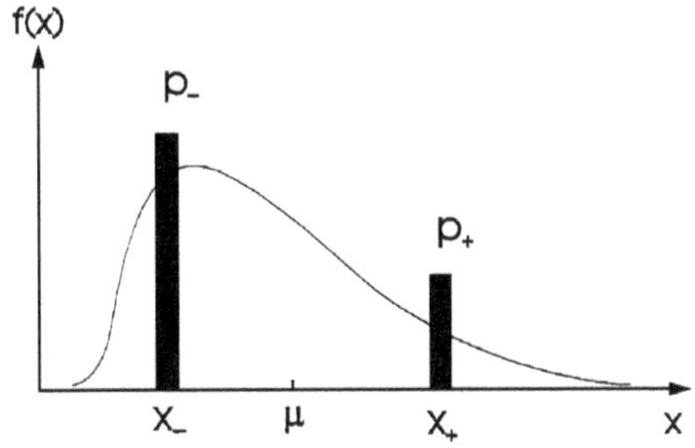

FIGURE 2.2 Schematic diagram of Rosenblueth's PE method in univariate case.

$$p_- = 1 - p_+ \tag{2.70d}$$

where

$$z_+ = \frac{\gamma}{2} + \sqrt{1 + \left(\frac{\gamma}{2}\right)^2} \tag{2.70e}$$

$$z_- = z_+ - \gamma \tag{2.70f}$$

with γ being the skew coefficient of the stochastic variable X.

When the distribution of the random variable X is symmetric ($\gamma = 0$), $z_- = z_+ = 1$ and $p_- = p_+ = 0.5$. This implies that, with a symmetric random variable, the two points are located at one standard deviation to either side of the mean, with equal probability mass assigned at the two locations.

For problems involving N stochastic variables, the two points for each variable are computed according to Eqs. (2.70a–f) and permuted producing a total of 2^N possible points of evaluation in the parameter space. The rth moment of $W = g(\mathbf{X}) = g(X_1, X_2, \ldots, X_N)$ about the origin can be approximated as

$$E(W^r) \approx \sum p_{(\delta 1, \delta 2, \ldots, \delta N)} \, w^r_{(\delta 1, \delta 2, \ldots, \delta N)} \tag{2.71}$$

in which subscript δ_i is a sign indicator that can only be + or – for representing the stochastic variable X_i having the value of $x_{i+} = \mu_i + z_{i+}\sigma_i$ or $x_{i-} = \mu_i - z_{i-}\sigma_i$, respectively; $p_{(\delta 1, \delta 2, \ldots, \delta N)}$ can be determined as

$$p_{(\delta 1, \delta 2, \ldots, \delta N)} = \prod_{i=1}^{N} p_{i,\delta i} + \sum_{i=1}^{N-1}\left(\sum_{j=i+1}^{N} \delta_i \delta_j a_{ij}\right) \tag{2.72}$$

in which

$$a_{ij} = \frac{\rho_{ij}/2^N}{\sqrt{\prod_{i=1}^{N}\left[1 + \left(\frac{\gamma_i}{2}\right)^2\right]}} \tag{2.73}$$

where ρ_{ij} is the correlation coefficient between stochastic variables X_i and X_j. The number of terms in the summation of Eq. (2.71) is 2^N which corresponds to the total number of possible combinations of + and – for all N stochastic variables.

For each term of the summation in Eq. (2.71), the model has to be evaluated once at the corresponding point in the parameter space. This indicates a potential drawback of Rosenblueth's PE method as it is applied to practical problems. When N is small, the method is practical for performing uncertainty analysis. However, for moderate or large N, the number of required function evaluations of $g(\mathbf{X})$ could be too numerous to implement practically, even on the computer. To circumvent this shortcoming, Harr (1989) developed an alternative PE method that reduces the 2^N function evaluations required by Rosenblueth's method down to $2N$.

Harr's Probabilistic Point Estimation Method. *Harr's point estimation method* (Harr, 1989) reduces the required function evaluations by PE methods. The method utilizes the first two moments (that is, the mean and variance) of the random variables involved and their correlations. Skew coefficients of the random variables are ignored by the method. Hence, the method is appropriate for treating random variables that are symmetric. For problems involving only a single random variable, Harr's PE method is identical to Rosenblueth's method with zero skew coefficient. The theoretical basis of Harr's PE method is built on the orthogonal transformation using eigenvalue-eigenvector decomposition.

Orthogonal transformation is an important tool for treating problems involving correlated random variables. The procedure maps correlated random variables from their original space to a new domain in which they become uncorrelated. Hence, the analysis is greatly simplified. Consider N multivariate random variables $\mathbf{X} = (X_1, X_2, \ldots, X_N)^T$ having a mean vector $\boldsymbol{\mu} = (\mu_1, \mu_2, \ldots, \mu_N)^T$ and covariance matrix $\mathbf{C}(\mathbf{X})$. The vector of correlated standardized random variables $\mathbf{X}' = \mathbf{D}^{-1/2}(\mathbf{X} - \boldsymbol{\mu})$; that is, $\mathbf{X}' = (X'_1, X'_2, \ldots, X'_N)^T$ with $X'_i = (X_i - \mu_i)/\sigma_i$ for $i = 1, 2, \ldots, N$, would have mean vector of zero, $\mathbf{0}$, and the covariance matrix equal to the correlation matrix $\mathbf{R}(\mathbf{X})$.

Orthogonal transformation can be made by several ways (Young and Gregory, 1973; Golub and Van Loan, 1989). One frequently used approach is the *eigenvalue-eigenvector factorization* by which the correlation matrix $\mathbf{R}(\mathbf{X})$ is decomposed as

$$\mathbf{R}(\mathbf{X}) = \mathbf{C}(\mathbf{X}') = \mathbf{V}\,\boldsymbol{\Lambda}\,\mathbf{V}^T \tag{2.74}$$

in which $\mathbf{V}$ is an $N \times N$ eigenvector matrix consisting of N eigenvectors as $\mathbf{V} = (\mathbf{v}_1, \mathbf{v}_2, \ldots, \mathbf{v}_N)$ with $\mathbf{v}_i$ being the ith eigenvector and $\boldsymbol{\Lambda} = \text{diag}(\lambda_1, \lambda_2, \ldots, \lambda_N)$ is a diagonal eigenvalue matrix. Frequently, the eigenvectors $\mathbf{v}$'s are normalized such that its norm is equal to unity; that is, $\mathbf{v}^T\,\mathbf{v} = 1$. Using the eigenvector matrix $\mathbf{V}$, the following transformation can be made

$$\mathbf{U} = \mathbf{V}^T\mathbf{X}' \tag{2.75}$$

The resulting transformed stochastic variables $\mathbf{U}$ have the mean $\mathbf{0}$ and covariance matrix $\boldsymbol{\Lambda}$ indicating that $\mathbf{U}$ are uncorrelated because their covariance matrix $\mathbf{C}(\mathbf{U})$ is a diagonal matrix $\boldsymbol{\Lambda}$. Hence, each new stochastic variable U_i has the standard deviation equal to $\sqrt{\lambda_i}$, for all $i = 1, 2, \ldots, N$.

For multivariate problems involving N stochastic variables, Harr's PE method selects its points for function evaluation which are located at the intersections of the N eigenvector axes and a hypersphere of radius $\sqrt{N}$ centered at the origin in the transformed $\mathbf{U}$-space. Referring to Fig. 2.3, the corresponding points in the original parameter space can be obtained as

$$\mathbf{x}_{i\pm} = \boldsymbol{\mu} \pm \sqrt{N}\,\mathbf{D}^{1/2}\,\mathbf{v}_i \qquad i = 1, 2, \ldots, N \tag{2.76}$$

in which $\mathbf{x}_{i\pm}$ is the vector of coordinates of the N stochastic variables in the parameter space on the ith eigenvector $\mathbf{v}_i$.

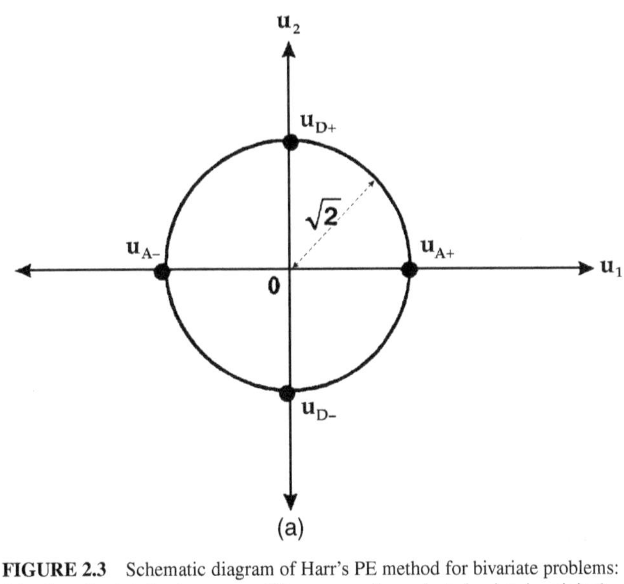

FIGURE 2.3 Schematic diagram of Harr's PE method for bivariate problems: (*a*) selected points in eigenspace; (*b*) corresponding selected points in original standardized space.

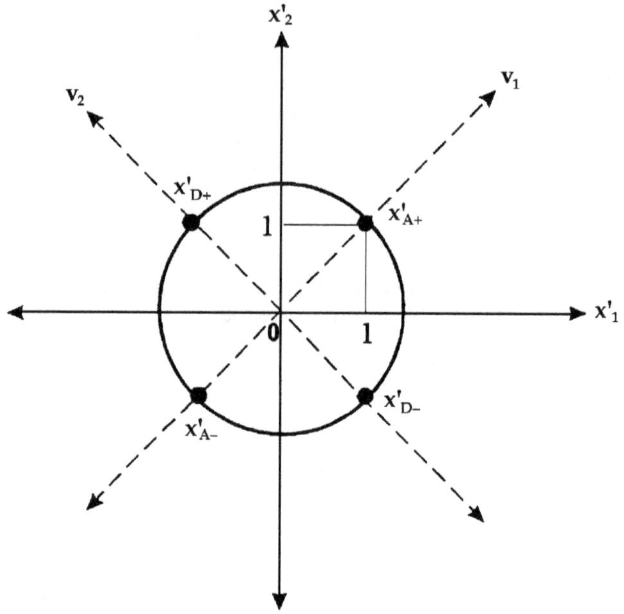

FIGURE 2.3 Schematic diagram of Harr's PE method for bivariate problems: (*a*) selected points in eigenspace; (*b*) corresponding selected points in original standardized space.

Based on the $2N$ points determined by Eq. (2.76), the function values at each of the $2N$ points can be computed. From that the rth moment of the function $W = g(\mathbf{X})$ can be calculated according to

$$\overline{w}_i^r = \frac{w_{i+}^r + w_{i-}^r}{2} = \frac{g^r(\mathbf{x}_{i+}) + g^r(\mathbf{x}_{i-})}{2} \qquad i = 1, 2, \ldots, N;\ r = 1, 2, \ldots \tag{2.77}$$

$$E[W^r] = \mu'_r(W) = \frac{\sum_{i=1}^{N} \lambda_i \overline{w}_i^r}{N} \qquad r = 1, 2, \ldots \tag{2.78}$$

General Remarks on Probabilistic Point Estimation Methods. As pointed out earlier, Rosenblueth's method requires 2^N model evaluations, which could be computationally prohibitive when N is large. Furthermore, the determination of the points in the parameter space for model evaluation and their corresponding probability masses (or weights) is indeterminate when the number of stochastic basic variables, N, is larger than 2. This implies that the solution to the Rosenblueth method is not unique. The simple equations, Eqs. (2.72 to 2.73), sometimes could produce negative probabilities for some selected points when a problem involves asymmetric correlated random variables. Panchalingam and Harr (1994) developed a more complicated formula to reduce, but not eliminate, the problem. Recently, the nonuniqueness difficulty of the Rosenblueth method has been circumvented (Tsai and Franceschini, 2002).

Note that Harr's method presented earlier is applicable to problems involving multivariate normal random variables. It selects points for model evaluation in the eigenspace by which eigenvalues are used as the weighting factors for estimating statistical moments of model output. Other variations, such as using a covariance matrix in orthogonal transform (He and Sällfors, 1994) and selecting points in standardized normal space (Chang et al., 1995), have been shown to produce a more accurate estimation of model output moments. In dealing with nonnormal multivariate random variables, proper normal transformation techniques, such as those described in Sections 2.4.4, can be applied.

In addition to the two probabilistic point estimation methods described herein, a computationally practical three-point PE method was proposed by Li (1992) and Zoppou and Li (1993) that allows incorporation of the first four moments of individual stochastic basic variables and the correlations among the stochastic basic variables.

2.4 LOAD-RESISTANCE INTERFERENCE RELIABILITY ANALYSIS

In numerous hydrologic and hydraulic problems, uncertainties in data and in theory, including design and analysis procedures, warrant a probabilistic treatment of the problems. The failure associated with a hydraulic structure is the result of the combined effect from inherent randomness of external load and various uncertainties involved in the analysis, design, construction, and operational procedures.

Failure of an engineering system occurs when the load L (external forces or demands) on the system exceeds the resistance R (strength, capacity, or supply) of the system. The definitions of reliability and failure probability, Eqs. (2.1) and (2.2), are applicable to component reliability, as well as total system reliability. In hydraulic and hydrologic analyses, the resistance and load are frequently functions of a number of stochastic variables; that is, $L = g(\mathbf{X}_L)$ and $R = h(\mathbf{X}_R)$. Accordingly, the reliability is a function of the stochastic variables involved as

$$p_s = P\,[g(\mathbf{X}_L) \leq h(\mathbf{X}_R)] \tag{2.79}$$

Computations of reliability using Eq. (2.79) do not consider the time-dependence of loads as if $\mathbf{X}_L$ and $\mathbf{X}_R$ are stationary random variables. This is referred to as the *static reliability model,* which is generally applied when the performance of the system subject to a single worst load event is of concern.

However, a hydraulic structure is expected to serve its designed function over an expected period of time. Engineers are frequently interested in knowing the reliability of the structure over its expected service life. In such circumstances, elements of service duration, randomness of occurrence of loads, and possible change of resistance characteristics over time must be considered. Reliability models incorporating these elements are called *time-dependent reliability models* (Kapur and Lamberson, 1977; Tung and Mays, 1980b; Wen, 1987).

2.4.1 Reliability Performance Measures

In reliability analysis, Eq. (2.79) is often expressed in terms of a performance function $W(\mathbf{X}) = W(\mathbf{X}_L, \mathbf{X}_R)$ as

$$p_s = P\,[W(\mathbf{X}_L, \mathbf{X}_R) \geq 0] = P\,[W(\mathbf{X}) \geq 0] \tag{2.80}$$

In reliability analysis, the system state is divided into a safe (satisfactory) set defined by $W(\mathbf{X}) \geq 0$ and a failure (unsatisfactory) set defined by $W(\mathbf{X}) < 0$. The boundary that separates the safe set and failure set is a surface defined by $W(\mathbf{X}) = 0$ and this surface is called the *failure surface* or *limit-state surface.* Because the performance function $W(\mathbf{X})$ defines the condition of the system, it is sometimes called the *system state function.*

The performance function $W(\mathbf{X})$ can be expressed differently as (Tung and Yen, 1993)

$$W_1(\mathbf{X}) = R - L = h(\mathbf{X}_R) - g(\mathbf{X}_L) \tag{2.81}$$

$$W_2(\mathbf{X}) = (R/L) - 1 = [h(\mathbf{X}_R)/g(\mathbf{X}_L)] - 1 \tag{2.82}$$

$$W_3(\mathbf{X}) = \ln\,(R/L) = \ln\,[h(\mathbf{X}_R)] - \ln\,[g(\mathbf{X}_L)] \tag{2.83}$$

Equation (2.81) is identical to the notion of safety margin, whereas Eqs. (2.82) and (2.83) are based on safety factor representations.

The *reliability index* β is another frequently used reliability indicator, which is defined as the reciprocal of the coefficient of variation of the performance function $W(\mathbf{X})$ as

$$\beta = \mu_W / \sigma_W \tag{2.84}$$

in which μ_W and σ_W are the mean and standard deviation of the performance function, respectively. From Eq. (2.84), assuming an appropriate PDF for the random performance function $W(\mathbf{X})$, the reliability can then be computed as

$$p_s = 1 - F_W(0) = 1 - F_{W'}(-\beta) \tag{2.85}$$

in which $F_W(\cdot)$ is the CDF of the performance variable W and W' is the standardized performance variable defined as $W' = (W - \mu_W)/\sigma_W$. The expressions of reliability p_s for eight different distributions of $W(\mathbf{X})$ are given in Table 2.5. In practice, the normal distribution is commonly used for $W(\mathbf{X})$, in which case the reliability can be computed as

$$p_s = 1 - \Phi(-\beta) = \Phi(\beta) \tag{2.86}$$

in which $\Phi(\cdot)$ is the standard normal CDF.

2.4.2 Direct Integration Method

From Eqs. (2.1) and (2.2), the computation of reliability requires knowledge of probability distributions of load and resistance, or the performance function, W. In terms of the joint PDF of the load and resistance, the reliability can be expressed as

$$p_s = \int_{r_1}^{r_2} \left[\int_{\ell_1}^{r} f_{R,L}(r,\ell)\, d\ell \right] dr = \int_{\ell_1}^{\ell_2} \left[\int_{\ell}^{r_2} f_{R,L}(r,\ell)\, dr \right] d\ell \tag{2.87}$$

in which r and ℓ are dummy arguments for the resistance and load, respectively; (r_1, r_2) and (ℓ_1, ℓ_2) are the lower and upper bounds for the resistance and load, respectively. This computation of reliability is called the *load-resistance interference.*

When the load and resistance are independent, Eq. (2.87) can be reduced to

$$p_s = E_R[F_L(R)] = 1 - E_L[F_R(L)] \tag{2.88}$$

in which $E_R[F_L(R)]$ is the expectation of the CDF of load over the possible range of resistance. A schematic diagram illustrating the load-resistance interference, when load and resistance are independent random variables, is shown in Fig. 2.4.

The method of direct integration requires the PDFs of the load and resistance or the performance function be known or derived. This information is seldom available in practice, especially for the joint PDF, because of the complexity of hydrologic and hydraulic models used in design. Explicit solution of direct integration can be obtained for only a few PDFs. Yen et al. (1986) listed the solution of the failure probability $p_f = 1 - p_s$ for eight different distributions of the performance function W. For most PDFs numerical integration may be necessary. When using numerical integration (including Monte Carlo simulation described in Sec. 2.5), difficulty may be encountered when one deals with a multivariate problem.

2.4.3 Mean-Value First-Order Second-Moment (MFOSM) Method

The *MFOSM method* for reliability analysis employs the FOVE method to estimate the statistical moments of the performance function $W(\mathbf{X})$. Once the mean and standard deviation of $W(\mathbf{X})$ are obtained, the reliability index β_{MFOSM} is computed as

TABLE 2.5 Reliability Formulas for Selected Distributions

Distribution of W	PDF, $f_W(w)$	Mean, μ_W	Coefficient of variation, Ω_W	Reliability $p_s = P(W \geq 0)$
Normal	$\frac{1}{\sigma\sqrt{2\pi}}\left[e^{-\frac{1}{2}\left(\frac{w-\mu}{\sigma}\right)^2}\right]; -\infty < w < \infty$	μ	σ/μ	$\Phi\left(\frac{\mu}{\sigma}\right)$
Lognormal	$\frac{1}{\sqrt{2\pi}w\sigma_{\ln W}}\left[e^{-\frac{1}{2}\left(\frac{\ln(w)-\mu_{\ln W}}{\sigma_{\ln W}}\right)^2}\right], w \geq 0$	$\mu_W = e^{\mu_{\ln W} + (\sigma^2_{\ln W}/2)}$	$\sqrt{e^{\sigma^2_{\ln W}} - 1}$	$\Phi\left(\frac{\mu_{\ln W}}{\sigma_{\ln W}}\right)$
Exponential	$\beta e^{-\beta(w - w_0)}, w \geq w_0$	$\frac{1}{\beta} + w_0$	$\frac{1}{1 + \beta w_0}$	$e^{-\beta(w - w_0)}$
Gamma	$\frac{\beta^\alpha (w-\xi)^{\alpha-1} e^{-\beta(w-\xi)}}{\Gamma(\alpha)}, w \geq \xi$	$\frac{\alpha}{\beta} + \xi$	$\frac{\sqrt{\alpha}}{\alpha + \beta\xi}$	$= 1 - I[-\lambda a, k]$
Beta	$\frac{1}{B(\alpha,\beta)} \frac{(w-a)^{\alpha-1}(b-w)^{\beta-1}}{(b-a)^{\alpha+\beta+1}}, a \leq w \leq b$	$a + \frac{\alpha}{\alpha+\beta}(b-a)$	$\sqrt{\frac{\alpha\beta}{\alpha+\beta+1}} \frac{(b-a)}{(\alpha+\beta)\mu_W}$	$1 - \frac{B_u(\alpha,\beta)}{B(\alpha,\beta)}; u = \frac{a}{a-b}$
Triangular	$\frac{2}{b-a}\left(\frac{w-a}{m-a}\right)$ for $a \leq w \leq m$	$\frac{a+m+b}{3}$	$\left[\frac{1}{2} - \frac{ab + am + bm}{6\mu_W^2}\right]^{1/2}$	$1 - \frac{(w-a)^2}{(b-a)(m-a)}$ for $a \leq w \leq m$
	$\frac{2}{b-a}\left(\frac{b-w}{b-m}\right)$ for $m \leq w \leq b$			$\frac{(b-w)^2}{(b-a)(b-m)}$ for $m \leq w \leq b$
Uniform	$\frac{1}{b-a}, a \leq w \leq b$	$\frac{a+b}{2}$	$\frac{1}{\sqrt{3}} \frac{b-a}{b+a}$	$\frac{b-w}{b-a}$

Source: Yen et al., 1986.

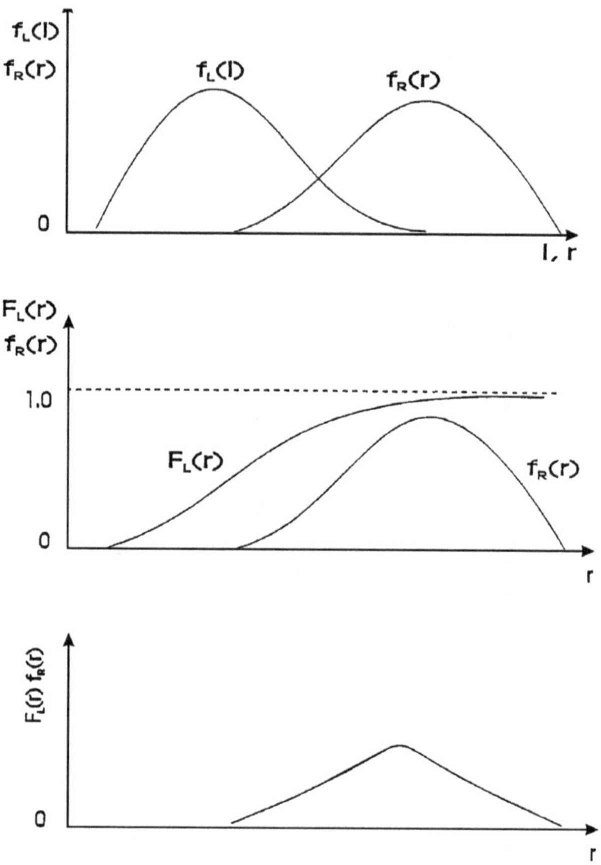

FIGURE 2.4 Schematic diagram of load-resistance interference.

$$\beta_{\text{MFOSM}} = \frac{W(\boldsymbol{\mu})}{\sqrt{\mathbf{s}^T\, \mathbf{C}(\mathbf{X})\, \mathbf{s}}} \tag{2.89}$$

from which the reliability is computed according to Eqs. (2.85) or (2.86).

Yen and Ang (1971), Ang (1973), and Cheng, et al. (1986b) indicated that, provided $p_s < 0.99$, reliability is not greatly influenced by the choice of distribution for W, and the assumption of a normal distribution is quite satisfactory. However, for reliability higher than this value (for example, $p_s = 0.999$), the shape of the tails of a distribution becomes critical. In such cases accurate assessment of the distribution of $W(\mathbf{X})$ should be made to evaluate the reliability or failure probability. The MFOSM method has been used widely in various hydraulic structural and facility designs (see Table2.8).

The application of the MFOSM method is simple and straightforward. However, it possesses certain weaknesses in addition to the difficulties with accurate estimation of extreme failure probabilities as mentioned above. These weaknesses include:

1. Inappropriate choice of the expansion point: In reliability computation, one should be concerned with those points in the parameter space that fall on the failure surface or limit-state surface.

2. Inability to handle distributions with large skew coefficient: This is mainly due to the fact that the MFOSM method incorporates only the first two moments of the random variables involved.

3. Generally poor estimation of the mean and variance of highly nonlinear functions: This is because the first-order approximation of a highly nonlinear function is not accurate.

4. Sensitivity of the computed failure probability to the formulation of the performance function *W*.

5. Inability to incorporate available information on probability distributions of the stochastic variables affecting the load and resistance.

From the above arguments, the general rule of thumb is not to rely on the result of the MFOSM method if any of the following conditions exist: (1) high accuracy requirements for the estimated reliability or risk, (2) high nonlinearity of the performance function, and (3) many skewed random variables are involved in the performance function.

2.4.4 Advanced First-Order Second-Moment (AFOSM) Method

The main thrust of the *AFOSM method* is to reduce the error of the MFOSM method associated with the nonlinearity and noninvariability of the performance function, while keeping the simplicity of the first-order approximation. In the AFOSM method, the expansion point $\mathbf{x}_* = (\mathbf{x}_{L*}, \mathbf{x}_{R*})$ is located on the failure surface defined by the limit-state equation $W(\mathbf{x}) = 0$ which defines the boundary that separates the system performance from being unsatisfactory (unsafe) or being satisfactory (safe), that is,

$$W(\mathbf{x}) \begin{cases} > 0 & \text{system performance is satisfactory (or safe region)} \\ = 0 & \text{failure surface} \\ < 0 & \text{system performance is unsatisfactory (or failure region)} \end{cases}$$

Among all the possible values of $\mathbf{x}$ that fall on the limit-state surface $W(\mathbf{x}) = 0$, one is more concerned with the combination of stochastic variables that would yield the lowest reliability or highest risk. The point on the failure surface with the lowest reliability is the one having the shortest distance to the point where the means of the stochastic variables are located. This point is called the *design point* by Hasofer and Lind (1974) or the *most probable failure point* (Shinozuka, 1983).

Consider that $\mathbf{X} = (X_1, X_2, \ldots, X_N)^T$ are N uncorrelated stochastic variables having a vector mean $\boldsymbol{\mu}$ and covariance matrix $\mathbf{D}$. The original stochastic variables X can be standardized into $\mathbf{X}'$ according to Eq. (2.27). Hence, the design point in $\mathbf{x}'$-space is the one that has the shortest distance from the failure surface $W(\mathbf{x}') = 0$ to the origin $\mathbf{x}' = \mathbf{0}$. Such a point can be found by solving

$$\text{minimize } \|\mathbf{x}'\| = (\mathbf{x}'^T \mathbf{x}')^{1/2} \tag{2.90a}$$

subject to

$$W(\mathbf{x}') = 0. \tag{2.90b}$$

This constrained nonlinear minimization problem can be solved by the Lagrangian multiplier method (see Sec. 1.3.2) and the design point $\mathbf{x}'_*$ can be obtained as

$$\mathbf{x}'_* = -\xi_* \|\mathbf{x}'_*\| \nabla_{x'} W(\mathbf{x}'_*) \tag{2.91}$$

Furthermore, from Eq. (2.91) the distance between the origin $\mathbf{x}' = \mathbf{0}$ to the design point $\mathbf{x}'_*$ can be obtained as

$$\mathbf{x}'_* = -\frac{\nabla_{x'} W(\mathbf{x}'_*)}{\|\nabla_{x'} W(\mathbf{x}'_*)\|} \|\mathbf{x}'_*\| = -\|\mathbf{x}'_*\| \boldsymbol{\alpha}_* \tag{2.92}$$

in which $\boldsymbol{\alpha}_* = \nabla_{\mathbf{x}'} W(\mathbf{x}'_*)/\|\nabla_{\mathbf{x}'} W(\mathbf{x}'_*)\|$ is a unit vector emanating from the design point $\mathbf{x}'_*$ and pointing toward the origin (see Fig. 2.5). The elements of $\boldsymbol{\alpha}_*$, α_{i*}, are called the *directional derivatives* representing the value of the cosine angle between the gradient vector $\nabla_{\mathbf{x}'} W(\mathbf{x}'_*)$ and axes of the standardized variables. Geometrically, Eq. (2.97) shows that the vector $\mathbf{x}'_*$ is perpendicular to the tangent hyperplane passing through the design point. The shortest distance can be expressed as

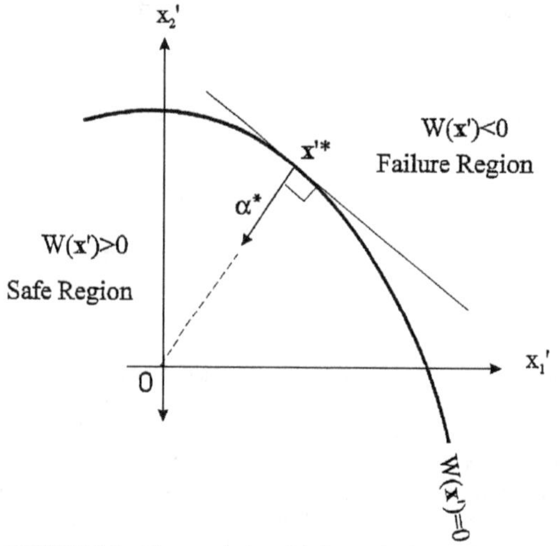

FIGURE 2.5 Characteristics of design point in standardized spaces.

$$|\mathbf{x}'_*| = -\boldsymbol{\alpha}_*^T\mathbf{x}'_* = \frac{-\sum_{i=1}^{N}\left(\frac{\partial W(\mathbf{x}')}{\partial x_i'}\right)_* x'_{i*}}{\sqrt{\sum_{j=1}^{N}\left(\frac{\partial W(\mathbf{x}')}{\partial x_i'}\right)_*^2}} \tag{2.93}$$

Recall that $X_i = \mu_i + \sigma_i X'_i$, for $i = 1, 2, \ldots, N$. By the chain rule in calculus, Eq. (2.93) can be written, in terms of the original variables **X**, as

$$|\mathbf{x}'_*| = \frac{\sum_{i=1}^{N}\left(\frac{\partial W(\mathbf{x})}{\partial x_i}\right)_* (\mu_i - x_{i*})}{\sqrt{\sum_{j=1}^{N}\left(\frac{\partial W(\mathbf{x})}{\partial x_i}\right)_*^2 \sigma_i^2}} \tag{2.94}$$

in which $\mathbf{x}_*$ is the point in **x**-space corresponding to the design point $\mathbf{x}'_*$ in **x**′-space which can be easily determined as $\mathbf{x}_* = \boldsymbol{\mu} + \mathbf{D}^{1/2}\mathbf{x}'_*$.

Refer to the first-order approximation of the performance function $W(\mathbf{X})$. Taking the expansion point $\mathbf{x}_o = \mathbf{x}_*$, the expected value and variance of the performance function $W(\mathbf{X})$ can be approximated as

$$\mu_W \approx \mathbf{s}_*^T(\boldsymbol{\mu} - \mathbf{x}_*) \tag{2.95}$$

$$\sigma_W^2 \approx \mathbf{s}_*^T\, \mathbf{C}(\mathbf{X})\, \mathbf{s}_* \tag{2.96}$$

in which $\mathbf{s}_* = (s_{1*}, s_{2*}, \ldots, s_{N*})^T$ is a vector of sensitivity coefficients of the performance function $W(\mathbf{X})$ evaluated at $\mathbf{x}_*$, and $\boldsymbol{\mu}$ and $\mathbf{C}(\mathbf{X})$ are the mean vector and covariance matrix of the stochastic variables, respectively. If the stochastic variables are uncorrelated, Eq. (2.96) reduces to

$$\sigma_W^2 = \sum_{i=1}^{N} s_{i*}^2\, \sigma_i^2 \tag{2.97}$$

in which σ_i is the standard deviation of the ith stochastic variable. The standard deviation of the performance function $W(\mathbf{X})$ can alternatively be expressed in terms of the directional derivatives as

$$\sigma_W = \sum_{i=1}^{N} \alpha_{i*}\, s_{i*}\, \sigma_i \tag{2.98}$$

where α_{i*} is the directional derivative for the ith stochastic variable at the expansion point $\mathbf{x}_*$

$$\alpha_{i*} = \frac{s_{i*}\,\sigma_i}{\sqrt{\sum_{j=1}^{N} s_{j*}^2\,\sigma_j^2}} \tag{2.99}$$

Equation (2.99) is identical to the one defined in Eq. (2.93). With the mean and standard deviation of the performance function $W(\mathbf{x})$ computed at $\mathbf{x}_*$, the AFOSM reliability index β_{AFOSM} can be determined as

$$\beta_{\text{AFOSM}} = \frac{\mu_W}{\sigma_W} = \frac{\sum_{i=1}^{N} s_{i*}\,(\mu_i - x_{i*})}{\sum_{i=1}^{N} \alpha_{i*}\, s_{i*}\, \sigma_i} \tag{2.100}$$

Equation (2.100) indicates that the AFOSM reliability index β_{AFOSM} is identical to the shortest distance from the origin to the design point in the standardized parameter space. β_{AFOSM} is also called the *Hasofer-Lind reliability index.*

Once β_{AFOSM} is computed, the reliability can be estimated as $p_s = \Phi(\beta_{\text{AFOSM}})$. Because $\beta_{\text{AFOSM}} = \|\mathbf{x}'_*\|$, the sensitivity of β_{AFOSM} with respect to the uncorrelated standardized random variables is

$$\nabla_{\mathbf{x}'}\beta_{\text{AFOSM}} = \nabla_{\mathbf{x}'}\|\mathbf{x}'_*\| = \frac{\mathbf{x}'_*}{|\mathbf{x}'_*|} = -\boldsymbol{\alpha}_* \tag{2.101}$$

Equation (2.101) shows that $-\alpha_{i*}$ represents the rate of change in β_{AFOSM} due to a one standard deviation change in stochastic variable X_i, that is,

$$-\alpha_{i*} = \left[\frac{\partial \beta_{\text{AFOSM}}}{\partial X'_i}\right]_{\mathbf{x}'_*} = \left[\frac{\partial \beta_{\text{AFOSM}}}{\partial X_i}\right]_{\mathbf{x}*} \sigma_i \tag{2.102}$$

It can also be easily shown that

$$\left(\frac{\partial p_s}{\partial X'_i}\right)_{\mathbf{x}'_*} = -\alpha_{i*}\,\Phi(\beta_{\text{AFOSM}}); \quad \left(\frac{\partial p_s}{\partial X_i}\right)_{\mathbf{x}*} = -\sigma_i\,\alpha_{i*}\,\Phi(\beta_{\text{AFOSM}}) \tag{2.103}$$

for $i = 1, 2, \ldots, N$.

In the case that $\mathbf{X}$ are independent normal stochastic variables, Hasofer and Lind (1974) proposed a recursive equation for determining $\mathbf{x}_*$,

$$\mathbf{x}_{(r+1)} = \boldsymbol{\mu} + \mathbf{D}\mathbf{s}_{(r)} \frac{[\mathbf{x}_{(r)} - \boldsymbol{\mu}]^T \mathbf{s}_{(r)} - W(\mathbf{x}_{(r)})}{\mathbf{s}_{(r)}^T\,\mathbf{D}\mathbf{s}_{(r)}} \qquad \text{for} \qquad r = 1, 2, \ldots \tag{2.104}$$

Based on Eq. (2.104), the Hasofer-Lind algorithm for the AFOSM reliability analysis can be outlined as follows:

Step 1 Select an initial trial solution $\mathbf{x}_{(r)}$.

Step 2 Compute $W(\mathbf{x}_{(r)})$ and the corresponding sensitivity coefficient vector $\mathbf{s}_{(r)}$.

Step 3 Revise solution point $\mathbf{x}_{(r+1)}$ according to Eq. (2.104).

Step 4 Check if $\mathbf{x}_{(r)}$ and $\mathbf{x}_{(r+1)}$ are sufficiently close. If yes, compute the reliability index β_{AFOSM} according to Eq. (2.103) and the corresponding reliability $p_s = \Phi(\beta_{\text{AFOSM}})$, then, stop otherwise, update the solution point by letting $\mathbf{x}_{(r)} = \mathbf{x}_{(r+1)}$ and return to Step 2.

Step 5 Compute the sensitivity of reliability with respect to changes in stochastic variables.

Due to the nature of nonlinear optimization, the algorithm AFOSM-HL does not necessarily coverge to the true design point associated with the minimum reliability index. Therefore, Madsen et al. (1986) suggest that different initial trial points be used and the smallest reliability index be selected to compute the reliability.

Ang and Tang (1984) also proposed an iterative procedure for finding the design point. The core of their updating procedure relies on the fact that, according to Eq. (2.100), the following relationship should be satisfied:

$$\sum_{i=1}^{N} s_{i*}(\mu_i - x_{i*} - \alpha_{i*}\beta_*\sigma_i) = 0 \tag{2.105}$$

Because variables X_i's are random and uncorrelated, Eq. (2.105) defines the failure point within the first-order context. Hence, Eq. (2.100) can be decomposed into

$$x_{i*} = \mu_i - \alpha_{i*}\beta_*\sigma_i \qquad \text{for} \qquad i = 1, 2, \ldots, N \tag{2.106}$$

Ang and Tang (1984) present the following iterative procedure to locate the design point $\mathbf{x}_*$ and the corresponding reliability index β_{AFOSM} under the condition that stochastic variables are independent normal random variables. The algorithm involves the following steps:

Step 1 Select an initial point $\mathbf{x}_{(r)}$ in the parameter space. For practicality, the point $\boldsymbol{\mu}$ is a viable starting point.

Step 2 At the selected point $\mathbf{x}_{(r)}$, compute the mean of the performance function $W(\mathbf{X})$ by

$$\mu_W = W(\mathbf{x}_{(r)}) + \mathbf{s}_{(r)}^T(\boldsymbol{\mu} - \mathbf{x}_{(r)}) \tag{2.107}$$

and the variance according to Eq. (2.91).

Step 3 Compute the corresponding reliability index β according to Eq. (2.84).

Step 4 Compute the values of directional derivative α_i for all $i = 1, 2, \ldots, N$, according to Eq. (2.99).

Step 5 Revise the location of expansion point $\mathbf{x}_{(r+1)}$ according to Eq. (2.112) using b obtained from Steps 4 and 5.

Step 6 Check if the revised expansion point $\mathbf{x}_{(r+1)}$ differs significantly from the previous trial expansion point $\mathbf{x}_{(r)}$. If yes, use the revised expansion point as the trial point by letting $\mathbf{x}_{(r)} = \mathbf{x}_{(r+1)}$ and go to Step 2 for another iteration. Otherwise, the iterative procedure is considered complete and the latest reliability index β_{AFOSM} is to be used in Eq. (2.86) to compute the reliability p_s.

Similar to the Hasofer-Lind algorithm, the above Ang-Tang algorithm does not guarantee convergence to the design point for all problems.

The task to determine the critical failure point $\mathbf{x}^*$ that minimizes the reliability is equivalent to minimizing the value of the reliability index β. Cheng et al. (1986a) proposed the following optimization model to identify the design point.

$$\text{minimize } |W(\mathbf{x})| \text{ subject to Eq. (2.106)} \tag{2.108}$$

The approach by Cheng et al. (1986a) is a viable tool when the iterative algorithms of Ang-Tang and Hasofer-Lind do not converge.

The AFOSM method described above is suitable for the case that all stochastic variables in the load and resistance functions are independent, normal random variables. In reality, stochastic variables in a performance function may be nonnormal and correlated. In the following, the treatment of stochastic variables that are nonnormal and correlated are discussed.

Treatment of Nonnormal Stochastic Variables. When nonnormal random variables are involved, it is advisable to transform them into equivalent normal variables. Rackwitz (1976) proposed an approach that transforms a nonnormal distribution into an equivalent normal distribution, so that the value of the CDF of the transformed equivalent normal distribution *is* the same as that of the original nonnormal distribution at the failure point $\mathbf{x}_*$; that is,

$$F_i(x_{i*}) = \Phi\left(\frac{x_{i*} - \mu_{i*N}}{\sigma_{i*N}}\right) \tag{2.109}$$

in which $F_i(x_{i*})$ is the CDF of the stochastic variable X_i having a value at x_{i*}; and μ_{i*N} and σ_{i*N} are the mean and standard deviation of the normal equivalent for the ith stochastic variable at $X_i = x_{i*}$. From Eq. (2.109), the following equations are obtained

$$\mu_{i*N} = x_{i*} - z_{i*}\,\sigma_{i*N} \tag{2.110}$$

$$\sigma_{i*N} = \frac{\Phi(z_{i*})}{f_i(x_{i*})} \tag{2.111}$$

in which $z_{i*} = \Phi^{-1}[F_i(x_{i*})]$ is the standard normal quantile.

To incorporate the normal transformation for nonnormal but uncorrelated stochastic variables, the iterative algorithms described previously for the AFOSM method can be modified. In the Hasofer-Lind algorithm, replace the updating equation in Step 3 by

$$\mathbf{x}_{(r+1)} = \boldsymbol{\mu}_N + \mathbf{D}_N\,\mathbf{s}_{(r)}\,\frac{(\mathbf{x}_{(r)} - \boldsymbol{\mu}_N)^T\,\mathbf{s}_{(r)} - W(\mathbf{x}_{(r)})}{\mathbf{s}_{(r)}^T\,\mathbf{D}_N\,\mathbf{s}_{(r)}} \tag{2.112}$$

In the Ang-Tang algorithm, compute the directional derivatives in Step 4 according to Eq. (2.104) substituting the σ_{i*N}'s for the corresponding standard deviations of nonnormal stochastic variable σ_i's. Then, in Step 5, use β and α_{iN} to revise the location of the expansion point $\mathbf{x}_{(r+1)}$ according to

$$x_{i(r+1)} = \mu_{iN} - \alpha_{iN}\,\beta\,\sigma_{iN} \qquad \text{for} \qquad i = 1, 2, \ldots, N \tag{2.113}$$

Treatment of Correlated Stochastic Variables. When some of the stochastic variables involved in the performance function are significantly correlated, transformation of correlated variables to uncorrelated ones is made before the step of normal transformation is taken. Consider that the stochastic variables in the performance function are multivariate normal random variables with the vector of mean $\boldsymbol{\mu}$ and a covariance matrix $\mathbf{C}(\mathbf{X})$. Following the orthogonal transformation described in Sec. 2.3.2, the directional derivatives of the performance function in the transformed, uncorrelated **Y**-space $\tilde{\alpha}$ are computed as

$$\tilde{\alpha}_i = \frac{\tilde{s}_i\,\tilde{\sigma}_i}{\sqrt{\sum_{j=1}^{N} \tilde{s}_j^2\,\tilde{\sigma}_j^2}} = \frac{\tilde{s}_i\,\sqrt{\lambda_i}}{\sqrt{\sum_{j=1}^{N} \tilde{s}_j^2\,\lambda_j}} \tag{2.114}$$

in which $\tilde{s}_{i*}$ is the sensitivity coefficient of the performance function $W(\mathbf{X})$ with respect to the transformed stochastic variable Y_i evaluated at $\mathbf{Y} = \mathbf{y}_*$ which could be obtained as

$$\tilde{s}_{i*} = \left(\frac{\partial W(\mathbf{X})}{\partial \mathbf{Y}_i}\right)_{y*} = \sigma_{i*}\,\mathbf{s}_*^T\,\mathbf{v}_i \tag{2.115}$$

in which $\mathbf{s}_*$ is the vector of sensitivity coefficients of the performance function with respect to the original stochastic variables evaluated at $\mathbf{X} = \mathbf{x}_*$; and $\mathbf{v}_i$ is the corresponding eigenvector in matrix $\mathbf{V}$ associated with the correlation matrix.

For the most general condition where stochastic variables are correlated and nonnormal, the following algorithm can be applied.

Step 1 Decompose the correlation matrix $\mathbf{R}(\mathbf{X})$ to find the eigenvector matrix $\mathbf{V}$ and eigenvalues λ.

Step 2 Select an initial point $\mathbf{x}_{(r)}$ in the original parameter space.

Step 3 At the selected point $\mathbf{x}_{(r)}$ compute the mean and variance of the performance function $W(\mathbf{X})$ according to Eqs. (2.107) and (2.96), respectively.

Step 4 Compute the corresponding reliability index β according to Eq. (2.84).

Step 5 Compute the mean μ_{iN} and standard deviation σ_{iN} of the normal equivalent using Eqs. (2.110) and (2.111) for the nonnormal stochastic variables.

Step 6 Compute sensitivity coefficients $\tilde{s}_i$ in the transformed domain according to Eq. (2.115) for all $i = 1, 2, \ldots, N$ with σ_i replaced by σ_{iN}.

Step 7 Compute the directional derivatives $\tilde{\alpha}$ according to Eq. (2.114).

Step 8 Using β and $\tilde{\alpha}_i$ obtained from Steps 4 and 7, compute the location of expansion point $\mathbf{y}_{(r+1)}$ in the transformed domain according to

$$y_{i,(r+1)} = -\tilde{\alpha}_i \beta \sqrt{\lambda_i} \qquad \text{for} \qquad i = 1, 2, \ldots, N \tag{2.116}$$

Step 9 Convert the obtained expansion point $\mathbf{y}_{(r+1)}$ in the transformed domain in Step 7 back to the original parameter space as

$$\mathbf{x}_{(r+1)} = \boldsymbol{\mu}_N + \mathbf{D}_N^{1/2}\, \mathbf{V}\mathbf{y}_{(r+1)} \tag{2.117}$$

in which $\boldsymbol{\mu}_N$ is the vector of means of the normal equivalent at solution point $\mathbf{x}$ and $\mathbf{D}_N$ is the diagonal matrix of variances of the normal equivalent.

Step 10 Check if the revised expansion point $\mathbf{x}_{(r+1)}$ differs significantly from the previous trial expansion point $\mathbf{x}_{(r)}$. If yes, use the revised expansion point as the new trial point by letting $\mathbf{x}_{(r)} = \mathbf{x}_{(r+1)}$; go to Step 3 for another iteration. Otherwise, the iterative procedure is considered complete and the latest reliability index β is used in Eq. (2.86) to compute the reliability p_s.

The use of the AFOSM reliability index removes the problem of lack of invariance associated with the MFOSM reliability index. This allows placing different designs on the same common ground for comparing their relative reliabilities using β_{AFOSM}. A design with a higher value of β_{AFOSM} would be associated with a higher reliability and lower failure probability. However, such conclusions are valid only if the curvatures of failure surfaces at the design points for different designs are similar.

One drawback of the preceding Ang-Tang algorithm is the potential inconsistency between the orthogonally transformed variables $\mathbf{U}$ and the normal-transformed space in computing the directional derivatives in Steps 6 and 7. This is because the eigenvalues and eigenvectors associated with $\mathbf{R}(\mathbf{X})$ will not be identical to those in the normal-transformed variables. To correct this inconsistency, a normal-transformation can be made in advance so that the algorithm is applied in the transformed space.

Based on the marginal PDFs of two stochastic variables X_i and X_j, and their correlation coefficient ρ_{ij}, Der Kiureghian and Liu (1986) apply Nataf's bivariate distribution model (Nataf, 1962)

$$\rho_{ij} = \int_{-\infty}^{\infty}\int_{-\infty}^{\infty} \left(\frac{x_i - \mu_i}{\sigma_i}\right)\left(\frac{x_j - \mu_j}{\sigma_j}\right) \Phi_{ij}(z_i, z_j, \rho^*_{ij})\, dz_i\, dz_j \tag{2.118}$$

in which Z_i and Z_j are the two correlated standard normal random variables having the equivalent marginal cumulative probabilities corresponding to a pair of random variables X_i and X_j in the original space; ρ^*_{ij} is the correlation coefficient between Z_i and Z_j; ρ_{ij} is the correlation coefficient between X_i and X_j; μ_i and σ_i are, respectively, the mean and standard deviation of X_i; Φ_{ij} is the bivariate standard normal PDF with zero means, unit standard deviations, and correlation coefficient ρ^*_{ij}.

Two conditions are inherently considered in the bivariate distribution model of Eq. (2.118): (1) the normal-transformation satisfies Eq. (2.109), and (2) the value of the correlation coefficient in the normal space lies between −1 and 1. For a pair of nonnormal stochastic variables X_i and X_j, with known means μ_i and μ_j, standard deviations σ_i and σ_j, and correlation coefficient ρ_{ij}, Eq. (2.118) can be applied to solve for ρ^*_{ij}. To avoid the required computation for solving ρ^*_{ij} in Eq. (2.118), Der Kiureghian and Liu (1986) developed a set of semiempirical formulas for ten marginal distributions commonly used in reliability computations as

$$\rho^*_{ij} = T_{ij} \cdot \rho_{ij} \tag{2.119}$$

in which T_{ij} is a transformation factor depending on the marginal distributions and correlation of the two random variables considered.

Recently, a practical third-order polynomial normal transform was examined by Chen and Tung (2003) in that the original nonnormal random variable X is related to the standard normal variable Z as $X = a_0 + a_1 Z + a_2 Z^2 + a_3 Z^3$. The values of polynomial coefficients can be determined by preserving statistical moments or quantiles of X. For multivariate problems, upon the determination of the polynomial coefficients for a pair of random variables, the correlation coefficient in the normal space can be easily obtained. From the numerical experiment conducted by Chen and Tung (2003), this simple technique is found to be quite accurate, and it is less restrictive in its usage for merely requiring the information of the first four statistical moments of a random variable rather than requiring a stronger assumption of the full distribution information. The technique is especially attractive when only samples of the random variables are available.

2.5 *RELIABILITY ANALYSIS: TIME-TO-FAILURE ANALYSIS*

In previous sections, evaluations of reliability are based on the analysis of interaction between loads on the system and the resistance of the system. A system would perform its intended function satisfactorily within a specified time period if its capacity exceeds the loading. Instead of considering detailed interactions of resistance and loading over time, a system or its components can be treated as a black box or a lumped-parameter system, and their performances are observed over time. This reduces the reliability analysis to a one-dimensional problem involving time as the only variable. In such cases, the *time-to-failure* (*TTF*) of a system or a component of the system is the random variable. It should be pointed out that the term *time* could be used in a more general sense. In some situations other physical scale measures, such as distance or length, may be appropriate for system performance evaluation.

The time-to-failure analysis is particularly suitable for assessing the reliability of systems and/or components that are repairable. For a system that is repairable after its failure, the time period it would take to have it repaired back to the operational state is uncertain. Therefore, the *time-to-repair* (*TTR*) is also a random variable.

For a repairable system or component, its service life can be extended indefinitely if repair work can restore the system as if it were new. Intuitively, the probability of a repairable system available for service is greater than that of a nonrepairable system. This section focuses on characteristics of failure, repair, and availability of repairable systems by time-to-failure analysis.

2.5.1 Failure Density, Failure Rate, and Mean-Time-To-Failure

Any system will fail eventually; it is just a matter of time. Due to the presence of many uncertainties that affect the operation of a physical system, the time that the system fails to satisfactorily perform its intended function is a random variable.

The *failure density function* is the probability distribution that governs the time occurrence of failure. The failure density function serves as the common thread in the reliability assessments in time-to-failure analysis. The reliability of a system or a component within a specified time interval [0, t] can be expressed as

$$p_s(t) = P(\text{TTF} > t) = \int_t^{\infty} f(\tau)\, d\tau \tag{2.120}$$

in which TTF is the random time-to-failure having $f(t)$ as the failure density function. The reliability $p_s(t)$ represents the probability that the system experiences no failure within [0,t]. The failure probability or *unreliability* can be expressed as $p_f(t) = 1 - p_s(t)$. Note that unreliability $p_f(t)$ is the probability that a component or a system would experience its first failure within the time interval (0,t). As the age of the system t increases, the reliability $p_s(t)$ decreases, whereas the unreliability $p_f(t)$ increases. Conversely, the failure density function can be obtained from reliability or unreliability as

$$f(t) = -\frac{d\,[p_s(t)]}{dt} = \frac{d\,[p_f(t)]}{dt} \tag{2.121}$$

The time-to-failure is a continuous, nonnegative random variable by nature. Many distribution functions described in Sec. 2.2.5 are appropriate for modeling the stochastic nature of time-to-failure. Among them, the exponential distribution, Eq. (2.42), is perhaps the most widely used. Besides its mathematical simplicity, the exponential distribution has been found, both phenomenologically and empirically, to adequately describe the time-to-failure distribution for components, equipment, and systems involving components with mixtures of life distributions.

The *failure rate* is defined as the number of failures occurring per unit time in a time interval (t, $t + \Delta t$) per unit of the remaining population at time t. The *instantaneous failure rate* $m(t)$ can be obtained as

$$m(t) = \lim_{\Delta t \to 0}\left[\frac{N_F(\Delta t)/\Delta t}{N(t)}\right] = \frac{f(t)}{p_s(t)} \tag{2.122}$$

where $N_F(\Delta t)$ is the number of failures in a time interval (t, $t + \Delta t$); and $N(t)$ is the number of failures from the beginning up to time t. This instantaneous failure rate is also called the *hazard function* or *force of mortality function* (Pieruschka, 1963).

The failure rate for many systems or components has a bathtub shape as shown in Fig. 2.6 in that three distinct life periods can be identified. They are *early life* (or *infant mortality*) *period, useful life period,* and *wearout life period.* In the early life period, quality failures and stress-related failures dominate, with little contribution from wearout failures. During the useful life period, all three types of failures contribute to the potential failure of the system or component, and the overall failure rate remains more or less constant over time. In the later part of life, the overall failure rate increases with age, because wearout failures and stress-related failures are the main contributors, and wearout becomes a more and more dominating factor for the failure of the system with age.

The reliability can be directly computed from the failure rate as

$$p_s(t) = \exp\left(\int_0^t m(\tau)\, d\tau\right) \tag{2.123}$$

Substituting Eq. (2.123) into Eq. (2.122), the failure density function $f(t)$ can be expressed, in terms of failure rate, as

$$f(t) = m(t)\left[\exp\left(\int_0^{-t} m(\tau)\, d\tau\right)\right] \tag{2.124}$$

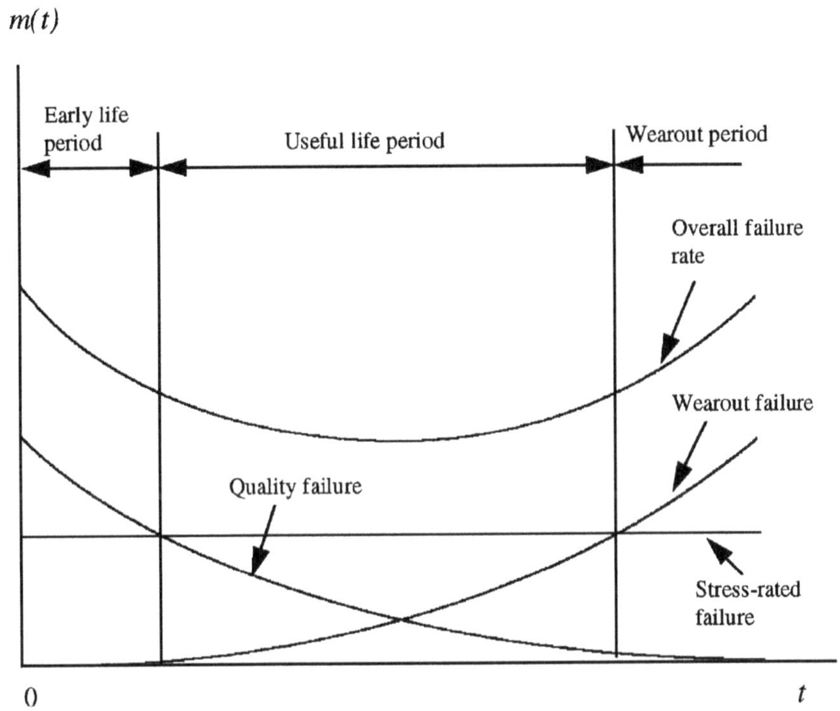

FIGURE 2.6 Shape of typical hazard function and its components.

In general, the reliability of a system or a component is strongly dependent on its age. This can be mathematically expressed by the conditional probability as

$$p_s(\xi|t) = \frac{p_s(t+\xi)}{p_s(t)} \tag{2.125}$$

in which t is the age of the system and up to that point the system has not failed, $p_s(\xi|t)$ is the reliability over a new mission period ξ, having successfully operated over a period of t.

A commonly used reliability measure of system performance is the *mean-time-to-failure* (*MTTF*) which is the expected time-to-failure. The MTTF can be mathematically defined as

$$\text{MTTF} = E(\text{TTF}) = \int_0^\infty \tau f(\tau)\, d\tau \tag{2.126}$$

The mean-time-to-failure for some failure density functions is given in Table 2.6.

2.5.2 Repairable Systems

For repairable water resources systems, such as pipe networks, pump stations, and storm runoff drainage structures, failed components within the system can be repaired or replaced so that the system can be put back into service. The time required to have the failed system repaired is uncertain, and consequently, the total time required to restore the system from its failure to operational state is a random variable.

TABLE 2.6 Mean-Time-to-Failure for Some Failure Density Functions

Distribution	PDF, $f(t)$	Reliability, $p_s(t)$	Failure rate, $m(t)$	MTTF
Normal	$\frac{1}{\sqrt{2\pi}\sigma_T}\exp\left[-\frac{1}{2}\left(\frac{t-\mu_r}{\sigma_T}\right)^2\right]$	$\Phi\left(\frac{t-\mu_T}{\sigma_T}\right)$	$\frac{f(t)}{\Phi\left(\frac{t-\mu_T}{\sigma_T}\right)}$	μ_T
Lognormal	$\frac{1}{\sqrt{2\pi}\sigma_{\ln T}t}\exp\left[-\frac{1}{2}\left(\frac{\ln(t)-\mu_{\ln T}}{\sigma_{\ln T}}\right)^2\right]$	$\Phi\left(\frac{\ln(t)-\mu_{\ln T}}{\sigma_{\ln T}}\right)$	$\frac{f(t)}{\Phi\left(\frac{\ln(t)-\mu_{\ln T}}{\sigma_{\ln T}}\right)}$	$\exp\left(\mu_{\ln T}+\frac{\sigma_{\ln T}^2}{2}\right)$
Exponential	$\beta e^{-\beta t}$	$e^{-\beta t}$	β	$\frac{1}{\beta}$
Rayleigh	$\frac{t}{\beta^2}\exp\left[-\frac{1}{2}\left(\frac{t}{\beta}\right)^2\right], \beta > 0$	$\exp\left[-\frac{1}{2}\left(\frac{t}{\beta}\right)^2\right]$	$\frac{t}{\beta^2}$	$1.253\,\beta$
Gamma	$\frac{\beta}{\Gamma(\alpha)}(\beta t)^{\alpha-1}e^{-\beta t}$	$\int_t^\infty f(\tau)\,d\tau$	$\frac{f(t)}{p_s(t)}$	$\frac{\alpha}{\beta}$
Gumbel	$e^{\pm y - e^{\pm y}};\ y=\frac{t-t_o}{\beta}$	$1-e^{-e^{\pm y}}$	$\frac{f(t)}{p_s(t)}$	$x_o \pm 0.577\,\beta$
Weibull	$\frac{\alpha}{\beta}\left(\frac{t-t_o}{\beta}\right)^{\alpha-1}e^{-\left(\frac{t-t_o}{\beta}\right)^\alpha}$	$e^{\left(\frac{t-t_o}{\beta}\right)^\alpha}$	$\frac{\alpha(t-t_o)^{\alpha-1}}{\beta^\alpha}$	$t_o+\beta\Gamma\left(1+\frac{1}{\alpha}\right)$
Uniform	$\frac{1}{b-a}$	$\frac{b-t}{b-a}$	$\frac{1}{b-t}$	$\frac{a+b}{2}$

Like the time-to-failure, the random time-to-repair (TTR) has the *repair density function* $g(t)$ describing the random characteristics of the time required to repair a failed system when failure occurs at time zero. The *repair probability* $G(t)$ is the probability that the failed system can be restored within a given time period $[0, t]$,

$$G(t) = P[\text{TTR} \le t] = \int_0^t g(\tau)\,d\tau \tag{2.127}$$

The repair probability $G(t)$ is one measure of maintainability (Kapur, 1988).

The *repair rate* $r(t)$, similar to the failure rate, is the conditional probability that the system is repaired per unit time, given that the system failed at time zero and is still not repaired at time t. The quantity $r(t)dt$ is the probability that the system is repaired during the time interval $(t, t + dt)$ given that the system fails at time t. The relation between the repair density function, repair rate, and repair probability is

$$r(t) = \frac{g(t)}{1 - G(t)} \tag{2.128}$$

Given a repair rate $r(t)$, the repair density function and the maintainability can be determined, respectively, as

$$g(t) = r(t)\exp\left(-\int_0^t r(\tau)\,d\tau\right) \tag{2.129}$$

$$G(t) = 1 - \exp\left(-\int_0^t r(\tau)\, d\tau\right) \tag{2.130}$$

The *mean-time-to-repair* (*MTTR*) is the expected value of time-to-repair of a failed system; that is,

$$\text{MTTR} = \int_0^{\infty} t g(t)\, dt \tag{2.131}$$

The MTTR measures the elapsed time required to perform the maintenance operation and is used to estimate the downtime of a system. It is also a measure for maintainability of a system. Various types of maintainability measures are derivable from the repair density function (Kraus, 1988).

The MTTR is a proper measure of the mean life span of a nonrepairable system. For a repairable system, a more representative indicator for the fail-repair cycle is the *mean-time-between-failure* (*MTBF*) which is the sum of MTTR and MTTR; that is,

$$\text{MTBF} = \text{MTTF} + \text{MTTR} \tag{2.132}$$

The *mean-time-between-repair* (*MTBR*) is the expected value of the time between two consecutive repairs, and it is equal to MTBF.

2.5.3 Availability and Unavailability

A repairable system experiences a repetition of repair-to-failure and failure-to-repair processes during its service life. Hence, the probability that a system is in operating condition at any given time t for a repairable system is different from that of a nonrepairable system. The term *availability* $A(t)$ is generally used for repairable systems, to indicate the probability that the system is in operating condition at any given time t. On the other hand, reliability $p_s(t)$ is appropriate for nonrepairable systems, indicating the probability that the system has been *continuously* in its operating state starting from time zero up to time t.

Availability can also be interpreted as the percentage of time that the system is in operating condition within a specified time period. In general, the availability and reliability of a system satisfies the following inequality relationship

$$0 \le p_s(t) \le A(t) \le 1 \tag{2.133}$$

with the equality holding for nonrepairable systems. The reliability of a system decreases monotonically to zero as the age of the system increases, whereas the availability decreases but converges to a positive probability. This is shown in Fig. 2.7.

The complement to availability is the *unavailability* $U(t)$, which is the probability that a system is in the failed condition at time t, given it is in operating condition at time zero. In other words, unavailability is the percentage of time that the system is not available for the intended service in time period $(0,t)$, given it is operational at time zero. Availability, unavailability, and unreliability satisfy the following relationships:

$$A(t) + U(t) = 1.0 \tag{2.134}$$

$$0 \le U(t) \le p_f(t) \le 1 \tag{2.135}$$

For a nonrepairable system, the unavailability is equal to the unreliability; that is, $U(t) = p_f(t)$.

Determination of availability or unavailability of a system requires full accounting of the failure and repair processes. The basic elements that describe such processes are the failure density function $f(t)$ and the repair density function $g(t)$. Consider a system with a constant failure rate λ and a constant repair rate η. The availability and unavailability, respectively, are

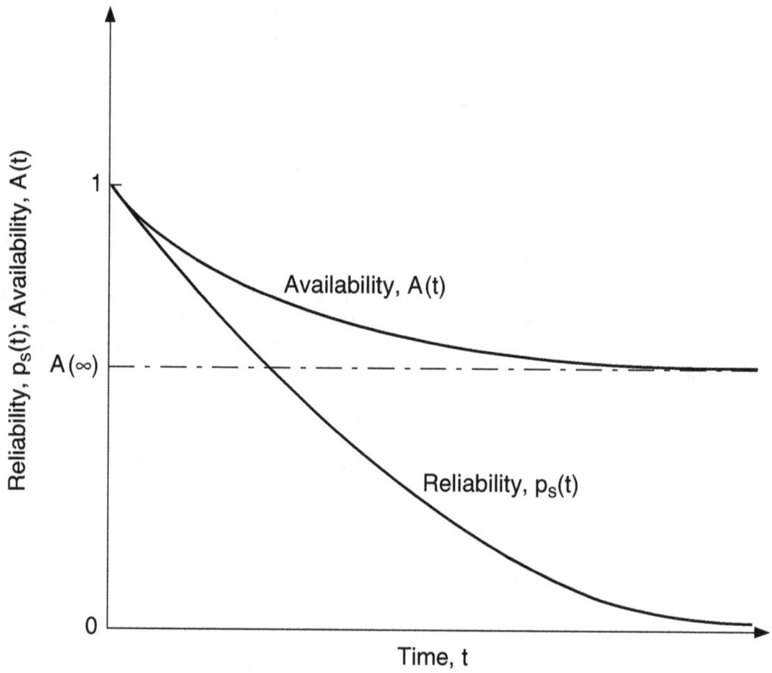

FIGURE 2.7 Variation of reliability and availability with time.

$$A(t) = \frac{\eta}{\lambda + \eta} + \frac{\lambda}{\lambda + \eta} e^{-(\lambda + \eta)t} \tag{2.136}$$

$$U(t) = \frac{\lambda}{\lambda + \eta}[1 - e^{-(\lambda + \eta)t}] \tag{2.137}$$

As the time approaches infinity ($t \to \infty$) the system reaches its stationary condition. Then, the stationary availability $A(\infty)$ and unavailability $U(\infty)$ are

$$A(\infty) = \frac{\eta}{\lambda + \eta} = \frac{1/\lambda}{1/\lambda + 1/\eta} = \frac{\text{MTTF}}{\text{MTTF} + \text{MTTR}} \tag{2.138}$$

$$U(\infty) = \frac{\lambda}{\lambda + \eta} = \frac{1/\eta}{1/\lambda + 1/\eta} = \frac{\text{MTTR}}{\text{MTTF} + \text{MTTR}} \tag{2.139}$$

Other properties for a system with constant failure rate and repair rate are summarized in Table 2.7.

2.6 MONTE CARLO SIMULATION

Simulation is a process of replicating the real world based on a set of assumptions and conceived models of reality (Ang and Tang, 1984). Because the purpose of a simulation model is to duplicate reality, it is a useful tool for evaluating the effect of different designs on system performance. The Monte Carlo procedure is a numerical simulation to reproduce random variables preserving the specified distributional properties.

TABLE 2.7 Summary of Constant Rate Model

Repairable systems	Nonrepairable systems
Failure process	
$m(t) = \lambda$	$m(t) = \lambda$
$p_s(t) = e^{-\lambda t}$	$p_s(t) = e^{-\lambda t}$
$p_f(t) = 1 - e^{-\lambda t}$	$p_f(t) = 1 - e^{-\lambda t}$
$f(t) = \lambda e^{-\lambda t}$	$f(t) = \lambda e^{-\lambda t}$
$\text{MTTF} = 1/\lambda$	$\text{MTTF} = 1/\lambda$
Repair process	
$r(t) = \eta$	$r(t) = 0$
$G(t) = 1 - e^{-\eta t}$	$G(t) = 0$
$g(t) = \eta e^{-\eta t}$	$g(t) = 0$
$\text{MTTR} = 1/\eta$	$\text{MTTR} = \infty$
Dynamic behavior of whole process	
$U(t) = \dfrac{\lambda}{\lambda+\eta}[1 - e^{-(\lambda+\eta)t}]$	$U(t) = 1 - e^{-\lambda t} = p_f(t)$
$A(t) = \dfrac{\eta}{\lambda+\eta} + \dfrac{\lambda}{\lambda+\eta}[1 - e^{-(\lambda+\eta)t}]$	$A(t) = e^{-\lambda t} = p_s(t)$
$w(t) = \dfrac{\lambda\eta}{\lambda+\eta} + \dfrac{\lambda^2}{\lambda+\eta}[1 - e^{-(\lambda+\eta)t}]$	$w(t) = f(t) = \lambda e^{-\lambda t}$
$\gamma(t) = \dfrac{\lambda\eta}{\lambda+\eta}[1 - e^{-(\lambda+\eta)t}]$	$\gamma(t) = 0$
$W(0, t) = \dfrac{\lambda\eta}{\lambda+\eta}t + \dfrac{\lambda^2}{(\lambda+\eta)^2}[1 - e^{-(\lambda+\eta)t}]$	$W(0, t) = p_f(t)$
$\Gamma(0,t) = \dfrac{\lambda\eta}{\lambda+\eta}t - \dfrac{\lambda\eta}{(\lambda+\eta)^2}[1 - e^{-(\lambda+\eta)t}]$	$\Gamma(0, t) = 0$
Stationary values of whole process	
$U(\infty) = \dfrac{\lambda}{\lambda+\eta} = \dfrac{\text{MTTR}}{\text{MTTF}+\text{MTTR}}$	$U(\infty) = 1$
$A(\infty) = \dfrac{\eta}{\lambda+\eta} = \dfrac{\text{MTTF}}{\text{MTTF}+\text{MTTR}}$	$A(\infty) = 0$
$w(\infty) = \dfrac{\lambda\eta}{\lambda+\eta} = \dfrac{1}{\text{MTTF}+\text{MTTR}}$	$w(\infty) = 0$
$\gamma(\infty) = \dfrac{\lambda\eta}{\lambda+\eta} = w(\infty)$	$\gamma(\infty) = 0$

Source: Adapted from Henley and Kumamoto, 1981.

2.6.1 Generation of Random Numbers

The most commonly used techniques to generate a sequence of pseudorandom numbers are those that employ some form of recursive computation. In principle, such recursive formulas are based on calculating the residual modulo of some integers of a linear transformation. The process of producing a random number sequence is completely deterministic. However, the generated sequence would appear to be uniformly distributed and independent.

Congruential methods for generating M random numbers are based on the fundamental congruence relationship which can be expressed as (Lehmer, 1951)

$$X_i = \{aX_{i-1} + c\} \pmod{m} \qquad i = 1, 2, \ldots, M \tag{2.140}$$

in which a is the multiplier, c is the increment, and m is an integer-valued modulus. Several forms of congruential algorithms are based on Eq. (2.140). Applying Eq. (2.140) to generate a random number sequence requires the specification of a, c, and m along with X_0, called the *seed.* Once the sequence of random number Xs is generated, the random number from the unit interval [0, 1] can be obtained as

$$U_i = \frac{X_i}{m} \qquad i = 1, 2, \ldots, M \tag{2.141}$$

The process of generating uniform random numbers is the building block in Monte Carlo simulations.

2.6.2 Classifications of Generation Algorithms

Algorithms to generate random variates from a specified distribution can be categorized into three types: *CDF-inverse method, acceptance-rejection (AR) method,* and *variable transformation method.* The CDF-inverse method is based on the fact that the CDF $F_X(x)$ of a random variable X is a nondecreasing function with respect to its value and $0 \le F_X(x) \le 1$.

For the great majority of continuous probability distributions applied in water resources engineering analysis, $F_X(x)$ is a strictly increasing function of x. Hence, there exists a unique relationship between $F_X(x)$ and u; that is, $u = F_X(x)$, as shown in Fig. 2.8. It can be shown that, if U is a standard uniform random variable defined over the unit interval [0, 1], the following relation holds,

$$X = F_X^{-1}(U) \tag{2.142}$$

To efficiently apply the CDF-inverse method for generating random numbers, an explicit expression between X and U is required. When analytical forms of the CDF inverse are not available, the CDF-inverse method requires solving

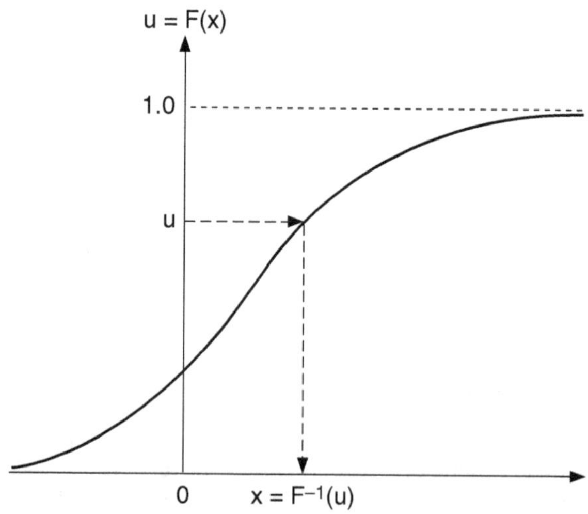

FIGURE 2.8 CDF-inverse method for random variate generation.

$$u = \int_{-\infty}^{x} f_X(t)\, dt \tag{2.143}$$

for x from the known u which could be inefficient.

An alternative approach is the *acceptance-rejection approach.* Its basic idea is to replace the original $f_X(x)$ by an appropriate PDF $h_X(x)$, from which random numbers can be easily and efficiently produced. The generated random number from $h_X(x)$ then is subject to testing before it is accepted as a random number from the original $f_X(x)$. This approach for generating random numbers has become popular.

The *variable transformation method* generates variates for the random variable of interest based on its known statistical relationship with other random variables whose variates can easily be produced. Therefore, the variable transformation method is sometimes effective for generating random variates from a complicated distribution based on variates produced from simple distributions.

2.6.3 Generation of Random Variates for Some Commonly Used Distributions

Several books have been written for generating univariate random numbers (Rubinstein, 1981; Dagpunar, 1988; Law and Kelton, 1991). A number of computer programs are available in the public domain. For example, STATLIB is a system for distribution of statistical software which is available through the e-mail address: stalib@lib.stat.cmu.edu. Press et al. (1989) present computer programs for Monte Carlo simulation which are available at low cost. There is other proprietary computer software frequently used for random number generation. An example of such software is the International Mathematical and Statistical Library (IMSL), which is available on most mainframe computer systems.

To generate multivariate random variates, the joint PDF is required. Compared with univariate random variate generators, algorithms for multivariate random variates are much more restricted to a few joint distributions, such as multivariate normal, multivariate lognormal (Parrish, 1990), multivariate gamma (Ronning, 1977), and others (Johnson, 1987). If the multivariate random variables involved are correlated with a mixture of marginal distributions, the joint PDF is difficult to formulate. Chang et al. (1994) proposed a practical multivariate Monte Carlo simulation procedure for problems involving mixtures of nonnormal random variables by utilizing the semiempirical formulas derived by Der Kiureghian and Liu (1985) and Liu and Der Kiureghian (1986). Other alternatives such as the recently developed polynomial normal transform technique (Chen and Tung, 2003) can be applied.

2.6.4 Monte Carlo Integration

In reliability analysis, computations of system and/or component reliability and other related quantities involve integration operations. A simple example is the time-to-failure analysis in which the reliability of a system within a time interval $(0, t)$ is obtained from Eq. (2.120). A more complex example of reliability computation is by the load-resistance interference in that the reliability is computed by Eq. (2.80).

For cases of integration on one or two dimensions and the integrands are well-behaved (e.g., no discontinuity), conventional numerical integration methods, such as trapezoidal approximation or Simpson's rule, are efficient and accurate. Gould and Tobochnik (1988) show that, in general, if the error for one-dimensional integration is $O(k^{-a})$, the error with a d-dimensional integration would be $O(k^{-a/d})$. This indicates that the accuracy of conventional numerical integration schemes decreases rapidly as the dimension of integration increases. For multiple integrals, the Monte Carlo method becomes a more suitable numerical technique for integration.

Consider the following integration problem

$$G = \int_{a}^{b} g(x)\, dx \qquad a \le x \le b \tag{2.144}$$

The *sample-mean method* is a simple Monte Carlo integration procedure based on the idea that the integral Eq. (2.144) can alternatively be expressed as

$$G = \int_a^b \left[\frac{g(x)}{f_X(x)} \right] f_X(x)\, dx \qquad a \le x \le b \tag{2.145}$$

in which $f_X(x) \ge 0$ is a PDF defined over $a \le x \le b$. The transformed integral given by Eq. (2.145) is equivalent to the computation of expectation of $g(X)/f_X(X)$, namely,

$$G = E_X\left[\frac{g(X)}{f_X(X)} \right] \tag{2.146}$$

with X being a random variable having a PDF $f_X(x)$ defined over $a \le x \le b$. The estimation of $E_X[g(X)/f_X(X)]$ by the Monte Carlo method is

$$\hat{G} = \frac{1}{M} \sum_{i=1}^{M} \frac{g(x_i)}{f_X(x_i)} \tag{2.147}$$

in which x_i is the random variate generated according to $f_X(x)$ and M is the number of random variates produced.

The sample-mean Monte Carlo integration involves the following steps:

1. Select $f_X(x)$ defined over the region of the integral from which M random variates are generated.
2. Compute $g(x_i)/f_X(x_i)$ for $i = 1, 2, \ldots, M$.
3. Calculate the sample average based on Eq. (2.147) as the estimate for G.

2.6.5 Variance Reduction Techniques

Because the Monte Carlo integration is a sampling procedure, the results obtained inevitably involve sampling errors which decrease as the sample size increases. Increasing sample size, for achieving higher precision, generally means an increase in computer time for generating random variates and data processing. *Variance reduction techniques* aim at obtaining high precision for the Monte Carlo simulation results without having to substantially increase the required sample size. Several variance reduction techniques are briefly described as follows.

Importance Sampling Technique. The *importance sampling technique* concentrates the distribution of sampling points in the part of the domain that is most "important" for the task rather than spreading the sampling points out evenly (Marshall, 1956). Rubinstein (1981) shows that the error associated with the sample-mean method can be considerably reduced if $f_X(x)$ is chosen to have a similar shape as $|g(x)|$. However, in practical implementations of this technique, consideration must be given to the tradeoff between the desired error reduction and the difficulties of sampling from $f_X(x)$, especially when $|g(x)|$ is not well behaved.

Antithetic-Variates Technique. Consider that Θ_1 and Θ_2 are two unbiased estimators of a parameter θ. The two estimators can be combined to form another estimator in the following fashion

$$\Theta_a = \frac{1}{2}(\Theta_1 + \Theta_2) \tag{2.148}$$

The new estimator Θ_a is also unbiased and has a variance as

$$\mathrm{Var}(\Theta_a) = \frac{1}{4}[\mathrm{Var}(\Theta_1) + \mathrm{Var}(\Theta_2) + 2\mathrm{cov}(\Theta_1, \Theta_2)] \tag{2.149}$$

From Eq. (2.149) one realizes that the variance associated with Θ_a could be reduced if Monte Carlo simulation can generate random variates that result in strongly negative correlation between Θ_1 and Θ_2.

In a Monte Carlo simulation, the estimators Θ_1 and Θ_2 are functions of generated random variates which, in turn, are related to the standard uniform random variates. Therefore, Θ_1 and Θ_2 are functions of the two standard uniform random variables U_1 and U_2. The objective to produce negative $\text{cov}[\Theta_1(U_1), \Theta_2(U_2)]$ can be achieved by producing U_1 and U_2 which are negatively correlated. A simple approach that generates negatively correlated uniform random variates with minimal computation is to let $U_1 = 1 - U_2$.

Correlated Sampling Technique. Correlated sampling techniques are especially effective for variance reduction when the primary objective of the simulation study is to evaluate small changes in system performance or to compare the difference in system performance between two specific designs (Rubinstein, 1981; Ang and Tang, 1984). For example, consider two designs *A* and *B* for the same system involving a vector of *N* random variables $\mathbf{X} = (X_1, X_2, \ldots, X_N)$ which could be correlated with a joint PDF $f_X(\mathbf{x})$ or independent, each with a marginal PDF, $f_i(x_i)$, $i = 1, 2, \ldots, N$. The performance of the system under the two designs can be expressed as

$$\Theta_A = g(\mathbf{a}, \mathbf{X}) \qquad \Theta_B = g(\mathbf{b}, \mathbf{X}) \tag{2.150}$$

in which $g(\;)$ is a function defining the system performance; and **a** and **b** are vectors of design parameters for designs *A* and *B*, respectively. Because the two performance measures Θ_A and Θ_B are dependent on the same random variables through the same performance function $g(\;)$, their estimators would be positively correlated. In this case, independently generating two sets of random variates, according to their probability laws for designs *A* and *B*, would still result in positive correlation between $\hat{\Theta}_A$ and $\hat{\Theta}_B$. However, an increase in correlation between $\hat{\Theta}_A$ and $\hat{\Theta}_B$ leading to further reduction in $\text{var}(\hat{\Delta}_\Theta)$ can be achieved by using a common set of standard uniform random variates for both designs *A* and *B*.

Stratified Sampling Technique. *Stratified sampling* is a well-established area in statistical sampling (Cochran, 1966). In spirit, it is similar to importance sampling. The basic difference between the two techniques is that the stratified sampling technique takes more observations at regions that are more "important" whereas importance sampling chooses an optimal PDF. Variance reduction by the stratified sampling technique is achieved through taking more samples at important subregions.

Latin Hypercubic Sampling (LHS) Technique. The *Latin hypercubic sampling technique* is a special method under the realm of stratified sampling that selects random samples of each random variable over its range in a stratified manner (McKay 1988). The technique divides the plausible range of each random variable into *K* equal-probability intervals for each stochastic variable involved. Within each interval, a random variate is generated resulting in *K* random variates for each stochastic variable. Then, the generated random sequences are randomly permuted for all random variables. The integral represented by Eq. (2.144) is estimated by

$$\hat{G} = \frac{1}{K} \sum_{k=1}^{K} g(\mathbf{x}_k) \tag{2.151}$$

Applications of the LHS techniques to uncertainty analysis of water resource engineering problems can be found in Table 2.8.

2.7 SYSTEM RELIABILITY

Most systems involve many subsystems and components whose performances affect in turn the performance of the system as a whole. The reliability of the entire system is affected not only by the reliability of individual subsystems and components, but also the interaction and configuration of the

TABLE 2.8 Example Applications of Reliability Analysis in Hydrosystem Design

Method	Hydrosystem type	References
Direct Integration	Levees	Plate & Ihringer (1986)
	Dams	Meon (1992)
	Urban drainage components	Adams & Papa (2000)
MFOSM	Storm sewers	Tang & Yen (1972); Tang et al. (1975); Yen & Tang (1976); Yen & Jun (1984);
	Culverts	Mays (1979); Yen et al. (1980); Tung & Mays (1980)
	Levees	Tung & Mays (1981a, b); Lee & Mays (1986)
	Bridges	Tung & Mays (1986)
	Open channels	Huang (1986); Yeh et al. (1993)
AFOSM	Storm sewers	Melching & Yen (1986)
	Dams	Cheng et al. (1982, 1993)
	Sea dikes & storm barriers	Vrijling (1987, 1993)
	Freeboard	Cheng et al (1986a)
	Bridge scour	Yen & Melching (1986)
	Runoff modeling	Melching et al (1990); Melching (1992)
	Groundwater	Sitar et al. (1987); Jang & Sitar (1990)
Monte Carlo	Bridge scour	Johnson (1992); Chang et al. (1993)
	Water distribution network	Bao & Mays (1990)
	Open channel	Mizumura & Ouazar (1992)
LHS	Sediment transport	Yeh & Tung (1993); Chang et al (1993)
	Water quality modeling	Sohrabi et al (2003); Melching & Bauwens (2001)
	Hydrologic modeling	Yu et al (2001); Christiaens & Feyen (2002)
Probabilistic PE	Open channels	Chang et al. (1993); Tung & Yeh (1993)
	Culverts	Lian & Yen (2003)
	Soil erosion	Chaves & Nearning (1991)
	Particle settling	Tsai & Franceschini (2002)
	Alluvial fan modeling	Zhao & Mays (1996)
Time-to-failure analysis	Water distribution systems	Mays (1989); Culinane et al (1992)
	Dams	Tang & Yen (1991, 1993); Kuo et al (1992)
Fault tree	Dams	Cheng et al (1982)
	Sea dikes	Vrijling (1987, 1993)
	Bridge scour	Yen & Melching (1991)
System reliability	Storm sewers	Yen et al (1976)
	Water distribution network	Tung (1985); Mays (1989); Duan et al (1990); Bouchart & Goulter (1992)
	Dams	Cheng et al (1982); Yen & Melching (1991)
	Storm surge barriers	Vrijling (1993)

subsystems and components. Furthermore, water resources systems involve multiple failure modes; that is, there are several potential modes of failure in which the occurrence of any or a combination of such failure modes constitutes the system failure (Ang and Tang, 1984). Due to the fact that different failure modes might be defined over the same stochastic variables space, the failure modes are correlated in general.

For a complex system involving many subsystems, components, and contributing stochastic variables, it is generally difficult, if not impossible, to directly assess the reliability of the system.

Instead, the system is divided into subsystems, components, and modes of operation, and the reliability of each is quantified.

2.7.1 Overview of System Reliability Evaluation

Evaluation of system reliability requires knowing what constitutes the system being in a failed or satisfactory state. Such knowledge is essential for system classification and dictates the methods to be used for system reliability determination.

From the reliability computation viewpoint, the classification of the system primarily depends on how the system performance is affected by its components or its modes of performance. A multiple-component system that requires all of its components to perform satisfactorily for satisfactory performance of the entire system is called a *series system.* Similarly, a single-component system involving several modes of operation is also viewed as a series system, if satisfactory performance of the system requires satisfactory performance of all its different modes of operation.

A second basic type of system is the *parallel system.* A parallel system is characterized by the property that the system would serve its intended purpose as long as at least one of its components or modes of operation performs satisfactorily.

For most real-life problems, systems are *complex systems* in which the components are arranged as a mixture of in-series and in-parallel, or in the form of a loop. In dealing with a complex system, the general approach is to reduce the system configuration, based on its component arrangement or modes of operation, to a simpler system for which the analysis can be performed easily. However, this goal may not be achieved for all cases, necessitating the development of a special procedure.

2.7.2 Reliability of Simple Systems

Series Systems. A series system requires that all of its components or modes of operation perform satisfactorily to ensure a satisfactory operation of the entire system. In the context of load-resistance interference, the failure probability of a series system involving M components or modes of operation can be expressed as

$$p_{f,\text{sys}} = P\left[\bigcup_{i=1}^{M} (W_i < 0)\right] = P\left[\bigcup_{i=1}^{M} (Z_i < -\beta_i)\right] \tag{2.152}$$

in which W_i is the performance function for the ith component or mode of operation.

In the framework of time-to-failure analysis, the reliability of a series system, when the performance of individual components is independent of each other, is

$$p_{s,\text{sys}}(t) = P\left(\bigcap_{i=1}^{M} F_i'\right) = \prod_{i=1}^{M} p_{s,i}(t) \tag{2.153}$$

in which $p_{s,i}(t)$ is the reliability of the ith component over the time interval $[0, t]$. Similarly, the availability of a series system involving M independent components is

$$A_{\text{sys}}(t) = \prod_{i=1}^{M} A_i(t) \tag{2.154}$$

in which $A_{\text{sys}}(t)$ and $A_i(t)$ are the availabilities of the system and the ith component, respectively, at time t.

For the special case of an exponential failure density function, the reliability of a series system involving M independent components is

$$p_{s,\text{sys}}(t) = \prod_{i=1}^{M} e^{-\lambda_i t} = \exp\left[-\left(\sum_{i=1}^{M} \lambda_i\right)t\right] \tag{2.155}$$

where λ_i is the constant failure rate of the ith component. Assuming an exponential repair function for each independent component, the availability for a series system is

$$A_{sys}(t) = \prod_{i=1}^{M} \left[\frac{\eta_i}{\lambda_i + \eta_i} + \frac{\lambda_i}{\lambda_i + \eta_i} e^{-(\lambda_i + \eta_i)t} \right] \tag{2.156}$$

in which η_i is the constant repair rate for the ith component. The stationary system availability is

$$A_{sys}(\infty) = \prod_{i=1}^{M} \left(\frac{\eta_i}{\lambda_i + \eta_i} \right) = \prod_{i=1}^{M} \left(\frac{\text{MTTF}_i}{\text{MTTR}_i + \text{MTTF}_i} \right) \tag{2.157}$$

in which MTTR_i and MTTF_i are, respectively, the mean-time-to-repair and mean-time-to-failure of the ith component.

Parallel Systems. For a parallel system, the entire system will perform satisfactorily if one or more of its components or modes of operation functions satisfactorily. In the framework of load-resistance interference for different modes of operation, the failure probability of a parallel system is

$$p_{f,sys} = P\left[\bigcap_{i=1}^{M} (W_i < 0) \right] = P\left[\bigcap_{i=1}^{M} (Z_i < -\beta_i) \right] \tag{2.158}$$

In the framework of time-to-failure analysis, the failure probability of a parallel system involving M independent components can be computed as

$$p_{f,sys}(t) = \prod_{i=1}^{M} p_{f,i}(t) \tag{2.159}$$

If each component has an exponential failure density function with parameter λ_i, for $i = 1, 2, \ldots, M$, the system failure probability is

$$p_{f,sys}(t) = \prod_{i=1}^{M} (1 - e^{-\lambda_i t}) \tag{2.160}$$

In a case where all components have an identical failure rate, that is, $\lambda_1 = \lambda_2 = \cdots = \lambda_M = \lambda$, the MTTF of the system is

$$\text{MTTF} = \frac{1}{\lambda} \sum_{i=1}^{M} \frac{1}{i} \tag{2.161}$$

The unavailability of a parallel system involving M independent components is

$$U_{sys}(t) = \prod_{i=1}^{M} U_i(t) \tag{2.162}$$

Under the condition of independent, exponential repair functions for the M components, the unavailability of a parallel system is

$$U_{sys}(t) = \prod_{i=1}^{M} \frac{\lambda_i}{\lambda_i + \eta_i} [1 - e^{-(\lambda_i + \eta_i)t}] \tag{2.163}$$

and the stationary system unavailability is

$$U_{sys}(\infty) = \prod_{i=1}^{M} \frac{\lambda_i}{\lambda_i + \eta_i} = \prod_{i=1}^{M} \left(\frac{\text{MTTR}_i}{\text{MTTR}_i + \text{MTTF}_i} \right) \tag{2.164}$$

***K*-Out-of-*M* Parallel Systems.** This is a parallel system of M components for which the system will function if K ($K < M$) or more components function. This type of system is also called a *partially redundant system.* The general reliability formula for this system is rather cumbersome. For components having identical reliability functions, that is, $p_{s,i}(t) = p_s(t)$, the system reliability, when component performances are independent, is

$$p_{s,\mathrm{sys}}(t) = \sum_{j=K}^{M} {}_MC_j\,[p_s(t)]^j\,[1 - p_s(t)]^{M-j} \tag{2.165}$$

in which ${}_MC_j$ is the binomial coefficient. Furthermore, if the failure density function is an exponential distribution, the system reliability can be expressed as

$$p_{s,\mathrm{sys}}(t) = \sum_{j=K}^{M} {}_MC_j\,[e^{-\lambda t}]^j\,[1 - e^{-\lambda t}]^{M-j} \tag{2.166}$$

The availability of the system can be obtained from substituting component availability for component reliability in Eq. (2.166).

Standby Redundant Systems. A *standby redundant system* is a parallel system in which only one component or subsystem is in operation. It is a special case of a K-out-of-M system with $K = 1$. If the operating component fails, then another component is operated. This type of system is different from the parallel system described previously where all the components are operating because standby units do not operate. The reliability for a system with M components out of which $M - 1$ units are on standby, is the probability that at most $M - 1$ components fail. Applying the Poisson distribution, this probability can be expressed by

$$p_{s,\mathrm{sys}}(t) = \sum_{i=0}^{M-1} \frac{(\lambda t)^i\, e^{-\lambda t}}{i!} \tag{2.167}$$

The above equation is valid under the following assumptions: The switch functions perfectly, the units are identical, the component failure rates are constant, the standby units are as good as new, and the unit failures are statistically independent. The mean-time-to-failure is

$$\mathrm{MTTF} = \int_0^\infty p_{s,\mathrm{sys}}(t)\,dt = \int_0^\infty \sum_{i=0}^{M-1} \frac{(\lambda t)^i e^{-\lambda t}}{i!}\,dt = \frac{M}{\lambda} \tag{2.168}$$

Because the system's operation is the result of a relay of a series of components, the system MTTF is the sum of the MTTF of the individual components.

2.7.3 Methods for Computing Reliability of Complex Systems

Many practical water resources systems, such as water distribution systems, have neither series nor parallel configuration. Evaluation of reliability for such complex systems is generally difficult. For some systems, it is possible to combine components into groups in such a manner that it would appear as in-series or in-parallel. For other systems, special techniques need to be developed that require a certain degree of insight and ingenuity from engineers. This section describes some of the potentially useful techniques for water resources system reliability evaluation.

State Enumeration Method. The *state enumeration method* lists all possible mutually exclusive states of the system components that define the state of the entire system. In general, for a system containing M components in which each can be classified into N operating states, there will be N^M possible states for the entire system. Once all the possible system states are enumerated, the states that result in successful system operation are identified, and the probability of the occurrence of each successful state is computed. The last step is to sum all of the successful state probabilities that yield the system reliability.

Path Enumeration Method. This is a very effective method for system reliability evaluation. A *path* is defined as a set of components or modes of operation that leads to a certain state of the system. In system reliability analysis, the system states of interest are those of failed state and operational state. Under this category, the *tie-set analysis* and *cut-set analysis* are the two well-known techniques.

Cut-Set Analysis. The *cut-set* is defined as a set of system components or modes of operation which, when failed, causes failure of the system. Cut-set analysis is powerful for evaluating system reliability for two reasons: (1) It can be easily programmed on digital computers for fast and efficient solution of any general system configuration, especially in the form of a network, and (2) the cut-sets are directly related to the modes of system failure and therefore identify the distinct and discrete ways in which a system may fail.

The cut-set method utilizes the minimum cut-set for calculating the system failure probability. The *minimum cut-set* is a set of system components that, when all fail, causes failure of the system, but when any one component of the set does not fail, the system does not fail. A minimum cut-set implies that all components of the cut-set must be in the failure state to cause system failure. Therefore, the components or modes of operation involved in the minimum cut-set are effectively connected in parallel, and each minimum cut-set is connected in series. Consequently, the failure probability of a system can be expressed as

$$p_{f,\text{sys}} = P\left[\bigcup_{i=1}^{I} C_i\right] = P\left[\bigcup_{i=1}^{I}\left(\bigcap_{j=1}^{J_i} F_{ij}\right)\right] \tag{2.169}$$

in which C_i is the ith of the total I minimum cut-sets, J_i is the total number of components or modes of operation in the ith minimum cut-set, and F_{ij} represents the failure event associated with the jth component or mode of operation in the ith minimum cut-set.

Tie-Set Analysis. As the complement of a cut-set, a *tie-set* is a minimal path of the system in which system components or modes of operation are arranged in series. Consequently, a tie-set fails if any of its components or modes of operation fail. All tie-sets are effectively connected in parallel; that is, the system will be in the operating state if any of its tie-sets are functioning. The main disadvantage of the tie-set method is that failure modes are not directly identified. Direct identification of failure modes is sometimes essential if a limited amount of a resource is available to place emphasis on a few dominant failure modes.

Conditional Probability Approach. The *conditional probability approach* starts with a selection of key components and modes of operation whose states (operational or failed) would decompose the entire system into simple series and/or parallel subsystems for which the reliability or failure probability of subsystems can be easily evaluated. Then, the reliability of the entire system is obtained by combining those of the subsystems using the conditional probability rule:

$$p_{s,\text{sys}} = p_{s|F_i'} \cdot p_{s,i} + p_{s|F_i} \cdot p_{f,i} \tag{2.170}$$

in which $p_{s|F_i'}$ and $p_{s|F_i}$ are the conditional system reliabilities given that component i is operational F_i', and failed F_i, respectively; and $p_{s,i}$ and $p_{f,i}$ are the reliability and failure probabilities of the ith component, respectively.

Except for very simple and small systems, a nested conditional probability operation is inevitable. Efficient evaluation of system reliability of a complex system hinges entirely on the proper selection of key components which generally is a difficult task when the system is large. Furthermore, the method cannot be easily adopted to computerized problem solving.

Fault Tree Analysis. Conceptually, *fault-tree analysis* traces from a system failure backward, searching for possible causes of the failure. A *fault tree* is a logical diagram representing the consequence of component failures (basic or primary failures) on system failure (top failure or top event). The major objective of fault tree construction is to represent the system condition, which may cause system failure, in a symbolic manner. In other words, the fault tree consists of event sequences that lead to system failure. There are actually two types of building blocks: *gate symbols* and *event sym-*

bols. Gate symbols connect events according to their causal relation such that there may be one or more input events but only one output event. Figure 2.9 shows the two commonly used gate symbols. Three types of commonly used event symbols are also shown in Fig. 2.9. A simple fault tree for the failure of an existing dam is shown in Fig. 2.10.

Henley and Kumamoto (1981) present heuristic guidelines for constructing fault trees:

1. Replace abstract events by less abstract events.
2. Classify an event into more elementary events.
3. Identify distinct causes for an event.
4. Couple trigger event with "no protection actions."
5. Find cooperative causes for an event.
6. Pinpoint component failure events.
7. Develop component failure.

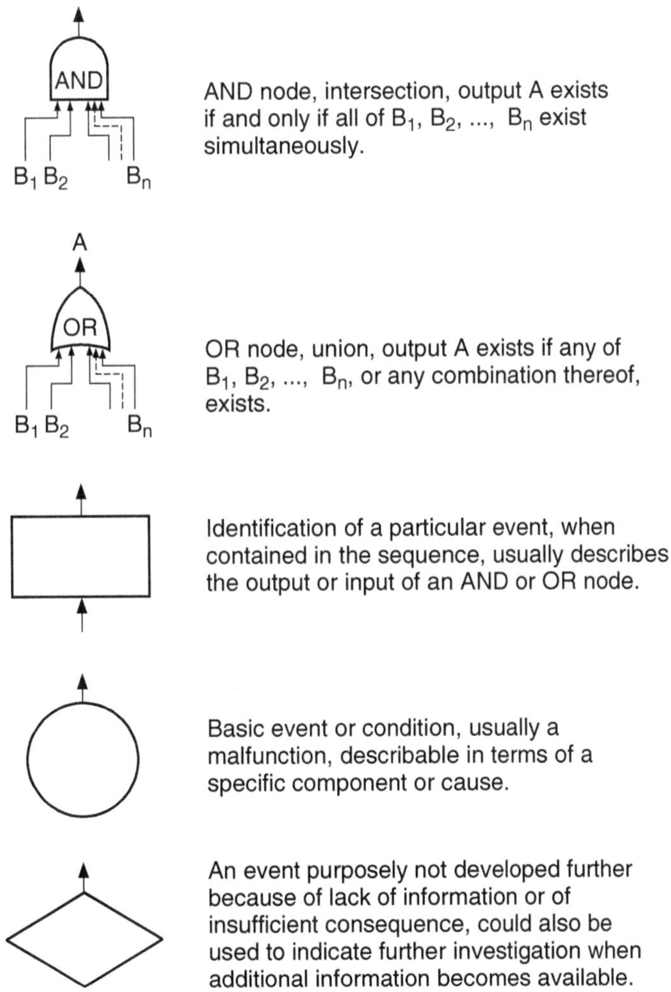

FIGURE 2.9 Some gate and event symbols used in a fault tree.

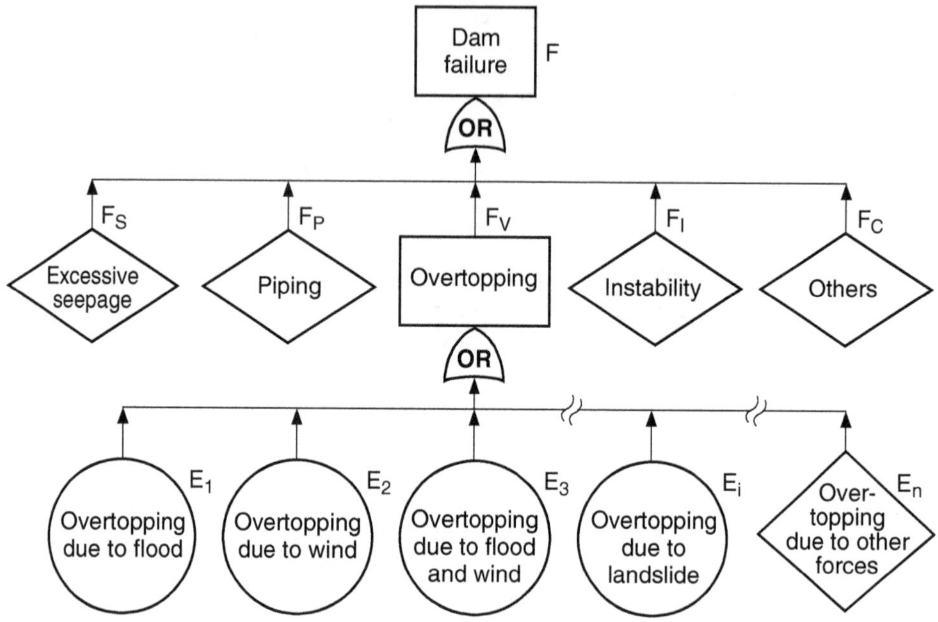

FIGURE 2.10 Simple fault tree for failure of existing dams. (*Cheng et al.*, 1982.)

To evaluate the fault tree, one should always start from the minimal cut-sets which, in essence, are *critical paths.* Basically, the fault tree evaluation comprises two distinct processes: (1) the determination of the logical combination of events that cause the top event (failure) expressed in minimal cut-sets and (2) the numerical evaluation of the failure probability. Several computer codes are available for generating cut-sets. For example, MOCUS (Fussell et al., 1974) was developed to obtain minimal cut-sets from fault trees.

The system availability $A_s(t)$ is the probability that the top event does not occur at time t, which is the probability of the system operating successfully when the top event is an OR combination of all system hazards. System reliability $p_{s,\text{sys}}(t)$ is the probability that the top event does not occur over time interval $(0, t)$. System reliability requires continuation of the nonexistence of the top event, and its value is less than or equal to the availability.

2.8 RISK-BASED DESIGN OF WATER RESOURCES SYSTEMS

All water resources engineering problems involve a number of interconnected and interrelated components. The analysis should take such interactions into account so that the overall behavior of the system is properly modeled. Water resources systems analysis using optimization techniques is described in Chap. 1 of this book. In this section, emphasis is placed on how to incorporate uncertainties for optimal water resources system designs.

Reliability analysis can be applied to the design of various hydraulic structures with or without considering *risk costs,* which are the costs associated with the failure of hydraulic structures or systems. Table 2.8 (on p. 2.44) lists some applications of reliability analysis of hydraulic structures that have appeared in the literature. The risk-based least-cost design of hydraulic structures promises to be, potentially, the most significant application of reliability analysis.

2.8.1 Risk-Based Design

The *risk-based design* of hydraulic structures integrates the procedures of economic, uncertainty, and reliability analyses in design practice. Engineers using a risk-based design procedure consider tradeoffs among various factors such as risk, economics, and other performance measures in hydraulic structure design. When risk-based design is embedded into an optimization framework, the combined procedure is called *optimal risk-based design.*

Basic Concept. Because the cost associated with the failure of a hydraulic structure cannot be predicted from year to year, a practical way to quantify it is to use an expected value on the annual basis. The *total annual expected cost* (*TAEC*) is the sum of the annual installation cost and annual expected damage cost which can be mathematically expressed as

$$\text{TAEC}(\mathbf{x}) = \text{FC}(\mathbf{x})*\text{CRF} + E(D|\mathbf{x}) \tag{2.171}$$

in which FC is the first or total installation cost, which is a function of the size and configuration $\mathbf{x}$ of the hydraulic structure considered; CRF is the *capital recovery factor,* which brings the present worth of the installation costs to an annual basis; and $E(D|\mathbf{x})$ is the annual expected damage cost associated with the structural failure.

Referring to Fig. 2.11, as the structural size increases, the annual installation cost increases and the annual expected damage cost associated with failure decreases. The optimal risk-based design determines the optimal structural size, configuration, and operation such that the annual total expected cost is minimized. Mathematically, the optimal risk-based design problem can be stated as:

$$\text{minimize TAEC}(\mathbf{x}) = \text{FC}(\mathbf{x})*\text{CRF} + E(D|\mathbf{x}) \tag{2.172a}$$

subject to

$$g_i(\mathbf{x}) = 0 \qquad i = 1, 2, \ldots, m \tag{2.172b}$$

where $g_i(\mathbf{x}) = 0$ are constraints representing the design specifications that must be satisfied. The solution to Eqs. (2.172a) to (2.172b) could be obtained through the use of appropriate optimization algorithms.

Evaluation of Annual Expected Damage Cost. In the optimal risk-based design of hydraulic structures, the thrust of the exercise is to evaluate $E(D|\mathbf{x})$ as the function of the PDFs of loading and resistance, damage function, and the types of uncertainty considered. The annual expected damage cost in the conventional risk-based hydraulic design can be computed as

$$\mathbf{E}_1(D|\mathbf{x}) = \int_{q_c^*}^{\infty} D(q|q_c^*)\, f(q)\, dq \tag{2.173}$$

where q_c^* is the deterministic flow capacity of a hydraulic structure subject to random loads following a PDF $f(q)$; and $D(q|q_c^*)$ is the damage function corresponding to the magnitude of the load q and the hydraulic structural capacity q_c^*. Due to the complexity of the damage function and the form of the PDF of loads, the integration of Eq. (2.173), in most real-life applications, is carried out numerically.

The conventional risk-based hydraulic design considers only the inherent hydrologic uncertainty due to the random occurrence of loads. It does not consider hydraulic and economic uncertainties. Furthermore, the probability distribution of the load to the water resources system is assumed known, which is generally not the case in reality.

Incorporation of Hydraulic Uncertainty. As described in Sec. 2.1.2, uncertainties also exist in the process of determining the flow-carrying capacity of the hydraulic structure. In other words, q_c is a quantity subject to uncertainty. From the uncertainty analysis of q_c, the statistical properties of q_c can be estimated. Hence, to incorporate the uncertainty features of q_c in the risk-based design, the annual expected damage can be calculated as

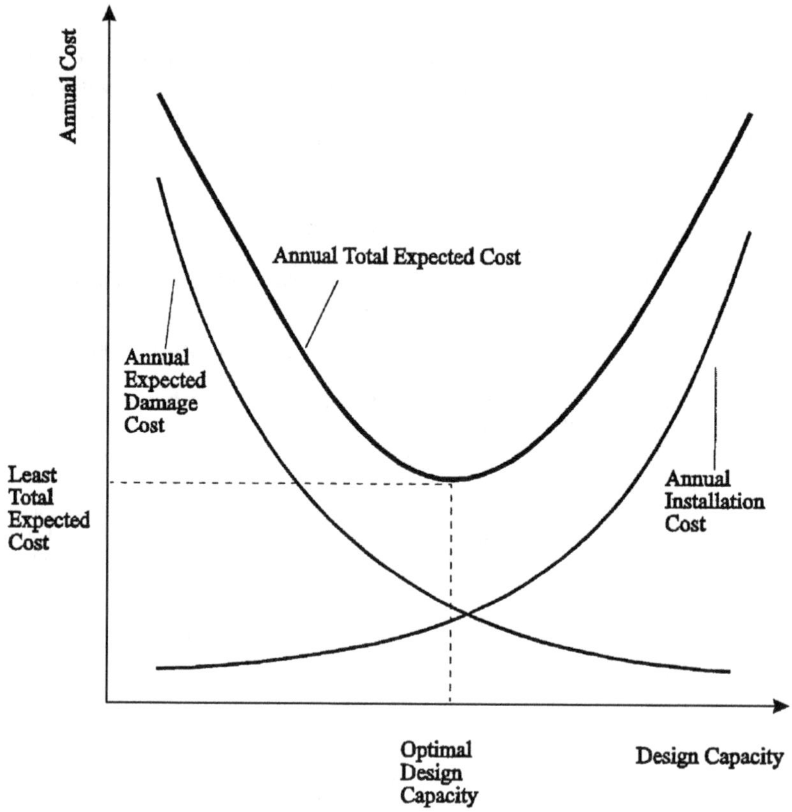

FIGURE 2.11 Optimal risk-based design concept.

$$E_2(D) = \int_0^{\infty} \left[\int_{q_c}^{\infty} D(q|q_c) f(q)\, dq \right] g(q_c) = \int_0^{\infty} E_1(D|q_c)\, g(q_c)\, dq_c \tag{2.174}$$

in which $g(q_c)$ is the PDF of random flow carrying capacity q_c.

Incorporation of Hydrologic Parameter Uncertainty. Because the occurrence of load is random by nature, the statistical properties such as the mean, standard deviation and skewness of the distribution calculated from a finite sample are also subject to sampling errors. Therefore, the estimated load of a specified return period q_T is also a random variable associated with its probability distribution (see Fig. 2.12), instead of being a single-valued quantity presented by its "average" as commonly done in practice. The expected damage corresponding to the load of a T-year return period can be expressed as

$$E(D_T|q_c^*) = \int_{q_c^*}^{\infty} D(q_T|q_c^*)\, h(q_T)\, dq_T \tag{2.175}$$

where $E(D_T|q_c^*)$ is the expected damage corresponding to a T-year load event given a known flow capacity of the hydraulic structure q_c^*; $h(q_T)$ is the sampling PDF of the estimated load with a T-year return period. Sampling distributions for a T-year event of some distributions in hydrologic frequency analysis have been developed by Stedinger (1983) and Chowdhury and Stedinger (1991).

To combine the inherent hydrologic uncertainty, represented by the PDF of the annual load event $f(q)$, and the hydrologic parameter uncertainty, represented by the sampling PDF for a load of a given return period $h(q_T)$, the annual expected damage cost can be written as

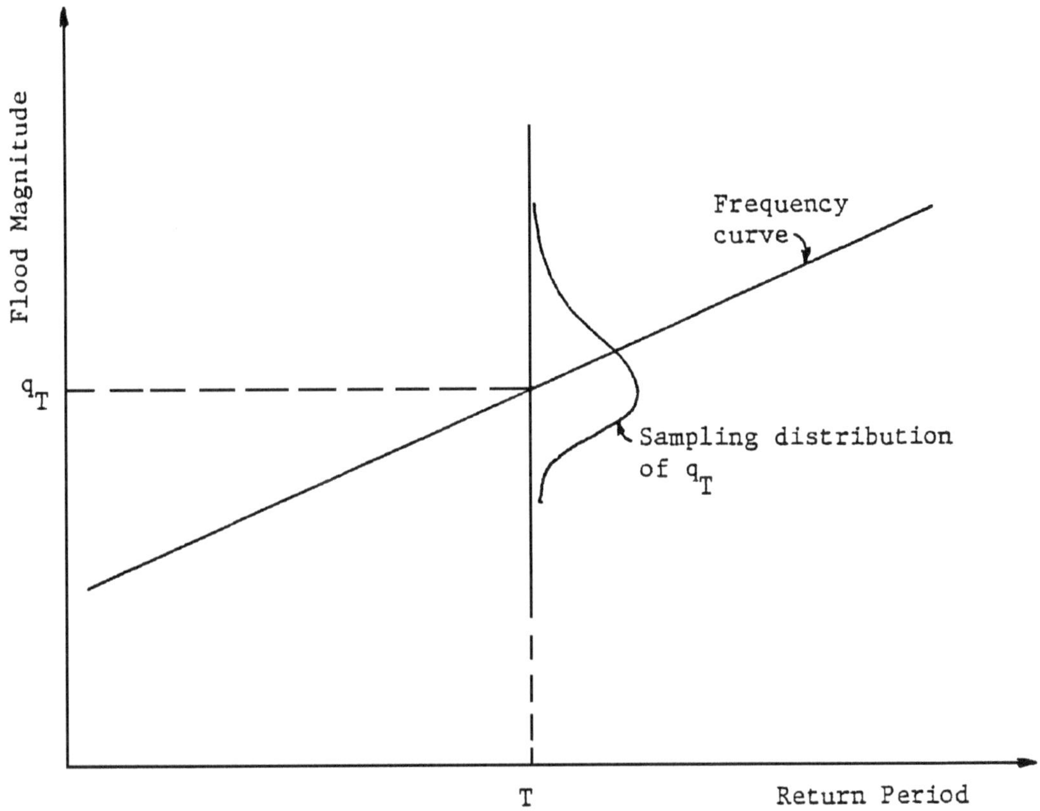

FIGURE 2.12 Sampling distribution associated with the flood estimator.

$$E_3(D|q_c^*) = \int_{q_c^*}^{\infty} \left[\int_{q_c^*}^{\infty} D(q_T|q_c^*)\, h(q_T|q)\, dq_T \right] f(q)\, dq \tag{2.176}$$

Incorporation of Combined Hydrologic and Hydraulic Uncertainties. To include hydrologic inherent and parameter uncertainties along with the hydraulic uncertainty associated with the flow carrying capacity, the annual expected damage cost can be written as

$$E_4(D) = \int_0^{\infty} \left[\int_{q_c}^{\infty} \left[\int_{q_c}^{\infty} D(q_T, q_c)\, h(q_T)\, dq_T \right] f(q)\, dq \right] g(q_c)\, dq_c$$
$$= \int_0^{\infty} E_3(D)\, g(q_c)\, dq_c \tag{2.177}$$

From the preceding formulations for computing annual expected damage in the risk-based design of hydraulic structures, one realizes that the mathematical complexity increases as more uncertainties are considered. However, to obtain an accurate estimation of annual expected damage associated with structural failure would require the consideration of all uncertainties, if such can be practically done. Otherwise, the annual expected damage would, in most cases, be underestimated, leading to inaccurate optimal design. Numerical investigations by Tung (1987) indicate that, without providing full account of uncertainties in the analysis, the resulting annual expected damage is significantly underestimated, even with a 75-year-long flood record.

Risk-Based Design Considering Intangible Factors. Besides the economic factors that can be quantified in monetary terms in the design of hydrosystems, there are other intangible factors that are noncommensurable and cannot be quantified. Some of the intangible factors might work against the principle of economic efficiency. Examples of intangible factors that are considered in the design and planning of hydrosystems may be potential loss of human lives, stability of water course, impacts on local society and environment, health hazards after floods, litigation potential, maintenance frequency of the systems, and others. The conventional optimal risk-based design yields the most economically efficient system, which may not be acceptable or feasible when other intangible factors are considered. Some work addressing the measure of risk in the case of intangible factors and acceptable risk issues can be found elsewhere (Jonkman et al., 2003; Vrijling et al, 1995).

As more intangible factors are considered in the risk-based design, it becomes a multiobjective or multicriteria decision making (MCDM) problem in which economic efficiency is but one of many factors to be considered simultaneously. Use of a multiple-criteria approach enhances more realistic decision-making and the design frequency so determined will be more acceptable in practice and defensible during litigation or negotiation with others. Tung et al. (1993) adopted the MCDM framework to incorporate intangible factors in risk-based design of highway drainage structures through which a more defensible extended risk-based design frequency can be determined from integrated consideration of tangible and intangible factors.

Risk-Based Design without Flood Damage Information. Conventional risk-based design and analysis of hydrosystems requires information with regard to various flood-related damages. Such information requires an extensive survey of type and value of various properties, economic and social activities, and other demographically related information in the regions that are affected by floods. For areas where flood-related damage data are unavailable, the conventional risk-based analysis cannot be implemented. In any design or analysis of a hydrosystem, one normally has to conduct a hydraulic simulation to delineate flood affected zone and other related flow characteristics, such as water depth and flow velocity. The hydraulic characteristics, combined with property survey data, would allow estimating flood damage for a specified flood event under consideration. In the situation where flood-related damage data are unavailable, the risk-based analysis of the relative economic merit of different flood defense systems can still be made by replacing the flood-related damage functions by relevant physical performance characteristics of the hydrosystems that are either required inputs for hydraulic modeling or can be easy extracted from model outputs. For examples, useful physical performance characteristics in urban drainage systems design and analysis could be pipe length (or street area) subject to surcharge, volume of surcharged water, maximum (or average) depth and velocity of overland flow. Although these performance characteristics may not completely reflect what the flood damages are, they are nevertheless provide a good indication of the potential seriousness of the flooding situation.

For a given design, the corresponding annual installation cost can be estimated. Also, the system responses under the different hydrologic loadings can be obtained by a proper hydraulic simulation model. Based on the annual project installation cost of the system and the expected hydraulic response of the system, a tradeoff analysis can be performed by examining the marginal improvement in hydraulic responses due to a one-unit increase in capital investment. Referring to Fig. 2.13 for a study to upgrade the level of protection for a urban drainage system in Hong Kong (Tung and So, 2003), it is observed that the annual expected surcharge volume decreases as the annual capital cost of the system increases due to increasing level of protection. The marginal cost (MC) corresponding to one reduction in surcharge volume can be written as $\mathrm{MC} = -\partial C/\partial S_v$ with C being the capital cost and S_v being the surcharge volume. As can be seen the value of MC starts very low from the existing system and increases to an annual capital cost around HK\$0.6M (which correspond to a 10-yr protection) beyond which the rate of increase in capital investment per unit reduction in surcharge volume becomes very high. From the trend of marginal cost, a decision-maker would be able to choose a sensible level of protection for project implementation.

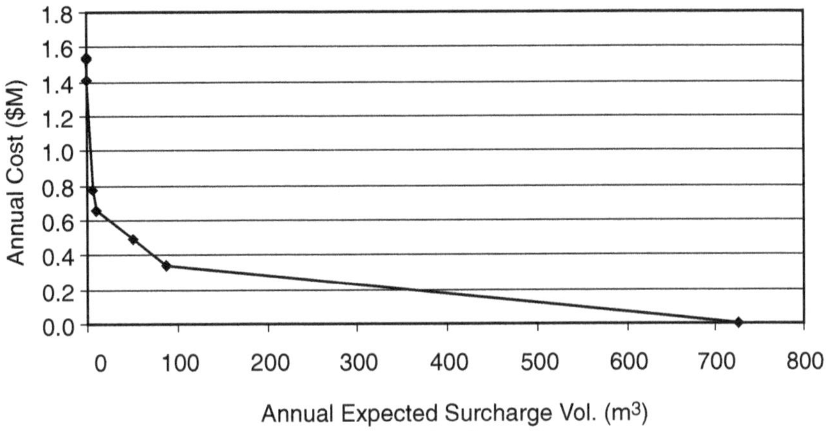

FIGURE 2.13 Annual project cost vs. annual expected surcharge volume.

2.8.2 Optimization of Water Resources Systems by Chance-Constrained Methods

In all fields of science and engineering, the decision-making process is dependent on several parameters describing system behavior and characteristics. More often than not, at least one of these system parameters cannot be assessed with certainty. In a system optimization model, if some of the coefficients in the constraints are random variables, compliance with the constraints, under a given set of solutions, cannot be ensured with certainty. Due to the random nature of the constraint coefficients, a certain likelihood always exists that constraints will be violated.

The basic idea of *chance-constrained methods* is to find the solution to an optimization problem such that the constraints will be complied with at a specified reliability. Chance-constrained formulations have been applied to groundwater management (Tung, 1986), reservoir operations (Loucks et al., 1981), waste load allocation (Tung and Hathhorn, 1990), and water distribution systems (Lansey et al., 1989). This section briefly describes the properties of chance-constrained models.

Refer to the general nonlinear optimization problem as stated in Eq. (1.17) and consider a constraint $g(\mathbf{x}) \leq b$ with $\mathbf{x}$ being a vector of decision variables. In general, decision variables $\mathbf{x}$ in an optimization model are controllable or have little uncertainty. Suppose that some of the parameters on the left-hand side of the constraint $g(\mathbf{x})$ and/or the right-hand side (RHS) coefficient $\mathbf{b}$ are subject to uncertainty. Under this circumstance, the chance-constrained formulation expresses the original constraint in a probabilistic format as

$$P[g(\mathbf{x}) \leq b] \geq \alpha \tag{2.178}$$

where $P[\ \ ]$ is the probability and α is the specified reliability for constraint compliance. Since the above chance-constrained formulation involves probability, it is not mathematically operational for algebraic solution. For this reason, the so-called *deterministic equivalent* must be derived. There are three cases in which the random elements in Eq. (2.178) could occur: (1) Only elements in $g(\mathbf{x})$ are random, (2) only the RHS b is random, and (3) both $g(\mathbf{x})$ and b are random.

The simplest case is case (2) where only the RHS coefficient b is random. The constraint can be rewritten as

$$P[g(\mathbf{x}) \leq B] \geq \alpha \tag{2.179}$$

where B is a random RHS coefficient associated with a CDF, $F_B(\ \ )$. The deterministic equivalent of the original chance-constraint Eq. (2.179) is

$$g(\mathbf{x}) \le b_{1-\alpha} \tag{2.180}$$

where $b_{1-\alpha}$ is the $(1-\alpha)$th quantile of the random RHS coefficient B satisfying $P[B \le b_{1-\alpha}] = 1-\alpha$. If the RHS coefficient B is a normal random variable with mean μ_B and standard deviation σ_B, Eq. (2.180) can be written as

$$g(\mathbf{x}) \le \mu_B + z_{1-\alpha}\,\sigma_B \tag{2.181}$$

with $z_{1-\alpha}$ being the $(1-\alpha)$th standard normal quantile.

In the third case, both the left-hand side and right-hand side of a constraint are random. They are denoted as $G(\mathbf{x})$ and B, respectively. The chance-constraint can then be expressed as

$$P[G(\mathbf{x}) - B \le 0] \ge \alpha \tag{2.182}$$

The deterministic equivalent of Eq. (2.182) can be derived as

$$F^{-1}_{G(\mathbf{x})-B}(\alpha) \ge 0 \tag{2.183}$$

where $F^{-1}_{G(\mathbf{x})-B}(\alpha)$ is the inverse of the CDF of random $G(\mathbf{x}) - B$ taken on the value of α.

As a special case, consider a linear programming (LP) formulation as

$$\text{Maximize } \mathbf{c}^T\mathbf{x}$$

subject to

$$\mathbf{Ax} \le \mathbf{b}$$

$$\mathbf{x} \ge \mathbf{0}$$

in which $\mathbf{A}$ is the technological coefficients matrix; $\mathbf{b}$ is the vector of RHS coefficients; and $\mathbf{c}$ is the vector of objective function coefficients. Suppose that the elements in $\mathbf{A}$ and $\mathbf{b}$ are subject to uncertainty. By imposing a reliability restriction α on the system constraints, the constraints in the above LP model can be transformed into the following chance-constrained formulation:

$$P[\mathbf{Ax} \le \mathbf{b}] \ge \alpha \tag{2.184}$$

In a chance-constrained LP model, the elements in $\mathbf{A}$, $\mathbf{b}$, and $\mathbf{c}$ can be random variables. When the objective function coefficient c_j's are random variables, it is common to replace them by their expected values. Consider the general case where elements A_{ij} and B_i are simultaneously random variables. Assume that random technological coefficients and random RHS coefficients are correlated within a constraint, and these coefficients are uncorrelated between constraints. The chance-constraint format can be written as

$$P\left[\sum_{j=1}^{n} A_{ij}x_j - B_i \le 0\right] \ge \alpha_i \tag{2.185}$$

The deterministic equivalent of Eq. (2.185) can be derived, assuming a normal distribution for the random terms in Eq. (2.185), as

$$\sum_{j=1}^{n} E[A_{ij}]x_j + \Phi^{-1}(\alpha_i)\sqrt{\mathbf{x}^T\mathbf{C}_i\mathbf{x} + 2\sum_{j=1}^{n} x_j \operatorname{cov}(A_{ij}, B_i) + \sigma^2_{B_i}} \le E(B_i), \quad i = 1, 2, \ldots, n \tag{2.186}$$

where $E[A_{ij}]$ is an expectation of the random technological coefficient A_{ij}, $\mathbf{C}_i$ is an $n \times n$ covariance matrix of n random technological coefficients $(A_{i1}, A_{i2}, \ldots, A_{in})$ in the ith constraint, and $\operatorname{cov}(A_{ij}, B_i)$ is the covariance between the random technological coefficient A_{ij} and the random RHS coefficient B_i for the ith constraint.

REFERENCES

Abramowitz, M. and I. A. Stegun, eds. (1972). *Handbook of Mathematical Functions with Formulas, Graphs and Mathematical Tables,* 9th ed., Dover, New York.

Adams, B. J. and F. Papa (2000). *Urban Stormwater Management Planning with Analytical Probabilistic Models,* John WileyNew York.

Ang, A. H.-S. (1973). "Structural Risk Analysis and Reliability-Based Design," *Journal of Structural Engineering Division,* ASCE, 99(9):1891–1910.

Ang, A. H.-S., and W. H. Tang (1984). *Probability Concepts in Engineering Planning and Design, Vol. II: Decision, Risk, and Reliability,* John Wiley, New York.

Bao, Y., and L. W. Mays (1990) "Model for Water Distribution System Reliability," *Journal of Hydraulic Engineering,* ASCE, 116(9):1119–1137.

Berthouex, P. M. (1975)."Modeling Concepts Considering Process Performance, Variability, and Uncertainty," in T. M. Keinath and M. P. Wanielista, eds., *Mathematical Modeling for Water Pollution Control Processes,* Ann Arbor Science, Ann Arbor, Mich., pp. 405–439.

Bouchart, F., and I. Goulter (1992). "Selecting Valve Location to Optimize Water Distribution Network Reliability," in J. T. Kuo and G. F. Lin, eds., *Stochastic Hydraulics '92, Proceedings 6th IAHR International Symposium, Taipei,* Water Resources Publications, Littleton, Colo., pp. 155–162.

Chang, C. H., J. C. Yang, and Y. K. Tung (1993)."Sensitivity and Uncertainty Analyses of a Sediment Transport Model: A Global Approach," *Journal of Stochastic Hydrology and Hydraulics,* 7(4): 299–314.

Chang, C. H., Y. K. Tung, and J. C. Yang (1994)."Monte Carlo Simulation for Correlated Variables with Marginal Distributions," *Journal of Hydraulic Engineering,* ASCE, 120(2): 313–331.

Chang, C. H., Y. K. Tung, and J.C. Yang (1995). Evaluation of Probabilistic Point Estimate Methods, *Applied Mathematical Modelling*, 19(2): 95-105.

Chaves, H. M. L. and Nearing, M. A. (1991)."Uncertainty Analysis of the WEPP Soil-Erosion Model," *Transactions of ASAE*, 34 (6): 2437-2445.

Chen, X. Y. and Y. K. Tung (2003). "Investigation of Polynomial Normal Transformation, *Journal of Structural Safety*, 25: 423–445.

Cheng, S. T., B. C. Yen, and W. H. Tang (1982). "Overtopping Risk for an Existing Dam," *Hydraulic Engineering Series,* no. 37, Department of Civil Engineering, University of Illinois at Urbana-Champaign, Ill.

Cheng, S. T., B. C. Yen, and W. H. Tang (1986b). "Wind Induced Overtopping Risk of Dams," in B. C. Yen, ed., *Stochastic and Risk Analysis in Hydraulic Engineering,* Water Resources Publications, Littleton, Colo., pp. 48–58.

Cheng, S. T., B. C. Yen, and W. H. Tang (1986b). "Sensitivity of Risk Evaluation to Coefficient of Variation," in B. C. Yen, ed., *Stochastic and Risk Analysis in Hydraulic Engineering,* Water Resources Publications, Littleton, Colo., pp. 266–273.

Cheng, S. T., B. C. Yen, and W. H. Tang (1993). "Stochastic Risk Modeling of Dam Overtopping," in B. C. Yen and Y. K. Tung, eds., *Reliability and Uncertainty Analyses in Hydraulic Design,* ASCE, pp. 123–132.

Chowdhury, J. U. and J. R. Stedinger (1991). "Confidence Interval for Design Floods with Estimated Skew Coefficient," *Journal of Hydraulic Engineering,* 117(7):811–831.

Christiaens, K. and J. Feyen (2002). Use of Sensitivity and Uncertainty Measures in Distributed Hydrological Modeling with an Application to the MIKE SHE Model, *Water Resources Research,* 38 (9): 1169.

Cochran, W. (1991) *Sampling Techniques,* 2d ed., Wiley, New York, 1966.

Cullinane, M. J., K. E. Lansey, and L. W. Mays (1992). "Optimization-Availability-Based Design of Water Distribution Networks," *Journal of Hydraulic Engineering,* ASCE, 118(3):420–441.

Dagpunar, J. (1988). *Principles of Random Variates Generation,* Oxford University Press, New York.

Der Kiureghian, A. and P. L. Liu (1985). "Structural Reliability under Incomplete Probability Information." *Journal of Engineering Mechanics,* ASCE 112(1):85–104.

Devore, J. L. (1987). *Probability and Statistics for Engineering and Sciences,* 2d ed., Brooks/Cole, Monterey, Calif.

Duan, N., L. W. Mays, and K. E. Lansey (1990). "Optimal Reliability-Based Design of Pumping and Distribution Systems," *Journal of Hydraulic Engineering,* ASCE, 116(2):249–268.

Epstein, B. (1948). "Some Application of the Mellin Transform in Statistics," *Annals of Mathematical Statistics,* 19:370–379.

Fisher, R. A. and L. H. C. Tippett (1928). "Limiting Forms of the Frequency Distribution of the Largest or Smallest Member of a Sample," *Proceedings of the Cambridge Philosophical Society,* 24: 180–190.

Fussell, J. B., E. B. Henry, and N. H. Marshall (1974). "MOCUS—A Computer Program to Obtain Minimal Cut Sets from Fault Trees," ANCR-1156.

Giffin, W. C. (1975) *Transform Techniques for Probability Modeling,* Academic New York.

Golub, G. H. and C. F. Van Loan (1989). *Matrix Computations,* 2d edition, Johns Hopkins University Press, Baltimore, Md. (1988).

Gould, H., and J. Tobochnik (1998). *An Introduction to Computer Simulation Methods: Applications to Physical Systems, Part 2,* Addison-Wesley, Reading, Mass.

Gumbel, E. J. (1958). *Statistics of Extremes,* Columbia University Press, New York.

Haan, C. T. (1977). *Statistical Methods in Hydrology,* Iowa State University Press.

Harr, M. E. (1989). "Probabilistic Estimates for Multivariate Analyses," *Applied Mathematical Modelling,* 13: 313–318.

Hasofer, A. M. and N. C. Lind (1974). "Exact and Invariant Second-Moment Code Format," *Journal of Engineering Mechanics Div.,* ASCE, 100(EM1):111–121.

He, J. and G. Sällfors (1994). "An Optimal Point Estimate Method for Uncertainty Studies," *Applied Mathematical Modelling,* 18(9): 494-499.

Henley, E. J. and H. Kumamoto (1981). *Reliability Engineering and Risk Assessment,* Prentice-Hall, Englewood Cliffs, N. J.

Huang, K. Z. (1986). "Reliability Analysis of Hydraulic Design of Open Channel," in B. C. Yen, ed., *Stochastic and Risk Analysis in Hydraulic Engineering,* Water Resources Publications, Littleton, Colo.

Interagency Advisory Committee on Water Data (1982). *Guidelines for Determining Flood Flow Frequency,* Bulletin 17B, U.S. Department of Interior, U.S. Geological Survey, Office of Water Data Coordination, Reston, Va.

Jang, Y. S. and N. Sitar (1990). "Reliability Analysis of Contaminant Transport Through Clay Liners," Department of Civil Engineering, Report no. UCB/GT-90/04.

Johnson, N. L. and S. Kotz (1976). *Distributions in Statistics: Continuous Multivariate Distributions,* John Wiley, New York.

Johnson, M. E. (1987). *Multivariate Statistical Simulation,* John Wiley, New York.

Johnson, P. A. (1992). "Reliability-Based Pier Scour Engineering," *Journal of Hydraulic Engineering,* ASCE, 118(10): 1344–1358.

Jonkman, S. N., P. H. A. J. M. van Gelder, and J. K. Vrijing (2003). "An Overview of Quantitative Risk Measures for Loss of Life and Economic Damage." *Journal of Hazardous Materials,* A99: 1-30.

Kapur, K. C. and L. R. Lamberson (1977). *Reliability in Engineering Designs,* John Wiley, New York.

Kapur, K. C. (1988). "Chapter 19—Mathematical and Statistical Methods and Models in Reliability and Life Studies," in W. G. Ireson and C. F. Crombs, Jr., eds., *Handbook of Reliability Engineering and Management,* McGraw-Hill, New York.

Kendall, M., A. Stuart, and J. K. Ord (1987). *Kendall's Advanced Theory of Statistics, Vol. 1: Distribution Theory,* 5th Edition, New York, Oxford University Press.

Kraus, J. W. (1988). "Chapter 15—Maintainability and Reliability," in W. G. Ireson and C. F. Coombs, Jr., eds., *Handbook of Reliability Engineering and Management,* McGraw-Hill, New York.

Kuo, J. T., B. C. Yen, J. I. Lin, G. P. Hwang, and S. K. Hsu (1992). "Dam Safety Inspection Scheduling for Shihmen Reservoir," in J. T. Kuo and G. F. Lin, eds., *Stochastic Hydraulics '92, Proceedings of the Six IAHR International Symposium, Taipei,* Water Resources Publications, Littleton, Colo., pp. 99–106.

Lansey, K. E., N. Duan, L. W. Mays, and Y. K. Tung (1989). "Water Distribution System Design under Uncertainties," *Journal of Water Resources Planning and Management,* 115(5): 630–645.

Law, A. M. and W. D. Kelton, (1986). *Simulation Modeling and Analysis,* McGraw-Hill, New York, 1991.

Lee, H. L. and L. W. Mays (1986). "Hydraulic Uncertainties in Flood Levee Capacity," *Journal of Hydraulic Engineering,* ASCE, 112(10): 928–934.

Lehmer, D. H. (1951). "Mathematical Methods in Large-Scale Computing Units," *Ann. Comp. Lab.,* Harvard University, 26: 141–146.

Li, K. S. (1992). "Point Estimate Method for Calculating Statistical Moments," *Journal of Engineering Mechanics,* ASCE, 118(7):1506-1511.

Lian, Y. Q. and B. C. Yen (2003). "Comparison of Risk Calculation Methods for a Culvert," *Journal of Hydraulic Engineering,* ASCE, 129 (2), February: 140–152.

Liu, P. L. and A. Der Kiureghian (1986). "Multivariate Distribution Models with Prescribed Marginals and Covariances," *Probabilistic Engineering Mechanics,* 1(2):105–112.

Loucks, D. P., J. R. Stedinger, and D. A. Haith (1986). *Water Resources Systems Planning and Analysis,* Prentice-Hall, Englewood Cliffs, N.J.

Madsen, H. O., S. Krenk, and N. C. Lind (1986). *Methods of Structural Safety,* Prentice-Hall, Englewood Cliffs, N.J..

Marshall, A. W. (1956). "The Use of Multi-Stage Sampling Schemes in Monte Carlo Computations," in M. A. Meyer, ed., *Symposium on Monte Carlo Methods,* John Wiley, New York.

Mays, L. W. (1979). "Optimal Design of Culverts under Uncertainties," *Journal of Hydraulic Division,* ASCE, 105(5):443–460, 1979.

Mays, L. W. ed. (1989). *Reliability Analysis of Water Distribution Systems,* ASCE, New York.

Mays, L. W. and Y. K. Tung (1992). *Hydrosystems Engineering and Management.* McGraw-Hill, New York.

McBean, E. A., J. Penel, and K.-L. Siu (1984). "Uncertainty Analysis of a Delineated Floodplain," *Canadian Journal of Civil Engineers,* 11:387–395.

McKay, M. D. (1988). "Sensitivity and Uncertainty Analysis Using a Statistical Sample of Input Values," in Y. Ronen, ed., *Uncertainty Analysis,* CRC Press, Boca Raton, Fla.

Melchers, R. E. (1987). *Structural Reliability Analysis and Prediction,* Ellis Horwood, Chichester, England.

Melching, C. S. (1992). "An Improved First-Order Reliability Approach for Assessing Uncertainties in Hydrologic Modeling," *Journal of Hydrology,* 132: 157–177.

Melching, C.S. and W. Bauwens (2001). "Uncertainty in Coupled Nonpoint Source and Stream Water-Quality Models," *J. of Water Resources Planning and Management,* ASCE, 127 (6): 403–413.

Melching, C. S., H. G. Wenzel, Jr., and B. C. Yen (1990). "A Reliability Estimation in Modeling Watershed Runoff with Uncertainties," *Water Resources Research,* 26: 2275–2286.

Meon, G. (1992). "Overtopping Probability of Dams under Flood Load," in J. T. Kuo and G. F. Lin, eds., *Stochastic Hydraulics '92, Proceedings of the Sixth IAHR International Symposium, Taipei,* Water Resources Publications, Littleton, Colo., pp. 99–106.

Mizumura, K. and D. Ouazar (1992)."Stochastic Characteristics of Open Channel Flow," in J. T. Kuo and G. F. Lin, eds., *Stochastic Hydraulics '92, Proceedings of the Sixth IAHR International Symposium, Taipei,* Water Resources Publications, Littleton, Colo., 1992, pp. 99–106.

Panchalingam, G. and M. E. Harr (1994). "Modelling of Many Correlated and Skewed Random Variables," *Applied Mathematical Modelling,* 18(11): 635-640.

Park, C. S. (1987). "The Mellin Transform in Probabilistic Cash Flow Modeling," *The Engineering Economist,* 32(2):115–134.

Parrish, R. S. (1990). "Generating Random Deviates from Multivariate Pearson Distributions." *Computational Statistics & Data Analysis,* 9:283–296.

Pieruschka, E. (1963). *Principles of Reliability,* Prentice-Hall, Englewood Cliffs, N.J.

Plate, E. J., and J. Ihringer (1986). "Failure Probability of Flood Levees on a Tidal River," in B. C. Yen, ed., *Stochastic and Risk Analysis in Hydraulic Engineering,* Water Resources Publications, Littleton, Colo., pp. 48–58.

Press, W. H., B. P. Flannery, S. A. Teukolsky, and W. T. Vetterling (1989). *Numerical Recipes—The Art of Scientific Computing (Fortran Version),* Cambridge University Press, New York.

Rackwitz, R. (1976). "Practical Probabilistic Approach to Design," *Bulletin* 112, Comité Européan du Beton, Paris.

Ronning, G. (1977) "A Simple Scheme for Generating Multivariate Gamma Distributions with Non-negative Covariance Matrix," *Technometrics,* 19(2):179–183.

Rosenblueth, E. (1975). "Point Estimates for Probability Moments," *Proceedings, National Academy of Science,* 72(10):3812–3814.

Rosenblueth, E. (1981). "Two-Point Estimates in Probabilities," *Applied Mathematical Modelling,* 5:329–335.

Rubinstein, R. Y. (1981). *Simulation and the Monte Carlo Method,* John Wiley, New York.

Shinozuka, M. (1983) "Basic Analysis of Structural Safety," *Journal of Structural Engineering Div.,* ASCE, 109(3):721–740.

Sitar, N., J. D. Cawlfield, and A. Der Kiureghian (1987). "First Order Reliability Approach to Stochastic Analysis of Subsurface Flow and Contaminant Transport," *Water Resources Research,* 23(5): 794–804.

Sohrabi, T. M., A. Shirmohammadi, T. W., Chu, H. Montas, A. P. Nejadhashemi (2003). "Uncertainty Analysis of Hydrologic and Water Quality Predictions for a Small Watershed Using SWAT2000," *Environmental Forensics,* 4 (4): 229–238.

Springer, M. D. (1979). *The Algebra of Random Variables,* John Wiley, New York.

Stedinger, J. R. (1983). "Confidence Intervals for Design Events," *Journal of Hydraulic Engineering,* 109(1):13–27.

Stedinger, J. R., R. M. Vogel, and E. Foufoula-Georgiou (1992). "Chapter 18: Frequency Analysis of Extreme Events," in D. R. Maidment, ed., *Handbook of Hydrology,* McGraw-Hill, New York.

Tang, W. H. and B. C. Yen (1972). "Hydrologic and Hydraulic Design under Uncertainties," *Proceedings, International Symposium on Uncertainties in Hydrologic and Water Resources Systems,* Tucson, Ariz., 2:868–882.

Tang, W. H., L. W. Mays, and B. C. Yen (1975) "Optimal Risk-Based Design of Storm Sewer Networks," *Journal of Environmental Engineering Division,* ASCE, 103(3):381–398.

Tang, W. H. and B. C. Yen (1991). "Dam-Safety Inspection Scheduling," *Journal of Hydraulic Engineering,* ASCE, 117(2):214–229.

Tang, W. H. and B. C. Yen (1993). "Probabilistic Inspection Scheduling for Dams," in B. C. Yen and Y. K. Tung, eds., *Reliability and Uncertainty Analyses in Hydraulic Design,* ASCE, pp. 107–122.

Tsai, C.W.S. and S. Franceschini, (2003). "Uncertainty Analysis Using an Improved Point Estimate Method." *AGU* 2002 *Fall Meeting*, San Francisco, CA.

Tung, Y. K. (1985). "Evaluation of Water Distribution Network Reliability," *Hydraulics and Hydrology in Small Computer Age, Proceedings of the Hydraulic Division Specialty Conference,* ASCE, Orlando, Fla., August 12–17, 1985.

Tung, Y. K. (1986). "Groundwater Management by Chance-Constrained Model," *Journal of Water Resources Planning and Management,* 112:1–19.

Tung, Y. K. (1987). "Effects of Uncertainties on Optimal Risk-Based Design of Hydraulic Structures," *Journal of Water Resources Planning and Management,* 113(5):709–722.

Tung, Y. K. (1990). "Mellin Transform Applied to Uncertainty Analysis in Hydrology/Hydraulics," *Journal of Hydraulic Engineering,* ASCE. 116(5):659–674.

Tung, Y. K. and W. E. Hathhorn (1990). "Stochastic Waste Load Allocation," *Journal of Ecological Modelling,* 51: 29–46.

Tung, Y. K., and L. W. Mays (1980) "Optimal Risk-Based Design of Hydraulic Structures," *Technical Report CRWR-171,* Center for Research in Water Resources, University of Texas, Austin, Tex.

Tung, Y. K., and L. W. Mays (1981a) "Risk Models for Levee Design," *Water Resources Research,* AGU, 17(4):833–841.

Tung, Y. K., and L. W. Mays (1981b) "Optimal Risk-Based Design of Flood Levee Systems," *Water Resources Research,* AGU, 17(4):843–852.

Tung, Y. K. and Y. W. So (2003). "Risk-based Urban Drainage System Improvement Without Considering Flood Damage." *Proceedings of XXX IAHR Congress (Theme-B)*, pp.533–540, Thessaloniki, Greece, 24–29 August.

Tung, Y. K., and K. C. Yeh (1993). "Evaluation of Safety of Hydraulic Structures Affected by Migrating Pit," *Journal of Stochastic Hydrology and Hydraulics,* 7(2):131–145.

Tung, Y. K. and B. C. Yen (1993). "Some Recent Progress in Uncertainty Analysis for Hydraulic Design," in B. C. Yen and Y. K. Tung, eds., *Reliability and Uncertainty Analyses in Hydraulic Design,* ASCE, pp. 17–34.

Vrijling, J. K. (1987). "Probabilistic Design of Water Retaining Structures," in L. Duckstein and E. J. Plate, eds., *Engineering Reliability and Risk in Water Resources,* Martinus Nijhoff, Dordrecht, The Netherlands, pp. 115–134.

Vrijling, J. K. (1993). "Development in Probabilistic Design of Flood Defenses in the Netherlands," in B. C. Yen and Y. K. Tung, eds., *Reliability and Uncertainty Analyses in Hydraulic Design,* ASCE, pp. 133–178.

Vrijling, J. K., W. van Hengel, and R. J. Houben (1995). "A Framework for Risk Evaluation." *Journal of Hazard Materials*, 43(3): 245–261.

Wen, Y. K. (1987). "Approximate Methods for Nonlinear Time-Variant Reliability Analysis," *Journal of Engineering Mechanics,* ASCE, 113(12):1826–1839.

Yeh, K. C. and Y. K. Tung (1993)."Uncertainty and Sensitivity of a Pit Migration Model," *Journal of Hydraulic Engineering,* ASCE, 119(2):262–281.

Yen, B. C. (1979). "Safety Factor in Hydrologic and Hydraulic Engineering Design," in E. A. McBean, K. W. Hipel and T. E. Unny, eds., *Reliability in Water Resources Management,* Water Resources Publications, Littleton, Colo., pp. 389–407.

Yen, B. C. and A. H.-S. Ang (1971). "Risk Analysis in Design of Hydraulic Projects," in C. L. Chiu, ed., *Stochastic Hydraulics, Proceedings of First International Symposium,* University of Pittsburgh, Pittsburgh, Pa., pp. 694–701.

Yen, B. C. and B. H. Jun (1984). "Risk Consideration in Design of Storm Drains," *Proceedings, 3d IAHR/IAW-PRC International Conference on Urban Storm Drainage,* Chalmers University of Technology, Goteborg, Sweden, 2:695–704.

Yen, B. C. and C. S. Melching (1991). "Reliability Analysis Methods for Sediment Problems," *Proceedings, 5th Federal Interagency Sediment Conference,* Las Vegas, 2:9.1–9.8, March.

Yen, B. C., and W. H. Tang (1976). "Risk-Safety Factor Relation for Storm Sewer Design," *Journal of Environmental Engineering Division,* ASCE, 102(2):509–516.

Yen, B. C., and Y. K. Tung (1993). "Some Recent Progress in Reliability Analysis for Hydraulic Design," in B. C. Yen and Y. K. Tung, eds., *Reliability and Uncertainty Analyses in Hydraulic Design,* ASCE, pp. 35–80.

Yen, B. C., S. T. Cheng, and C. S. Melching (1986). "First-Order Reliability Analysis," in B. C. Yen, ed., *Stochastic and Risk Analysis in Hydraulic Engineering,* Water Resources Publications, Littleton, Colo., 1986, pp. 1–36.

Yen, B. C., S. T. Cheng, and W. H. Tang (1980). "Reliability of Hydraulic Design of Culverts," *Proceedings, International Conference on Water Resources Development,* IAHR Asian Pacific Division 2d Congress, Taipei, Taiwan, 2:991–1001.

Yen, B. C., H. G. Wenzel, Jr., L. W. Mays, and W. H. Tang (1976). "Advanced Methodologies for Design of Storm Sewer Systems," Research Report, no. 112, Water Resources Center, University of Illinois at Urbana-Champaign, Ill., August.

Young, D. M. and R. T. Gregory (1973). *A Survey of Numerical Mathematics, Vol. II,* Addison-Wesley, Reading, Mass.

Yu, P. S., T. C. Yang, and S. J. Chen (2001). "Comparison of Uncertainty Analysis Methods for a Distributed Rainfall-Runoff Model," *Journal of Hydrology*, 244: 43–59.

Zhao, B. and L. W. Mays (1996). "Uncertainty and Risk Analyses for FEMA Alluvial-Fan Method," *Journal of Hydraulic Engineering*, ASCE, 122 (6): 325–332.

Zoppou, C. and K. S. Li (1993). "New Point Estimate Method for Water Resources Modeling," *Journal of Hydraulic Engineering*, ASCE, 119(11): 1300–1307.

CHAPTER 3
REGIONAL WATER SUPPLY PLANNING AND CAPACITY EXPANSION MODELS

Messele Z. Ejeta
California Department of Water Resources
Sacramento, California

Larry W. Mays
Department of Civil and Environmental Engineering
Arizona State University
Tempe, Arizona

3.1 INTRODUCTION

Water supply planning on a regional scale has attracted more interest in recent years. The major reason for this is that any effect on the quantity and quality of the water resources at a certain location in a region has a bearing on the quantity and quality of this resource at other locations in the region. To meet water supply needs in a given region, it becomes necessary to obtain water from different sources in the region or outside it. This entails not only blending different quality waters but also maintaining the quality of the water distributed within accepted standards, which is an important planning aspect for water supply agencies. Yeh et al. (2000) state that in a multisource, multiquality regional water distribution system, water agencies often find it necessary to impose blending requirements at certain control points in the system in order to secure the desired water quality downstream of the control points.

The desire for efficient water resources management has led to the development of various models for optimal management of water supply systems. Some such models that have been developed in the past consist of nonlinear programming (NLP) models with different sets of objective functions and constraints. These include combinations of nonlinear objective functions and constraints, nonlinear objective functions and a mix of linear and nonlinear constraints, nonlinear objective functions and linear constraints, and so on. Ocanas and Mays (1980, 1981a, b) developed a water reuse planning NLP model for determining the optimum allocation of water and reuse of wastewater on a regional basis for multiperiod planning, which was solved by using the large-scale generalized reduced gradient and successive linear programming methods. Schwartz and Mays (1983) also developed water reuse and wastewater treatment alternative planning models using a dynamic programming approach. Mehrez et al. (1992) developed a model for real-time control and operation of a multisource regional system with multiple customers, which involved both quality and quantity constraints.

A capacity expansion model for water energy systems was developed by Lall and Mays (1981) as a multisystem, multiperiod capacity expansion problem, which was solved using Bender's decomposition with a specialized algorithm. Matsumoto and Mays (1983) also developed a capacity expansion model for large-scale water energy systems as a multisystem, multiperiod capacity expansion problem. Bender's decomposition is applied to the original problem to decompose it into three subproblems: capacity, production, and distribution. Each subproblem was solved using a specialized algorithm (Matsumoto and Mays, 1983). Kim and Mays (1994) formulated the rehabilitation and replacement of water distribution system components as a mixed integer nonlinear programming (MINLP) problem and solved it by decomposing the MINLP problem into a mixed integer linear programming (MILP) master problem and an NLP subproblem. The NLP subproblem is solved using the generalized reduced gradient (GRG2) (Lasdon and Waren, 1986) computer code, and the MILP master problem is solved using ZOOM, a 0/1 mixed integer programming code.

Herein, two NLP models for regional water supply planning and an MINLP model for capacity expansion of water supply infrastructure are discussed. The NLP models are formulated with nonlinear objective functions to maximize net benefits. The first NLP model is a yearly static model, whereas the second one is a seasonal dynamic model. The revenue from water supplies from different sources, the cost of supplying the water from these sources, and the damage due to poor-quality water being delivered to users are considered in the objective functions of these models. These NLP models are applied to a regional water supply system located along the Rio Grande from Caballo Dam in New Mexico to El Paso County, Texas. The solution obtained for the seasonal model gives the "best" reservoir release policy for four seasons of a year and a maximum net benefit that is slightly higher than that obtained using the static NLP model. Furthermore, compared to the static model, this model generally maintained better-quality water in terms of total dissolved solids (TDS).

The MINLP model for the capacity expansion problem is formulated for water supply conveyance and treatment infrastructure over a long-term planning period. This model incorporates two sets of decision variables: (1) the optimum timing of the capacity expansion of water supply conveyance systems and water treatment plants, and (2) the water allocation policy that maximizes the net benefit. The solution methodology interfaces a simulated annealing (SA) heuristic search algorithm with the GRG2 nonlinear optimization code. The model is applied to El Paso County's water supply system assuming three water supply regions, two potential water sources (surface water and groundwater) for each region, and a planning period of 10 years. The optimum capacity expansion schedules for the water supply infrastructure components for optimum water supply allocations are given.

3.2 MODEL FORMULATIONS

3.2.1 Regional Water Supply Models

A general problem on a regional (watershed) scale is considered in which all the water users in the region are first identified. The general water supply customers considered include municipality, industry, hydropower, irrigation, ranching, recreation, aquaculture, environmental protection, and remediation, whereas general water sources considered include in-stream flow, groundwater pumping, precipitation, interbasin transfer, and return flow or drainage. The various equations used in the models are discussed in the following subsections.

Continuity Equations. A watershed is divided into a total of I reaches where the delineation of a reach is based on whether there is one or more diversions and/or one or more return flows at only one point in the reach (see Fig. 3.1). In each reach, a total of J activities is considered. This approach enables one to model the entire watershed on a reach-by-reach approach.

To develop the continuity equations, which are, in general, mass conservation equations, an arbitrary reach i in Fig. 3.1 is chosen. In Fig. 3.1, all the possible flows associated with reach i are indicated. To derive the first continuity equation, the point of diversion, or return, in reach i is considered. The equation is expressed as

$$\sum_{k=(i-n)}^{i-1} \sum_{j=1}^{J} (Q_{r_{kji}} - Q_{s_{kji}}) + Q_{ds_i} - \sum_{j=1}^{J} Q_{d_{ij}} - Q_{us\,(i+1)} = 0 \tag{3.1}$$

where $Q_{r_{kji}}$ = return flow from demand point j in reach k to reach i ($k \neq i$)

$Q_{s_{kji}}$ = seepage loss in return flow from demand point j in reach k to reach i ($k \neq i$)

$Q_{d_{ij}}$ = diverted flow from reach i for demand point j, where j represents any kind of demand for water ($i = 1, 2, \ldots, I$; $j = 1, 2, \ldots, J$)

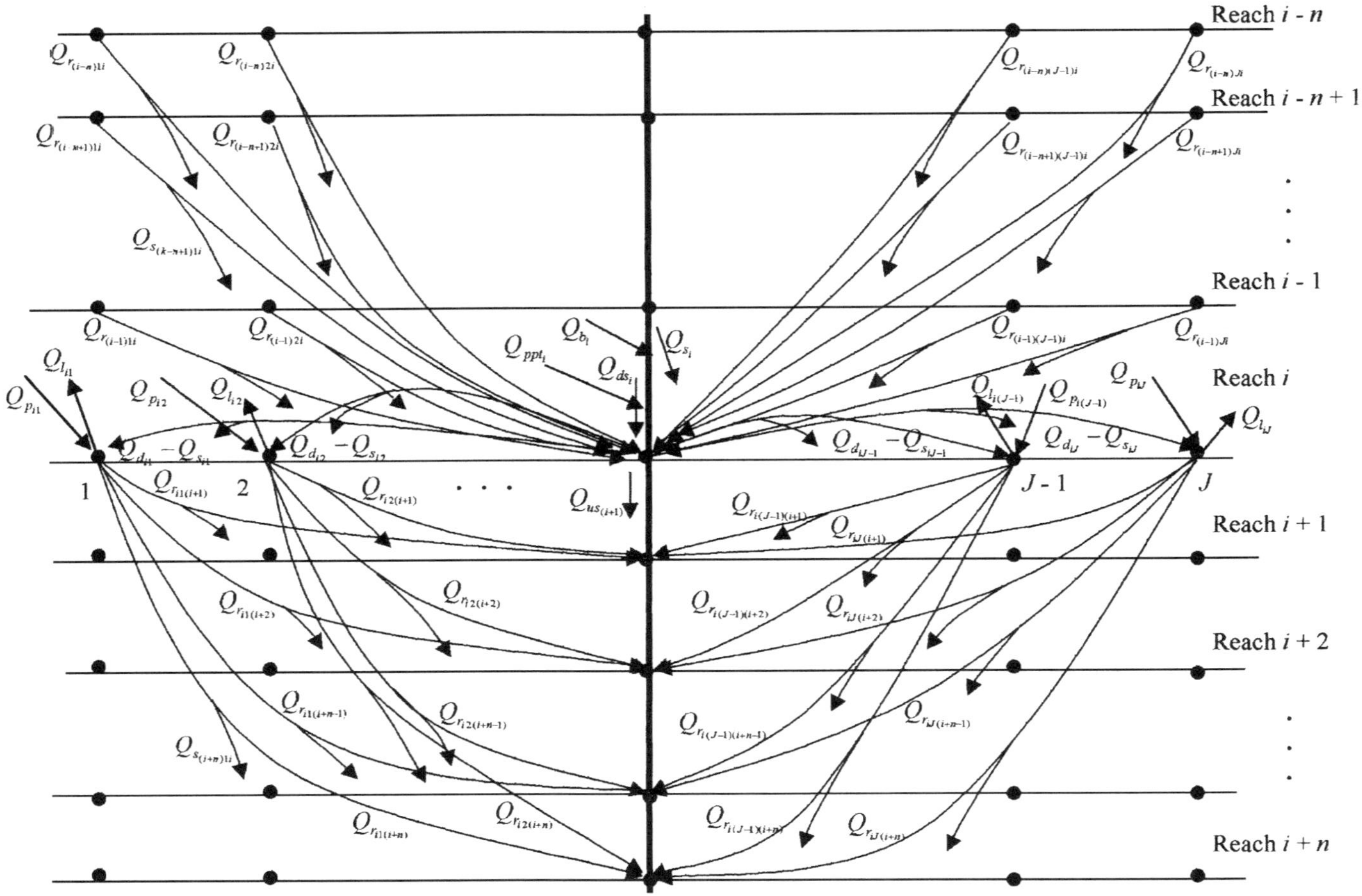

FIGURE 3.1 Schematic representation of the hydraulics of a typical watershed.

Q_{ds_i} = flow at downstream end of reach i, that is, in-stream flow into point of diversion in reach i, referred to herein as a *node*
Q_{us_i} = flow at upstream end of reach i
n = maximum number of reaches a return flow travels before draining into the main stream

The same equation holds true for the entire reaches with boundary conditions for the return flows near the upstream and the downstream ends of the watershed. Also, for each demand point, the following continuity equations can be derived.

$$(Q_{d_{ij}} - Q_{s_{ij}}) + Q_{p_{ij}} - Q_{l_{ij}} - \sum_{k=(i+1)}^{i+n} Q_{r_{ijk}} = 0 \qquad \text{for all } i, j \tag{3.2}$$

where $Q_{s_{ij}}$ = seepage loss in reach i on way to demand point j
$Q_{p_{ij}}$ = pumped flow from reach i for demand point j
$Q_{l_{ij}}$ = consumptive use in reach i at demand point j

Furthermore, within each reach, the following continuity equation is written.

$$Q_{us_i} + Q_{b_i} + Q_{ppt_i} - Q_{ds_i} - Q_{s_i} = 0 \qquad \text{for all } i \tag{3.3}$$

where Q_{b_i} = interbasin transfer to reach i
Q_{ppt_i} = flow into reach i accountable to net precipitation contribution
Q_{s_i} = in-stream seepage loss in the stream in reach i

Water Quality Equations. A typical quality parameter for water distribution on a watershed scale is the amount of TDS in the water supplied. This water quality parameter is taken as an example to develop the equations for water quality management as part of water supply system planning on a watershed scale. Generally, however, multiple water quality parameters may be considered because the quality equations developed will be similar.

The constraints for the point of diversion in reach i can be expressed as

$$\sum_{k=(i-n)}^{i=1} \sum_{j=1}^{J} C_{r_{kji}}(Q_{r_{kji}} - Q_{s_{kji}}) + C_{ds_i}Q_{ds_i} - C_{w_i}\left(\sum_{j=1}^{J} Q_{d_{ij}} + Q_{us\,(i+1)}\right) = 0 \tag{3.4}$$

where $C_{r_{kji}}$ = quality parameter in $Q_{r_{kji}}$
C_{ds_i} = quality parameter in Q_{ds_i}
C_{w_i} = weighted quality parameter in $Q_{d_{ij}}$

The flow out of the nodes will have the same quality assuming "perfect" mixing conditions (Yang et al., 1999). Thus, the following equalities also hold true at the diversion point in reach i.

$$C_{w_i} = C_{d_{ij}} = C_{us(i+1)} \tag{3.5}$$

Defining $C_{w_{ij}}$ as the weighted TDS concentration at any demand point j in reach i, the constraints for demand point j can be expressed as

$$C_{d_{ij}}(Q_{d_{ij}} - Q_{s_{ij}}) + C_{p_{ij}} Q_{p_{ij}} - C_{w_{ij}}\left(Q_{l_{ij}} + \sum_{k=(i+1)}^{i+n} Q_{r_{ijk}}\right) = 0 \qquad \text{for all } i, j \tag{3.6}$$

Similar equations can be also derived for the in-stream flow in each reach as

$$C_{us_i}Q_{us_i} + C_{b_i}Q_{b_i} + C_{ppt_i}Q_{ppt_i} - C_{ds_i}Q_{ds_i} - \frac{C_{us_i} + C_{ds_i}}{2} Q_{s_i} = 0 \qquad \text{for all } i \tag{3.7}$$

The average TDS concentration is used for the seepage loss in reach i assuming that the change in the concentration in the reach is approximately linear.

Flow Regulation. In Fig. 3.1, the flow system represents the general scenario in which diversions and pumpages in any reach and return flows from any reach to any other reach downstream are possible. However, this may not be the case in practice because, for instance, the number of demand points in one reach may not be equal to those in other reaches. Introducing coefficients with values of 0 or 1 for each of the arcs can regulate these flows, where 0 is used if there is no flow along the arc and 1 is used if there is a flow along the arc. The only exceptions that do not require the 0 or 1 coefficients are the in-stream flows $Q_{u_{si}}$ and $Q_{d_{si}}$, because the coefficient of these flows is always 1. The use of such coefficients has advantages in sensitivity analysis to the optimization model developed, which can be done by turning on and off all uncertain activities.

Let $K_{d_{ij}}$ be the coefficient of flow for $Q_{d_{ij}}$, $K_{p_{ij}}$ be the coefficient of flow for $Q_{p_{ij}}$, and so on. Then, the above continuity and TDS equations are modified by introducing these coefficients.

$$\sum_{k=(i-n)}^{i-1} \sum_{j=1}^{J} (K_{r_{kji}}Q_{r_{kji}} - K_{s_{kji}} + Q_{s_{kji}}) + Q_{ds_i} - \sum_{j=1}^{J} K_{d_{ij}}Q_{d_{ij}} - Q_{us\,(i+1)} = 0 \text{ for all } i \tag{3.8}$$

$$(K_{d_{ij}}Q_{d_{ij}} - K_{s_{ij}}Q_{s_{ij}}) + K_{p_{ij}}Q_{p_{ij}} - K_{l_{ij}}Q_{l_{ij}} - \sum_{k=(i+1)}^{i+n} K_{r_{ijk}}Q_{r_{ijk}} = 0 \text{ for all } i, j \tag{3.9}$$

$$Q_{us_i} + K_{b_i}Q_{b_i} + K_{ppt_i}Q_{ppt_i} - Q_{ds_i} - K_{s_i}Q_{s_i} = 0 \qquad \text{for all } i \tag{3.10}$$

$$\sum_{k=(i-n)}^{i-1} \sum_{j=1}^{J} C_{r_{kji}}(K_{r_{kji}}Q_{r_{kji}} - K_{s_{kji}}Q_{s_{kji}}) + C_{ds_i}Q_{ds_i} - C_{w_i}\left(\sum_{j=1}^{J} K_{d_{ij}}Q_{d_{ij}} + Q_{us\,(i+1)}\right) = 0 \qquad \text{for all } i \tag{3.11}$$

$$C_{d_{ij}}(K_{d_{ij}}Q_{d_{ij}} - K_{s_{ij}}Q_{s_{ij}}) + C_{d_{ij}}(K_{p_{ij}}Q_{p_{ij}}) - C_{w_{ij}}(K_{l_{ij}}Q_{l_{ij}} + \sum_{k=(i+1)}^{i+n} K_{r_{ijk}}Q_{r_{ijk}}) = 0 \qquad \text{for all } i, j \tag{3.12}$$

$$C_{us_i}Q_{us_i} + C_{b_i}(K_{b_i}Q_{b_i}) + C_{ppt_i}(K_{ppt_i}Q_{ppt_i}) - C_{ds_i}Q_{ds_i} - \frac{C_{us_i} + C_{ds_i}}{2\,K_{s_i}Q_{s_i}} = 0 \qquad \text{for all } i \tag{3.13}$$

$$C_{w_i} - C_{d_{ij}} = 0 \qquad \text{for all } i, j \tag{3.14}$$

$$C_{w_i} - C_{us\,(i+1)} = 0 \qquad \text{for all } i \tag{3.15}$$

$$C_{w\,(i-1)} - C_{ds_i} = 0 \qquad \text{for all } i \tag{3.16}$$

Additional Constraints. Apart from the mass balance constraints, some other physical and resource constraints may be prevalent. The capital expenditure available for pumping the water and the amount that can be pumped may be limited. Environmental regulations may require that the TDS level at any point be kept below some acceptable level. Such constraints generally can be given as follows.

$$K_{p_{ij}} Q_{p_{ij}} \leq A \tag{3.17}$$

$$C(K_{p_{ij}} Q_{p_{ij}}) \leq B \tag{3.18}$$

$$K_{w_{ij}} C_{w_{ij}} \leq C \tag{3.19}$$

where A = constant
B = constant or a function of amount of pumped water
C = constant or a function of quality parameter in supplied water
$C(K_{p_{ij}} Q_{p_{ij}})$ = cost of any pumping at demand point j in reach i

Objective Function. The objective function can be expressed in terms of the profits from the water resources allocated at each of the demand points, the costs associated with each allocation, and the damages caused due to an undesirable water quality level (such as unacceptable TDS level) at each user point. Defining $P(Q_{d_{ij}}, Q_{p_{ij}})$ as the profit as a function of the allocated supplies (both surface water and groundwater), $C(Q_{d_{ij}}, Q_{p_{ij}})$ as the cost of allocation, and $D(C_{d_{ij}}, C_{p_{ij}}) = D(C_{w_{ij}})$ as the damage, the objective is to maximize the net profit Z given as

$$Z = P(Q_{d_{ij}}, Q_{p_{ij}}) - C(Q_{d_{ij}}, Q_{p_{ij}}) - D(C_{w_{ij}}) \tag{3.20}$$

subject to constraint equations (3.8) to (3.19).

3.2.2 Capacity Expansion Model

Modeling Approach. In this subsection, a water supply system capacity expansion problem is formulated as an MINLP model. Costs for construction, operation, and expansion of the infrastructure are considered as continuous functions of water demands, and the times of expansion are modeled using 0 or 1 integer variables. Thus, for any given unit of the infrastructure, there exist $P - 2$ possible alternatives of expansion of the unit in any P time units of a planning period, such as 20 or 30 years. No expansion during the first or the last time unit in the planning period may be practically feasible, and, thus, we obtain $P - 2 + 1 = P - 1$ possible alternatives for P time units of a planning period. One of the alternatives that doesn't involve any expansion is to initially construct a capacity that is sufficient for the demand during the last time unit of the planning period. Figure 3.2 shows a pictorial representation of such alternatives for any single unit. The objective is therefore to determine the size of the water supply infrastructures and/or when to expand them within the time framework of the planning period at a minimum cost.

The number of possible combinations of construction times grows exponentially with the number of the components to be expanded. In general, for c components that are expandable during any one time unit from a total of p time units, the total possible alternatives will be p^c. For instance, for a certain water supply system with four supply ends, with each end requiring a supply line and a treatment facility, the total number of possible combinations for a planning period of 50 years becomes $(50 - 1)^8 \approx 3.32 \times 10^{13}$. Although a problem with a small number of combinations could be solved by the total enumeration approach, a problem of this size is practically not feasible to solve by this approach. Alternative approaches involving some heuristic search techniques may be employed to obtain a reasonable optimum solution.

For a mathematical formulation, the water supply system is divided into I reaches with J activities (diversion, pumpage, reservoir, etc.) in each reach (see Fig. 3.3). The optimal capacity expansion problem is formulated, in general, as a problem to minimize the sum of costs including construction

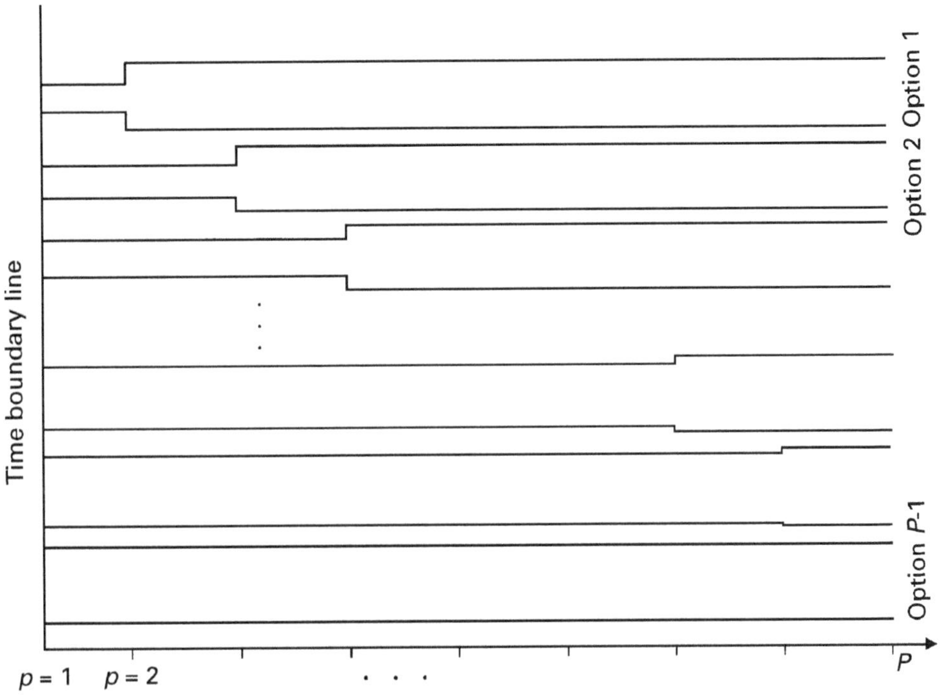

FIGURE 3.2 Different alternative times of capacity expansion of a single water supply unit.

costs, operation costs, and capacity expansion costs subject to constraints on water demand, capacity, mass conservation, and capital availability. The objective function is to minimize the sum of construction, operation, and expansion costs of water conveyance and treatment systems.

Construction Costs. The total construction costs for water supply lines for all activity points j in all reaches i for a supply quantity of Q can be written as

$$\mathrm{CC}_Q = \sum_{i=1}^{I} \sum_{j=1}^{J} \left[\sum_{p=2}^{P-1} X_{ijp} f_c(Q_{ijp}) + X_{ijP} f_c(Q_{ijP}) \right] \tag{3.21}$$

where CC_Q = total cost of construction of supply lines
Q_{ijp} = supply to activity point j in reach i during time p of planning period
$f_c(Q_{ijp})$ = construction cost function for supply line of Q_{ijp}
X_{ijp} = 0 or 1 integer associated with expansion time of capacity of supply line to activity point j in reach i during time p of planning period.

Similarly, the total construction cost for all activity points j in all reaches i can be written as

$$\mathrm{CC}_{\mathrm{CAP}} = \sum_{i=1}^{I} \sum_{j=1}^{J} \left[\sum_{p=2}^{P-1} Y_{ijp} f_c(\mathrm{CAP}_{ijp}) + Y_{ijP} f_c(\mathrm{CAP}_{ijP}) \right] \tag{3.22}$$

where $\mathrm{CC}_{\mathrm{CAP}}$ = cost of construction of treatment facilities
CAP_{ijp} = capacity of treatment facility at activity point j in reach i during time p of planning period

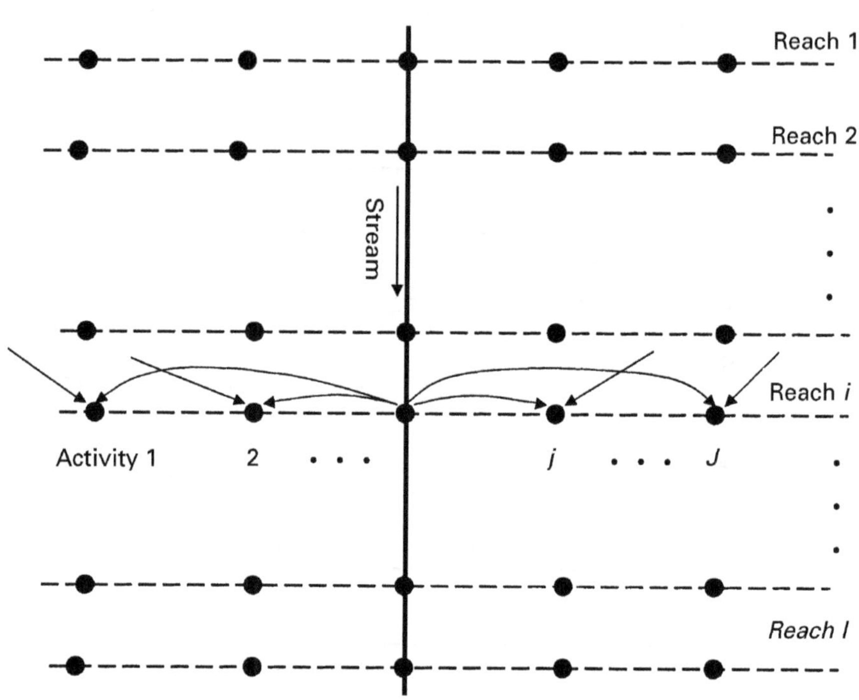

FIGURE 3.3 A water supply system divided into I reaches and J activities with both surface and groundwater supplies.

$f_c(\text{CAP}_{ijp})$ = construction cost for water treatment facility of capacity CAP_{ijp}

Y_{ijp} = 0 or 1 integer associated with expansion time of capacity of water treatment facility at activity point j in reach i during time p of planning period.

In both Eqs. (3.21) and (3.22), the last alternative, which is the "expansion" during time P, is written outside the summation over time to indicate that this case is practically different from the other cases in the sense that there is no capacity expansion involved if this alternative is selected. For such a case, the unit or facility that is constructed at the beginning of the planning period will be sufficient for all the time units in the planning period, that is, no capacity expansion is involved (see Fig. 3.2).

Operation Costs. The total operation cost function can be written for both supply lines and water treatment facilities as

$$\text{OC}_Q = \sum_{i=1}^{I} \sum_{j=1}^{J} \left\{ \sum_{p=2}^{P} X_{ijp} \left[f_o(Q_{ij(p-1)}) \left(\frac{(1+i)^{p-1}-1}{i(1+i)^{p-1}} \right) + f_o(Q_{ijP}) \left(\frac{(1+i)^{P-p}-1}{i(1+i)^{P}} \right) \right] \right\} \tag{3.23}$$

$$\text{OC}_{\text{CAP}} = \sum_{i=1}^{I} \sum_{j=1}^{J} \left\{ \sum_{p=2}^{P} Y_{ijp} \left[f_o(\text{CAP}_{ij(p-1)}) \left(\frac{(1+i)^{p-1}-1}{i(1+i)^{p-1}} \right) + f_o(\text{CAP}_{ijP}) \left(\frac{(1+i)^{P-p}-1}{i(1+i)^{P}} \right) \right] \right\} \tag{3.24}$$

where OC_Q = total cost of operation of supply lines
OC_{CAP} = total cost of operation of treatment facilities
f_o = annual operation function

$\left(\frac{(1+i)^{p-1}}{i(1+i)^{p-1}}\right)$ and $\left(\frac{(1+i)^{P-p}-1}{i(1+i)^{P}}\right)$ = discount factors to present-worth values

In both Eqs. (3.23) and (3.24), the first term in the sum over time is taken up to time $p-1$ to avoid any overlap between the operation of the old and the expanded units. In other words, if the expanded unit is ready for time unit p, the operation cost during this time unit should be the operation cost of the expanded unit, not that of the old unit; the operation of the old unit must have ceased at the end of time unit $p-1$.

Expansion Costs. The total expansion cost function can be written for both supply lines and water treatment facilities, respectively, as

$$EC_Q = \sum_{i=1}^{I}\sum_{j=1}^{J}\sum_{p=2}^{P-1} X_{ijp}\frac{f_e(Q_{ijP}-Q_{ijp})}{(1+i)^{p}} \tag{3.25}$$

$$EC_{CAP} = \sum_{i=1}^{I}\sum_{j=1}^{J}\sum_{p=2}^{P-1} Y_{ijp}\frac{f_e(CAP_{ijP}-CAP_{ijp})}{(1+i)^{p}} \tag{3.26}$$

where EC_Q = total cost of expansion of supply lines
EC_{CAP} = total cost of expansion of treatment facilities

$\frac{f_e(Q_{ijP}-Q_{ijp})}{(1+i)^{p}}$ = unit expansion cost discounted to present-worth value

Constraints. The constraints include water demand constraints, capacity constraints, mass conservation constraints, and capital availability constraints. For all *i, j,* and *p,* the water demand constraints and the capacity constraints can be given as

$$Q_{ijp} \geq Q_{ijp}^{\min} \tag{3.27}$$

and

$$CAP_{ijp} \geq CAP_{ijp}^{\min} \tag{3.28}$$

where $Q_{ijp}^{\min}$ is the minimum supply required for activity point *j* in reach *i* during time *p* of the planning period and $CAP_{ijp}^{\min}$ is the capacity required for minimum supply requirement at activity point *j* in reach *i* during time *p* of the planning period. The mass conservation constraints can be given as

$$\sum_{j=1}^{J}\sum_{i=1}^{I} Q_{ijp} \leq Q_p^{\text{tot}} \tag{3.29}$$

for all *p,* where Q_p^{tot} is the total available water supply during time *p* of the planning period.

Defining TCC, TOC, and TEC as the total capital available for construction, operation, and expansion, respectively, the capital availability constraints can be written for each category as

$$CC_Q + CC_{CAP} \leq TCC \tag{3.30}$$

$$\mathrm{OC}_Q + \mathrm{OC}_{\mathrm{CAP}} \le \mathrm{TOC} \tag{3.31}$$

$$\mathrm{EC}_Q + \mathrm{EC}_{\mathrm{CAP}} \le \mathrm{TEC} \tag{3.32}$$

In Eqs. (3.30) to (3.32), each of the cost terms may be substituted by the right-hand side expressions of Eqs. (3.21) to (3.26).

In addition to the above constraints, each of the integer variables for each activity and in each reach must add to 1 because any unit is expanded once. Thus,

$$\sum_{p=2}^{P} X_{ijp} = 1 \tag{3.33}$$

and

$$\sum_{p=2}^{P} Y_{ijp} = 1 \tag{3.34}$$

for all i and j.

Therefore, the general mathematical formulation of the MINLP capacity expansion problem can be formulated as follows. Minimize:

$$\mathrm{CC}_Q + \mathrm{CC}_{\mathrm{CAP}} + \mathrm{OC}_Q + \mathrm{OC}_{\mathrm{CAP}} + \mathrm{EC}_Q + \mathrm{EC}_{\mathrm{CAP}} \tag{3.35}$$

subject to constraint equations (3.27) to (3.34).

3.3 APPLICATIONS TO THE RIO GRANDE PROJECT AND THE CITY OF EL PASO WATER SUPPLY

3.3.1 Overview of the Rio Grande Project

The Rio Grande project studied stretches from Caballo Dam in New Mexico to the United States–Mexico boundary below El Paso, Texas, on the Rio Grande (see Fig. 3.4). The U.S. Bureau of Reclamation completed the construction of the project in 1916. The primary source of water is the San Juan Mountains of southern Colorado that drain to the Rio Grande, which flows south through New Mexico and Texas to the Gulf of Mexico (Engineering-Science, Inc., 1991). The project's potential water demands include

1. Municipal water supplies for Hatch, Las Cruces, Anthony, and El Paso
2. Irrigation water supplies for the irrigation areas at Percha, Leasburg, Mesilla, and El Paso District No. 1, which consists of two locations
3. Industrial water supplies for Las Cruces and El Paso
4. A 60,000-acre·ft annual water supply to Ciudad Juarez, a city in Mexico, under an international treaty

The competition for the demand of the Rio Grande project water is increasing to the extent where the total water of the project will fall short of the requirements of the different parties, which will have practically few alternative sources. The irrigation districts are currently the primary beneficiaries of the Rio Grande project water. Return flows from drainage from water diverted upstream increase the salinity of the streamflow downstream thereby increasing the damage from irrigation, municipal, and industrial water supplies downstream due to poor water quality. Regulations by the U.S. Environmental Protection Agency (EPA) and local agencies recommend that the TDS level be maintained below 1000 parts per million (ppm).

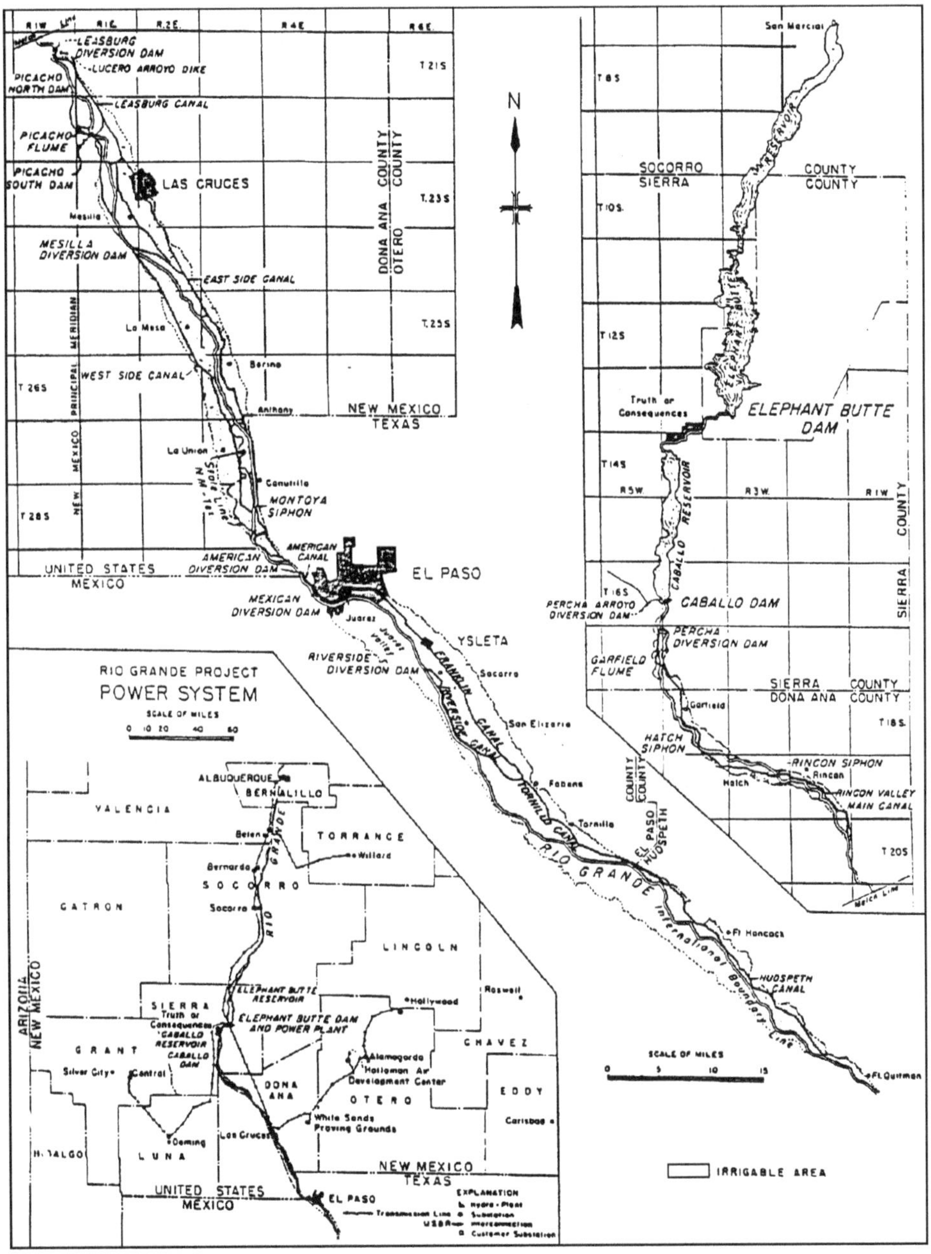

FIGURE 3.4 The Rio Grande project. (*Adapted from USBR project data, 1981.*)

The sources of project water are the Caballo Reservoir, which supplies an annual average of 790,000 acre·ft of water, and groundwater pumping at several locations. The growth of the population of the cities and towns and the maximum limit regulation on the amount of groundwater pumped pose a serious water shortage problem. In addition, the delivery of poor-quality water is also a problem because a portion of the water supply returns to the river, carrying high levels of TDS that affect the water quality of the mainstream flow system downstream. The drainage water first filtrates into the subsurface and then flows to the mainstream with a fairly uniform TDS content approximated at about 1500 ppm.

The five diversion dams for irrigation purposes include Percha, Leasburg, Mesilla, American, and Riverside. The first three diversion dams serve the Elephant Butte Irrigation District (EBID) which consists of the Percha, Leasburg, and Mesilla irrigation areas. The last two diversion dams serve the El Paso County Water Improvement District (EPCWID) No. 1. Percha Diversion Dam is located 2 mi downstream of Caballo Dam and diverts water to the Rincon Valley Main Canal to serve irrigation in the Rincon Valley. Leasburg Diversion Dam is located 62 mi north of El Paso at the head of the Mesilla Valley and diverts water into the Leasburg Canal to serve irrigation in the upper Mesilla Valley. Mesilla Diversion Dam is located 40 mi north of El Paso at the head of the Mesilla Valley and diverts water into the East Side Canal and West Side Canal for irrigation in the lower Mesilla Valley. The American Diversion Dam is located 2 mi northwest of El Paso, immediately above the location where the river becomes the international boundary line between the United States and Mexico. It diverts water into the American Canal for irrigation in the upper portion of the El Paso Valley. The Riverside Diversion Dam is located 15 mi southeast of El Paso and diverts water into the Riverside Canal to serve irrigation in the lower portion of the El Paso Valley.

The current number of estimated households in El Paso is 120,000 whereas that of Hatch, Las Cruces, and Anthony is 1000, 19,000, and 2000, respectively. Hatch, Las Cruces, and El Paso are located in the Rincon, Mesilla, and El Paso valleys, on the opposite sides of the irrigation areas of Percha, Leasburg, and El Paso, respectively. Anthony is located in the Mesilla Valley, below the Mesilla irrigation area.

Several studies have been completed on the Rio Grande project by different agencies and consulting firms, including those by Engineering-Science, Inc. (1991) and jointly by Boyle Engineering Corp. and Parsons Engineering Science, Inc. (1998). The general approach utilized in these studies was the identification and assessment of different management alternatives for the project using available data so that the alternative that resulted in the best management would be selected.

The study by Engineering-Science, Inc. (1991) assessed various opportunities for water storage facilities and the possibilities of additional water supplies by transmountain diversion and by salvage of present losses so that water shortages in the project region would be overcome. The study by Boyle Engineering Corp. and Parsons Engineering Science, Inc. (1998) was aimed at drain mitigation strategies so that acceptable water quality would be maintained for the downstream users in the project area. Based on the results of the simulation study, almost all the selected strategies reduced the problem of quality issues to some extent. However, none of these studies considered the formulation of the problem as a single optimal problem for which the best solution may be obtained.

3.3.2 The City of El Paso Water Supply System

The city of El Paso, Texas, is located along the Rio Grande at the boundary with Mexico (see Fig. 3.4). Most of the population in El Paso County is located in El Paso, which is one of the fastest-growing cities in the nation. The population of El Paso is estimated to become 1.16 million by 2040, whereas that of El Paso County is estimated to become 1.37 million (Boyle Engineering Corporation, 1991), as compared to the 1990 estimates of 529,723 and 752,188, respectively. By 2040, El Paso County's municipal and industrial water requirements are expected to equal 300,000 acre·ft per year, which corresponds to an average consumption rate of 196 gallons per capita per day (gcd). The city is facing a very high water supply shortage because of the apparent increase in demand with time and because its current primary source of groundwater supply is being withdrawn to its potential. It has a limited share of the water of the Rio Grande project. In 1990, the city owned approximately 2000 acres per year of water rights land from the Rio Grande project (Boyle Engineering Corporation, 1991).

The county's water supply system consists of seven major supply regions: Northwest, Northeast, Central, Lower Valley, East, Fort Bliss, and Hueco. Figure 3.5 shows these seven planning areas, and Table 3.1 gives the population projections for these planning areas for the time from 1980 to 2040, in 10-year intervals. Potential water supply sources and the potential quantity that can be obtained from each of these sources, which include alternative surface water supplies, alternative groundwater supplies, and other alternative sources, have been identified (Boyle Engineering Corporation, 1991). Boyle Engineering Corporation (1991) recommended attempting to overcome future water shortage problems by a more balanced approach of conserving and utilizing both surface water and groundwater supplies to meet the demands.

3.3.3 Static NLP Model Application

The NLP model developed is applied to the Rio Grande project by considering yearly and seasonal time frames. Groundwater and Caballo reservoir's average annual storage of 790,000 acre·ft are considered as the water sources, whereas water requirements for irrigation and municipal purposes are considered as the primary demands. The practical significance of this application can be summarized as follows.

1. All the concerned parties get a fair share of the Rio Grande water.
2. The water requirements under the international treaty are satisfied.
3. The minimum requirements for industrial purposes are satisfied.
4. The net profit from the irrigation and municipal water supplies is maximized.
5. The TDS damage is minimized.

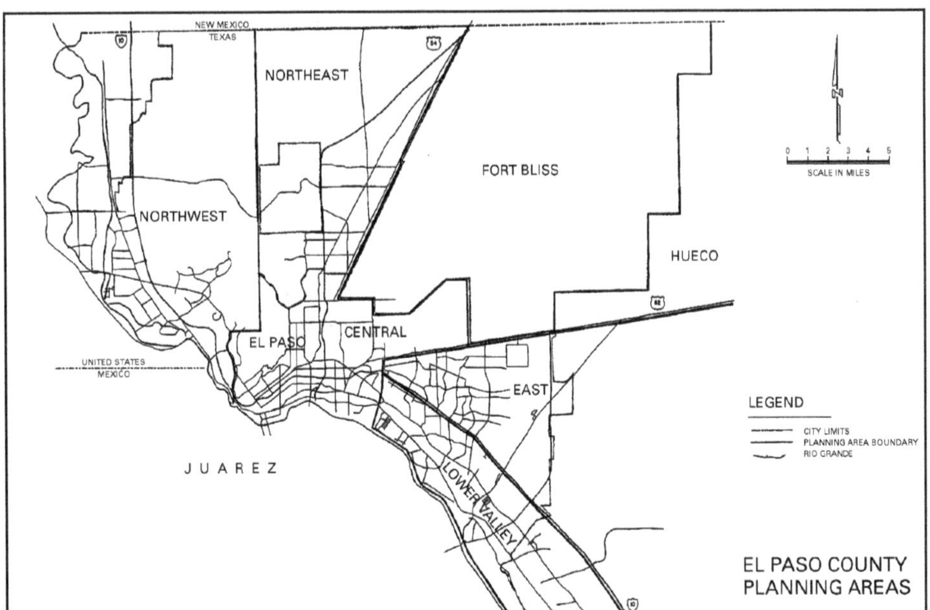

FIGURE 3.5 Major water supply regions in El Paso County. (*Boyle, 1991.*)

TABLE 3.1 Population Projections for the Seven Planning Areas of the City of El Paso

Planning area	Year						
	1980	1990	2000	2010	2020	2030	2040
Northwest	48,938	90,111	135,031	176,800	231,371	280,907	304,634
Northeast	75,398	88,940	106,866	123,696	138,897	154,365	159,162
Central	141,533	140,694	143,184	145,744	145,648	146,184	146,471
Lower Valley	115,807	152,177	192,046	244,025	305,063	370,283	406,870
East	72,023	110,610	141,711	179,014	220,213	267,535	301,026
Fort Bliss	26,200	26,661	26,700	26,700	26,700	26,700	26,700
Hueco	———	———	6,650	9,816	13,872	18,731	23,053
Total	479,899	609,193	752,188	905,795	1,081,764	1,264,705	1,367,916

Source: Boyle Engineering Corporation (1991a,b,c).

The potential income from the irrigation water supply is dependent on the location of two major irrigation districts within the project region. These include the EBID and the EPCWID No. 1. The average yearly return from the former irrigation district is estimated to be about \$266 per acre of land, which needs a total of 5 acre·ft of water per year. Of this water requirement, 3 acre·ft of water should be obtained from diverted surface water from the Rio Grande and 2 acre·ft of water should be pumped in the vicinity of the irrigation area. For EPCWID No. 1, the average return per acre of land irrigated is estimated to be about \$187 per year. All the water for irrigation purposes for this district, also 5 acre·ft per year per acre of land, should be obtained from surface water. Because of its adverse salinity effect on crop growth, groundwater use is not to be opted for in this irrigation district. Table 3.2 gives the incomes from different crops and the weighted average return per acre of irrigated land.

The return from domestic water supply can be derived using the concepts of econometrics. In general, the return is given as

$$R = aQ - \left(\frac{b}{N}\right)Q^2 \tag{3.36}$$

where R = return, \$/yr
Q = yearly domestic water supply, acre·ft
N = number of households in city or town
a and b = constants

Table 3.3 gives the values of a and b for different towns and cities in the project area.

The cost of supplying surface water from the river to the towns and cities is estimated at \$503.7 per acre·ft per year, whereas the cost of supplying water from the same source to irrigation districts is estimated at \$15 per acre·ft per year. The cost of supplying groundwater to the towns and cities is estimated at \$325 per acre·ft per year, and the cost of supplying groundwater to the irrigation districts is estimated at \$10 per acre·ft per year.

The economic damage that results from supplying saline water to domestic uses and irrigation districts is generally small, but appreciable. The damage results from the reduced lives of household appliances such as dishwashers, water heaters, food waste disposers, water softeners, and evaporative coolers. In the towns and cities that are present in the study area, the damage due to salinity is

TABLE 3.2 Return from Agricultural Activity

Irrigation district	Grains	Forage	Cotton	Chile	Pecans	Average
EBID						
Acreage	4,550	24,240	17,048	17,459	13,767	
Benefit, \$/acre·ft	60	350	126	278	300	266
EPCWID						
Acreage	6,676	11,331	22,404	1,464	5,047	
Benefit, \$/acre·ft	65	319	133	253	187	178

TABLE 3.3 Values of a and b in Eq. (3.36) for Different Cities in the Project Area (1999/2000 Estimates)

S. No.	Town/city name	a value	b value	No. of households	b/N value
1	Hatch	8,948	7,480	1,000	−7.479
2	Las Cruces	21,604	18,050	19,000	−0.950
3	Anthony	8,948	7,480	2,000	−3.739
4	El Paso	8,948	7,480	120,000	−0.062

estimated at 3 cents per household per 1 ppm of TDS per year. Therefore, the TDS damage generally can be expressed as

$$D = 0.03NC(Q) \tag{3.37}$$

where D = damage, \$/yr
N = number of households for which Q is supplied
$C(Q)$ = TDS concentration in Q

On the other hand, the damage to crops because of the salinity of the irrigation water occurs in the form of reduced yield of irrigated crops due to toxic and osmotic effects. In the project area, this is reflected by the reduced income per acre of irrigated land in EPCWID No. 1. The salinity damage to irrigated crops is estimated as the average difference between the income from agriculture in EBID and that in EPCWID's irrigation district, which is presumed to be due to the increase in the salinity in the latter irrigation district. Thus the damage is estimated to be 2.6 cents per the increased TDS over the base value of 450 ppm per acre·ft of water supplied per year.

As shown in Fig. 3.6, the entire project area is subdivided into a total of 17 reaches. In these reaches, the points of interest considered include diversion node, return node, two potential irrigation areas (one on each side), and municipal and industrial points of use. These points of use are currently taken for this project to be the maximum possible points of interest in each reach. Possible return flows to this node or reach from as far as four reaches upstream are considered, which is the maximum distance currently observed to be traveled by the drainage system in the project region (see Fig. 3.7).

At each demand point in a reach, water supply may be obtained from diverted water or from pumpage except for agricultural purposes in reaches 12 and 15 where no pumpage is desirable. Possible return flows from each demand point to the stream as far downstream as four reaches below the reach, in retrospect of the continuity at the node, are considered (see Fig. 3.8). The municipal and industrial pumpages are from deeper groundwater storage, which has a significantly lower TDS level of approximately 250 ppm. However, the pumped waters for irrigation purposes for the Percha, Leasburg, and

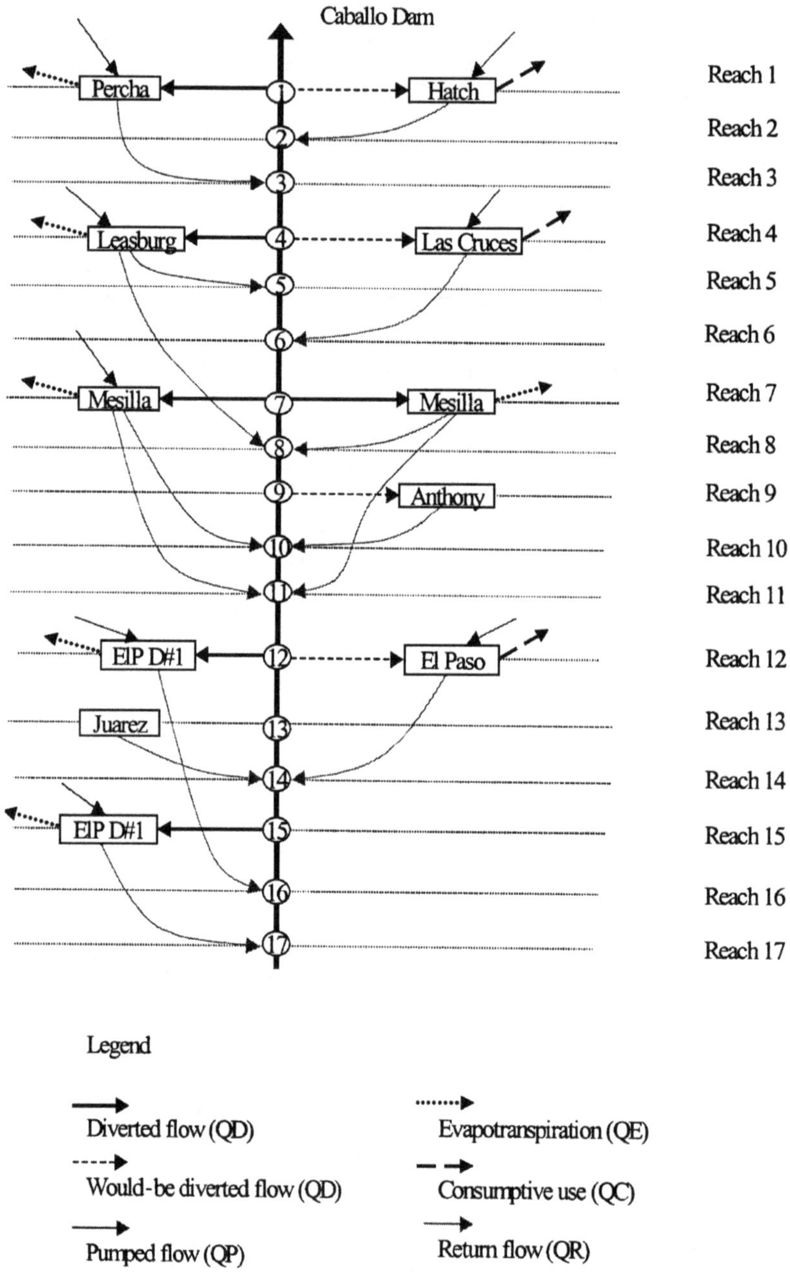

FIGURE 3.6 Schematic diagram of the Rio Grande project system.

Mesilla irrigation areas are from shallow groundwater that has a high TDS level, approximately 1500 ppm, and are assumed to be essentially drawn from the river. Before the diverted water reaches the point of use for irrigation purposes, a significant amount of seepage loss occurs. Although this water is a loss to the immediate purpose for which it is diverted, it will find its way back to the river as a drainage flow or percolate deep into the groundwater storage. For this project, it is assumed that about

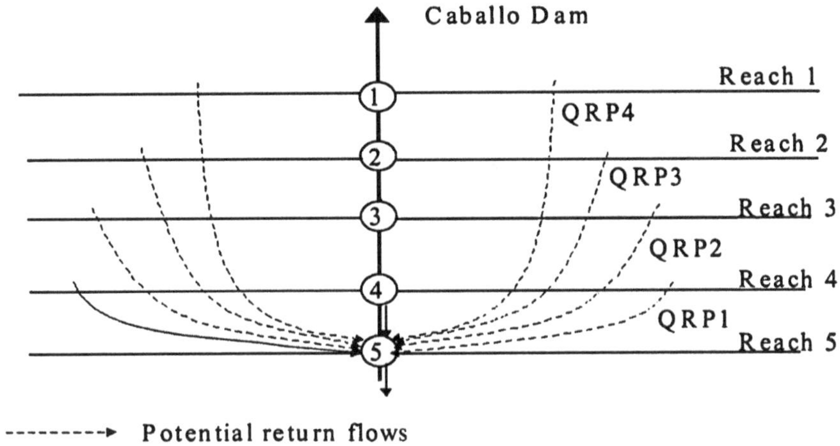

FIGURE 3.7 Schematic diagram showing continuity at a node.

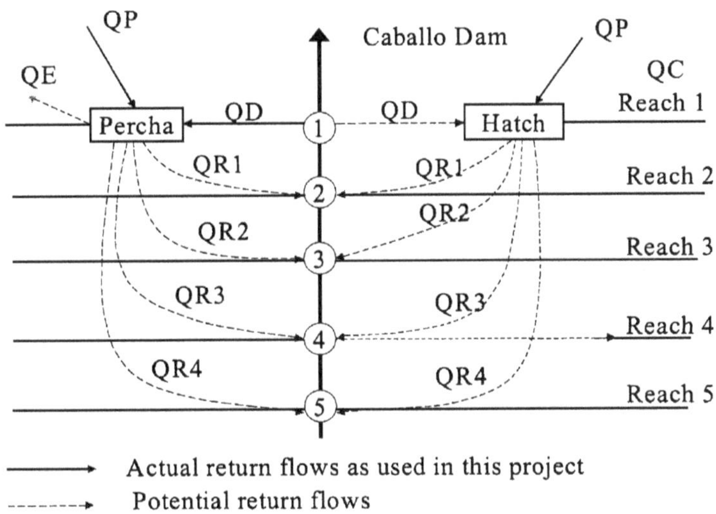

FIGURE 3.8 Schematic diagram showing continuity at a point of use.

40 percent of the water diverted for irrigation purposes will be lost and will become seepage flow, which will eventually drain back to the stream flow. The return flows from the municipalities take place through conduits and thus there is negligible seepage loss.

At each of these demand points, the consumptive use is accounted for by determining the amount of water that is consumed. At the agricultural areas, this constitutes the evapotranspiration, whereas at the municipal points of use, the consumptive losses constitute the consumptive uses for purposes such as drinking and cooking. At the industrial points of use, the consumptive losses constitute the amount that is used in the industrial by-product. To account for these effects, it is assumed that about 50 percent of the water used for irrigation and municipal purposes and about 70 percent of the water used for industrial purposes are "used up." The remaining water from each point of use is assumed to find its way back into the stream flow in the form of drainage.

3.3.4 Seasonal NLP Model Application

The seasonality aspect of the management of the Rio Grande project results from the fact that some of the activities occur only during certain seasons of the year. The irrigation activities occur during the spring, summer, and fall seasons. In addition, the magnitude of the urban water demands depends on the season. For each city and town, this demand is maximum during the summer season and minimum during the winter season. Because of these reasons, consideration of seasonal effects may prove to be more appropriate for this project. Therefore, the planning year is divided into four seasons and the problem is reformulated as a more complex, but more realistic, NLP problem. The seasons considered include December to February as season 1, March to May as season 2, June to August as season 3, and September to November as season 4. The approximate demand proportions by the different activities are given in Table 3.4. This approach surpasses the original problem formulation for a full year that was presented earlier. However, this approach tends to transform the prevailing management practice into a more realistic optimal model.

The general form of the objective function for the urban water demands is essentially the same as the one used in the static model except that the annual urban water demand is taken as the sum of the seasonal demands. For the cities and towns, the modified income (benefit) from the supply of demand is expressed as

$$a[Q(S_1) + Q(S_2) + Q(S_3) + Q(S_4)] - \frac{b}{N}[Q(S_1) + Q(S_2) + Q(S_3) + Q(S_4)]^2 \tag{3.38}$$

where $Q(S_k)$ denotes the demand during season k where $k = 1, 2, 3$, or 4. The other parameters are as defined earlier. It may be noted further that Eq. (3.38) is essentially the same as Eq. (3.36) except that in this case the total yearly demand is substituted by the sum of the four seasonal demands.

For the agricultural activities, it is assumed that the total income will be generated only during the harvesting season of September through November, that is, season 4. The total income taken is the same as that used in the static model, but the water demands for this purpose prevail during the last three seasons of the year to generate this income. Thus, the total income per acre of land is taken as \$266 in the agricultural areas in reaches 1, 4, and 7 and as \$187 in the agricultural areas in reaches 12 and 15. Similarly, the water supply costs and the TDS damages used in the objective function are taken as the sum of the seasonal costs and damages. The costs of supplying surface water and groundwater to the towns and cities are taken as \$503.7 and \$325 per acre·ft, respectively, for the four seasons of the year. These figures were taken in the static model as the annual costs of surface water and groundwater supplies, respectively. The cost of supplying water from the river to the irrigation districts is estimated at \$15 per acre·ft for the last three seasons of the year during which diversions to the irrigation districts prevail. Similarly, the cost of supplying groundwater to the irrigation districts is estimated at \$10 per acre·ft for the two seasons of the year during which pumpage

TABLE 3.4 Approximate Demand Proportions by Different Activities at the Rio Grande Project

Activity	Dec–Feb	Mar–May	Jun–Aug	Sep–Nov
Municipal (fractions)	0.15	0.25	0.35	0.25
Agricultural (unit proportions)				
Diversion	—	0.30	2.20	0.50
Pumpage	—	1.50	0.50	—
Subtotal	—	1.80	2.70	0.50

to the irrigation districts is allowed. Again, these figures were taken in the static model as the annual costs of surface water and groundwater supplies for irrigation, respectively.

The TDS damages are also broken down into seasons so that the seasonality effects are reflected. The seasonal TDS damages that result from both the urban and the irrigation supplies are approximated by taking the same proportionate values as the seasonal demands (see Table 3.4). Thus, the seasonal urban TDS damages per 1 ppm are 0.45, 0.75, 1.05, and 0.75 cents for the first, second, third, and fourth seasons, respectively. Similarly, the estimated irrigation TDS damages per ppm for the last three seasons are 0.936, 1.404, and 0.26 cents, respectively.

In the static model, it was assumed that the total drainage flow occurs during the same year of diversion for all agricultural, municipal, and industrial activities. However, in this model, it is assumed that the total drainage flow from the agricultural activities is spread over two seasons, with about 75 percent of the drainage occurring during the same season as the diversion season and the remaining 25 percent of the drainage occurring during the subsequent season. No diversion or pumping activity for irrigation exists during the first season. However, there is carryover drainage during this season from the diversion during the fourth season of the previous year. This is accounted for by assuming the same 75 percent return flow during the same season and the remaining 25 percent return flow during the first season of the following year. For practical purposes, the quantity of this drainage is taken to be the same as the previous year's drainage during the same season, that is, the first season. Because the drainage from the municipal and industrial activities takes place in conduits, it is assumed to occur during the same season of diversion and pumping.

To write the above distribution of drainage flows over two seasons in equation form, let $Q_{lm(kij)}$ be the return flow from reach i, that was diverted for purpose j, to l reaches downstream from the water diverted during season m and returning during season k. Then the two general drainage flow distributions over two seasons can be given as

$$Q_{lm\,(kij)} = 0.75\,(Q_{lm\,(kij)} + Q_{l\,(m+1)\,(kij)}) \tag{3.39}$$

$$Q_{l\,(m+1)\,(kij)} = 0.25\,(Q_{lm\,(kij)} + Q_{l\,(m+1)\,(kij)}) \tag{3.40}$$

To better illustrate the notations used here consider, for instance, $Q_{22(S2,R1,A1)}$, which is the drainage flow during season 2 that is returning to two reaches downstream of reach 1 (R1) from the water supplied during season 2 (S2) to agricultural activity 1 (A1). Figure 3.9 shows the general scheme of the seasonal flow system of the Rio Grande project for any given typical season. In general, there is one such scheme for every season.

3.3.5 Capacity Expansion Model Application

Solution Methodology. The MINLP model formulated consists of two sets of variables: discrete time variables and continuous cost variables associated with water demand and treatment, thus making analytical solution of the problem difficult. However, if the discrete binary integer variables are set to specific values, then the problem at that discrete point in time reduces to an NLP problem. The resulting NLP problem can be solved by an appropriate NLP computer program. Nonetheless, such a solution can be obtained only after the values of the integer variables have been fixed.

For small problems, it may be possible to fix the values of the binary discrete time variables and solve the reduced NLP at each of the discrete points, one at a time, using an NLP computer program. In such a case, the best solution will be obtained from these optimal solutions. For a large number of discrete points, in which case it would become very tedious to use this approach, the use of some form of heuristic search algorithm, such as SA, becomes a necessity. One such algorithm may be the use of an automated computer program that is capable of performing the following tasks:

1. Generate the values of the discrete time variables one at a time.
2. Use this information to reduce the MINLP to an NLP.
3. Obtain the optimum solution of the reduced NLP at each of the discrete points.

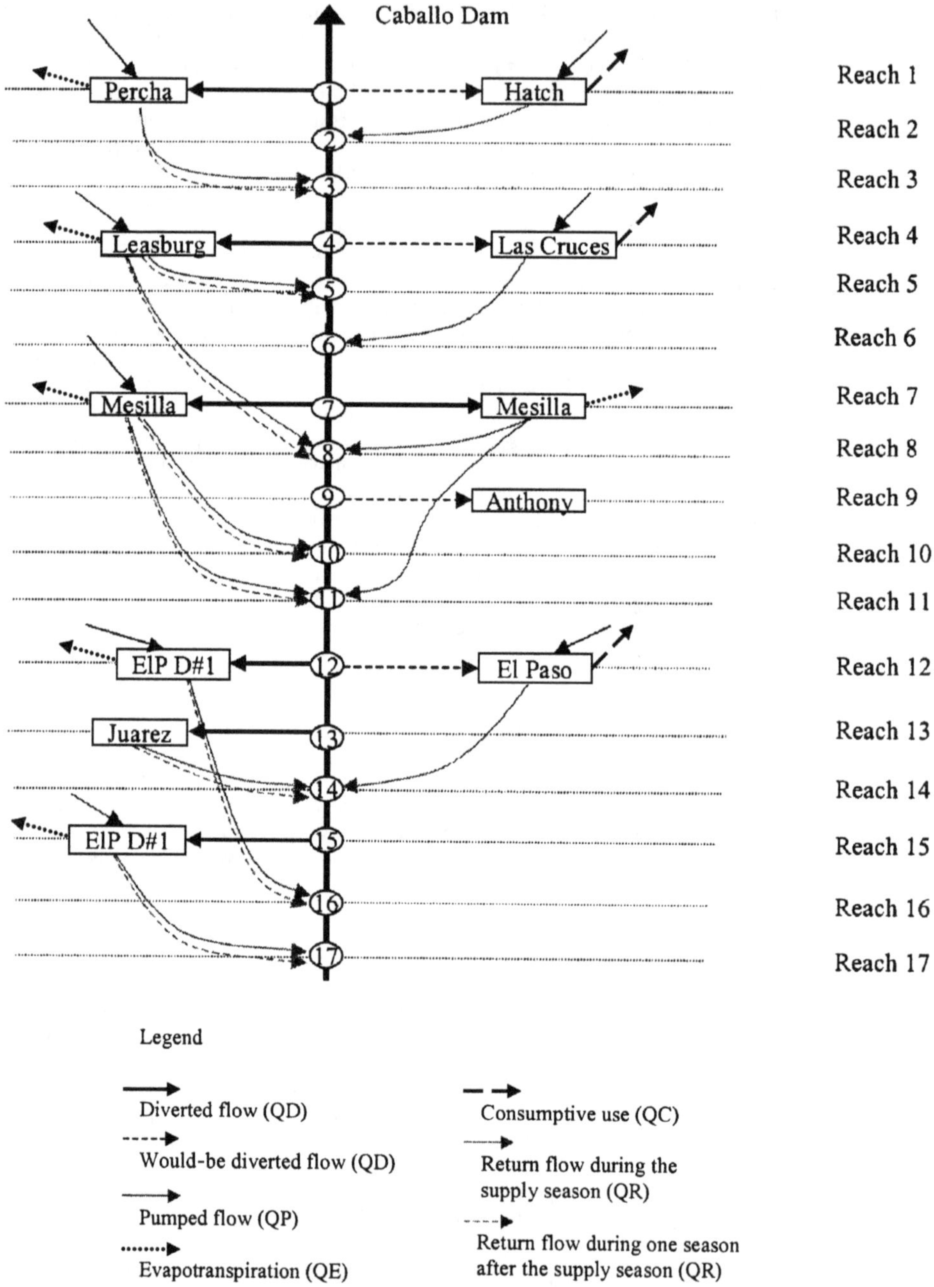

FIGURE 3.9 Schematic diagram of the Rio Grande project system showing the drainage for any given season in which diversion to agriculture exists.

4. Return the best optimum solution among those obtained for all the reduced NLPs.

This solution strategy is pursued in two phases. In the first phase the simulated annealing heuristic search algorithm is implemented to solve a combinatorial problem. In the second phase, the SA code is interfaced with GRG2 to solve the MINLP problem.

The general SA algorithm is given below (Wolsey, 1998, with modifications in the notations).

1. Get an initial solution $\mathbf{X}_o$.
2. Get an initial temperature T and a reduction factor α with $0 < \alpha < 1$.
3. While not yet frozen, do the following:
  3.1 Perform the following loop k times:
    3.1.1 Pick a random neighbor $\mathbf{Y}$ of $\mathbf{X}$.
    3.1.2 Let $\delta = f(\mathbf{Y}) - f(\mathbf{X})$.
    3.1.3 If $\delta \le 0$, set $\mathbf{X}_{i+1} = \mathbf{Y}$.
    3.1.4 If $\delta > 0$, set $\mathbf{X}_{i+1} = \mathbf{Y}$ with probability $e^{-\delta/T}$; otherwise, set $\mathbf{X}_{i+1} = \mathbf{X}_i$.
  3.2 Set $T \leftarrow \alpha T$ (reduce the temperature).
4. Return the best solution found.

The implementation of the SA heuristic search algorithm requires that a neighborhood and a search strategy be defined first. The starting point with a given set of time units is chosen randomly, and its neighbors are selected. For any given discrete point, the number of neighbors considered is a function of the total number of units to be expanded.

In the course of the neighborhood search, if the current point is at or adjacent to a boundary, that is, where one or more of the time lines are at their limits which include the beginning and the end of the planning period along each time line, then special attention is paid to the search mechanism. This is tackled by the search algorithm in such a way that whether the randomly selected neighbor is at or adjacent to any boundary line is checked before any move. If it is already at the boundary, then no move is allowed, that is, the new neighbor coincides with the center point. If it is already one unit away from the boundary and if the selected neighbor were to move two time units to the boundary, then no two time unit move is allowed; instead, it moves only one time unit to reach the boundary.

The Generalized Reduced Gradient (GRG2). GRG2 solves nonlinear optimization problems with the general form

$$\text{Minimize} \quad f(\mathbf{x}) \tag{3.41a}$$

subject to

$$\mathbf{g}(\mathbf{x}) = \mathbf{0} \tag{3.41b}$$

$$\underline{\mathbf{x}} \le \mathbf{x} \le \overline{\mathbf{x}} \tag{3.41c}$$

where $\mathbf{x}$ = n-dimensional column of vector variables
$f(\mathbf{x})$ = objective as a function of $\mathbf{x}$
$\mathbf{g}(\mathbf{x})$ = m constraints
$\underline{\mathbf{x}}$ = lower bound on $\mathbf{x}$
$\overline{\mathbf{x}}$ = upper bound on $\mathbf{x}$

The fundamental idea of the generalized reduced gradient is to express m variables, called basic variables, in terms of $n - m$ variables, called nonbasic variables. The objective function is thus expressed as a function of the nonbasic variables, which are bound between their upper and lower limits. The resulting formulation, called the reduced problem, is solved using the unconstrained NLP solution technique with modifications to account for the bounds for the nonbasic variables. Solving a sequence of reduced problems solves the original NLP problem.

Interfacing the Simulated Annealing Algorithm with GRG2. The SA search algorithm code was interfaced with the GRG2 code to solve the MINLP problem. Figure 3.10 shows the general flow-

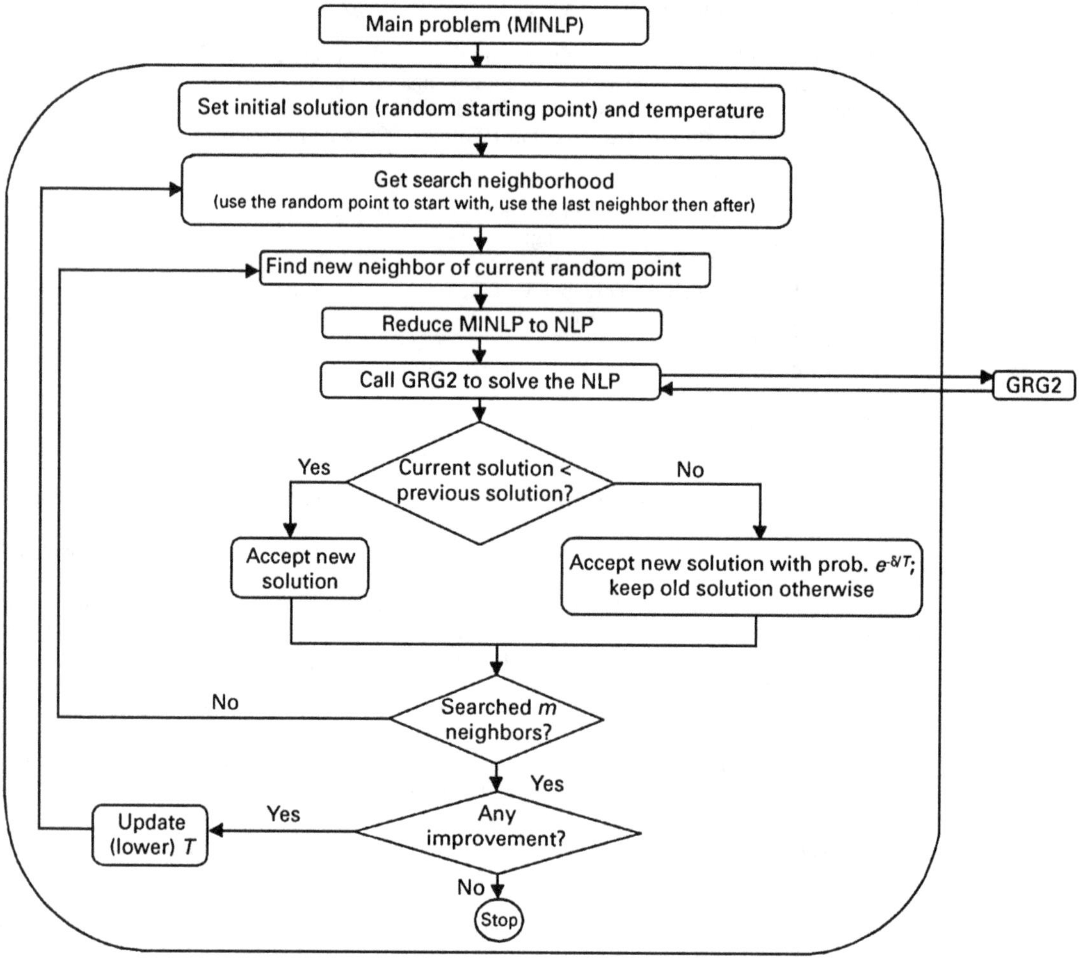

FIGURE 3.10 A flowchart showing the procedures used by the model.

chart of the computer program used to solve the MINLP problem. To use the GRG2 code, a user-supplied GCOMP subroutine was written. The main elements of this subroutine are the definitions of the objective function and the constraint functions, which are given as a list of functions whereby the index of the objective function is specified in the main program. Furthermore, two options exist for the user in selecting the methodology to calculate the gradient. GRG2's default option is to use the forward differencing procedure. Alternatively, partial derivatives of the functions given in the GCOMP subroutine can be determined by the user and another subroutine called PARSH can be provided to the model. For this application, GRG2's default option was used.

General Construction, Operation, and Expansion Cost Equations. In general, many capital facilities are usually constructed with a capacity that will satisfy the requirements over some years to come, instead of a capacity that satisfies immediate requirements. The main reason for this lies in the economies of scale available with a large plant that may be achieved in investment cost or operating cost (Hinomoto, 1974). Chenery (1952) proposed the following power function for an industrial facility.

$$C = \alpha K^{\beta} \tag{3.42}$$

where C = investment cost
K = design capacity
α and β = positive parameters determined by observed data

In this equation, if $K = 1$, then C equals α, which can be defined as the investment cost of a system with a capacity of one unit. Parameter β determines the manner in which investment cost changes capacity. The investment cost increases with capacity at an increasing or decreasing rate depending on whether β is greater than or less than 1, respectively.

The general equation for the construction of a water supply unit, including setup costs K_c, can be expressed as

$$\mathrm{CC}(Q_{ijp}) = K_c + \alpha_c Q_{ijp}{}^{\beta c} \tag{3.43}$$

where α_c and β_c are constants. Using this general form, the operation cost for a water supply facility during a planning period p is expressed as

$$\mathrm{OC}(Q_{ijp}) = K_o + \alpha_o Q_{ijp}{}^{\beta_o} \tag{3.44}$$

and for expansion of the water supply facility as

$$\mathrm{EC}(Q_{ijp}) = K_e + \alpha_e (Q_{ij\,(p+1)} - Q_{ijp})^{\beta_c} \tag{3.45}$$

where K_c, K_o, K_e α_o, β_o, α_e, and β_e are constants. The construction, operation, and expansion costs of a water treatment facility are expressed, respectively, as

$$\mathrm{CC}(\mathrm{CAP}_{ijp}) = C_c + \theta_c (\mathrm{CAP}_{ijp})^{\phi_c} \tag{3.46}$$

$$\mathrm{OC}(\mathrm{CAP}_{ijp}) = C_o + \theta_o (\mathrm{CAP}_{ijp})^{\phi_o} \tag{3.47}$$

$$\mathrm{EC}(\mathrm{CAP}_{ijp}) = C_e + \theta_e (\mathrm{CAP}_{ij\,(p+1)} - \mathrm{CAP}_{ijp})^{\phi_e} \tag{3.48}$$

where C, θ, and ϕ are constants with the subscripts c, o, and e denoting construction, operation, and expansion, respectively.

The EPA has developed empirical functions for the estimation of the three types of costs for water and wastewater conveyance systems and water and wastewater treatment plants. Clark et al. (2002) developed different cost equations for the construction of water supply distribution system components; for water supply distribution system rehabilitation; and for the construction of pumps, tanks, and reservoirs. The cost functions for water supply distribution system components for different pipe systems include:

1. Base installed pipe costs
2. Trenching and excavation costs
3. Embedment costs
4. Backfill and compaction costs
5. Valve fitting and hydrant costs
6. Dewatering costs
7. Sheeting and shoring costs
8. Horizontal boring costs
9. Pavement removal and replacement costs

10. Utility interference costs

11. Traffic control costs

12. Household service connection costs

The general cost equation for each of these different costs is given as

$$y = a + bx^c + du^e + fxu \tag{3.49}$$

where y = cost of a particular component, \$/ft
x = design parameter (for example, pipe diameter)
u = indicator variable

The variables *a, b, c, d, e,* and *f* are estimated using regression techniques. Table 3.5 gives the values of these parameters for selected pipe types.

The aggregated cost of construction can be determined in an additive manner by considering the construction cost components involved (Clark et al., 2002). Thus, considering the cost of using a ductile iron pipe (base installed, with push-on joints) and the cost of the major construction activities involved, including trenching and excavation, embedment, backfill, and compaction, and valve fitting and hydrant placement, the total aggregated cost can be estimated by

$$y_{\text{total}} = 50.74 + 0.33x^{1.72} + 0.32x^{0.67} + 0.26x^{1.46} - 0.062x^{0.73} + 0.02x^{1.79} + 0.16x \tag{3.50}$$

where y_{total} is the aggregated construction cost (\$/ft) and x is the pipe diameter (in). To use Eq. (3.50) in the model, the diameter is expressed in terms of the flow rate in the pipe because the flow rate, rather than the pipe diameter, is used in the model. Also, the total distance of construction must be determined. Using the equation of continuity and an average design velocity of 5 ft/s, the diameter is expressed in terms of Q as $x = 6.06Q^{0.5}$ from which Eq. (3.51) is obtained.

$$y_{\text{total}} = 50.74 + 7.32Q^{0.860} + 1.07Q^{0.335} + 3.61Q^{0.730} - 0.23Q^{0.365} + 0.50Q^{.895} + 0.97Q^{0.5} \tag{3.51}$$

Therefore, if L_{ij} is the length of construction from the diversion node in reach i to activity point j, the total construction cost $\text{CC}(Q_{ijp})$ for a water supply line in reach i for activity j during planning period p can be given as

$$\text{CC}(Q_{ijp}) = L_{ij}(50.74 + 7.32Q_{ijp}^{0.86} + 1.07Q_{ijp}^{0.335} + 3.61Q_{ijp}^{0.73} - 0.23Q_{ijp}^{0.365} + 0.50Q_{ijp}^{0.895} + 0.97Q_{ijp}^{0.5}) \tag{3.52}$$

Although Eq. (3.52) can be used as an estimate for construction costs, such costs may change with time. In fact, the original equations given in the sum of equations in Eq. (3.50) were derived based on the Producer Price Index (PPI) for 1997. However, such costs can be updated by the use of standard indices. For example, the PPI for 1997 is 358.5, whereas that for 2000 is 388.5, which yields a ratio of 1.08. Therefore, the above costs can be updated to 2000 by multiplying the 1997 costs by 1.08 (Clark et al., 2002).

The costs for pump, reservoir, and tank construction and operation can be given by an equation that is essentially similar to Eq (3.24) as (Clark et al., 2002):

$$y = a + bx^c + du^e + fxu \tag{3.53}$$

where y = construction cost (\$)
x = design variable
u = indicator variable

Parameters *a, b, c, d, e,* and *f* are determined from regression analysis. Table 3.6 gives values of these parameters for some selected activities.

Using the parameters given in Table 3.6, aggregated cost equations for construction, operation, and maintenance of pump stations and reservoirs are determined. Therefore, if a horizontal split case

TABLE 3.5 Parameters for Distribution System Component Cost Equations for Selected Pipe Types

Type of activity	Conditions	Pipe diameter range, in	Indicator variable *u* value	Parameter values						
				a	*b*	*c*	*d*	*e*	*f*	R^2
Ductile iron pipe	Base installed cost, with push-on joints	4–144	50, 52	−44.33	0.33	1.72	2.87	0.74	0.0	0.99
Trenching and excavation	Sandy gravel soil with 1:1 side slope	1. 4–8 2. 8–144	4, 6, 8, 10 ,12 4, 6, 8, 10, 12	−24.18 2.91	0.32 0.0018	0.67 1.90	16.66 0.13	0.38 1.77	0.0 0.0	0.99 0.98
Embedment	Concrete arch	4–144	—	7.11	0.26	1.46	0.0	0.0	0.0	0.99
Backfill and compaction	Sandy native soil with a 1:1 side slope	4–144	4, 6, 8, 10, 12	−0.094	−0.062	0.73	0.18	2.03	0.02	0.99
Valve fitting and hydrant	Medium frequency using ductile iron fitting	12–72	—	9.83	0.02	1.79	0.0	0.0	0.0	0.99
Horizontal boring	—	4–60	—	503.67	1.99	1.52	0.0	0.0	0.0	0.98

Source: Clarke et al. (2002).

TABLE 3.6 Parameters for Costs of Selected Pumps, Tanks, and Reservoirs

Type of activity	Conditions	Independent variable description and range	Indicator variable u values
Horizontal split case pumps	Includes suction discharge piping	Firm pumping capacity (0.5–200 mgd)	100, 200, or 300 ft of total dynamic head
Pump station site work	—	Firm pumping capacity (0.5–200 mgd) for extreme conditions	1 for average conditions and 2
Building installation for horizontal centrifrigal pump	—	Firm pumping capacity (0.5—200 mgd)	2 for a simple slab
Construction of electrical and instrumentation	—	Firm pumping capacity (0.5–200 mgd)	100, 200, or 300 ft of total dynamic head
Expansion for horizontal split case pump	Includes suction and discharge pumping	Additional pumping capacity (0.5–200 mgd)	100, 200, or 300 ft of total dynamic head
Expansion of site work	—	Additional pumping capacity (0.25–33 mgd) conditions	1 for average conditions and 2 for extreme
Expansion of electrical and instrumentation	—	Additional pumping capacity (0.25–12 mgd)	100, 200, or 300 ft of total dynamic head
Annual operation and maintenance for horizontal centrifugal pump installations	—	Firm pumping capacity (0.5–200 mgd)	100, 200 or 300 ft of total dynamic head

Source: Clarke et al. (2002).

TABLE 3.6 Parameters for Costs of Selected Pumps, Tanks, and Reservoirs (*Continued*)

Parameter values						
a	*b*	*c*	*d*	*e*	*f*	R^2
33,246.43	15,615.82	0.99	0.0	—	22.78	0.99
−14,695.02	25,947.34	0.42	0.28	1.00	0.0	0.99
225,500.9	8,629.99	0.84	−116,177.1	0.88	−1,449.77	0.99
24,953.27	40,018.32	0.62	0.072	2.20	33.93	0.91
12,766.78	23,136.77	1.06	0.18	1.81	0.0	0.82
−26,101.60	31,414.44	0.04	159.61	0.0	0.0	0.98
−31,272.00	7,902.90	0.93	9,500.48	0.34	9.92	0.90
10,094.42	1,369.70	0.97	0.08	1.69	135.00	0.99

pump with suction and discharge piping with 200 ft total dynamic head is selected, the aggregated construction cost can be computed as the sum of the base construction cost, the pump station site work, the building cost, and the construction cost for electricity and instrumentation. Considering these costs, we obtain the following equation:

$$\mathrm{CC}(Q_{ijp}) = 63{,}506.35 + 23{,}639.68Q_{ijp}^{0.95} + 37{,}931.97Q_{ijp}^{0.87} + 12{,}451.92Q_{ijp}^{0.84} + 52{,}454.49Q_{ijp}^{0.62} + 13{,}062.42Q_{ijp} \qquad (3.54)$$

The annual operation and maintenance cost for the selected type of horizontal centrifugal pump installations can also be determined by referring to Table 3.6. For a pump station of firm capacity of 0.5 to 200 mgd and total dynamic head of 200 ft, this cost is obtained as

$$\mathrm{OC}(Q_{ijp}) = 10{,}713.62 + 2091.67Q_{ijp}^{0.97} + 41{,}775.18Q_{ijp} \qquad (3.55)$$

The expansion cost for the selected type of pump can be determined as the aggregate of the expansion costs of the pump, the site work, the building, and the extension of electrical power and instrumentation. The parameters of these costs are obtained from Table 3.6.

$$\mathrm{EC}(Q_{ijp}) = 27{,}151.70 + 36{,}747.74Q_{ijp}^{1.06} + 31{,}967.71Q_{ijp}^{0.04} + 11{,}064.73Q_{ijp}^{1.07} + 11{,}859.66Q_{ijp}^{0.93} - 7935.95Q_{ijp} \qquad (3.56)$$

where $\mathrm{EC}(Q_{ijp})$ is the total construction cost for a supply line in reach i for activity j during time p of the planning period. Expansion costs for pipelines depend on the approach used in the process, which include pipe bursting, microtunneling, and horizontal directional drilling. For this example, pipe bursting, a method for replacing pipe by bursting from within while simultaneously pulling in a new pipe (Selvakumar et al., 2000), is selected. The liner pipe can be the same size or as much as two pipe sizes larger than the existing pipe (Selvakumar et al., 2000). Boyce and Bried (1998) estimate a range of \$7 to \$9 per inch diameter for installed costs for pipe replacement. Thus, an approximate cost equation for pipe replacement is obtained as

$$\mathrm{EC}(Q_{ijp}) = 8(12)\,(6.06Q^{0.5})L_{ij} = 581.76Q_{ijp}^{0.5}L_{ij} \qquad (3.57)$$

3.4 MODEL RESULTS

3.4.1 Static Model

The formulation given is prepared as the input code for the GAMS/MINOS solver (Brook et al., 1992), and the results shown in Tables 3.7 and 3.8 are obtained, which give a net objective value of \$434.12 million. A summary of the major breakdown of the costs and benefits of this project in the major water-use districts is given in Table 3.8, whereas Fig. 3.11 gives its pictorial illustration. The results clearly show that the damage from the TDS is more pronounced in the irrigation districts than in the cities and towns. The TDS damages to the cities and towns can be considered practically negligible when compared to the gross income from the cities and towns.

The solution of the problem depends on the initial values of some of the parameters in the model. As a result, different local maximum points were obtained depending on the initial conditions used. The most important parameter among these appeared to be the initial value of the river flow. Using different values of the initial river flow (from 400,000 to 850,000), the result in Table 3.9 was obtained. A comparison of the calculated TDS values at each of the nodes from 1 to 13 with 30-year available data at these nodes is given in Table 3.10 and Fig. 3.12.

The sensitivity analysis to the most important factor in this project, the average annual water supply, shows two important trends (see Fig. 3.13). First, even when there is more than enough

TABLE 3.7 "Best" Optimal Solution Obtained

Objective function value, million $	Water supply to the irrigation districts, acre·ft/yr			Water supply to the cities/towns, acre·ft/yr		
	Area name	Diversion	Pumpage	City/town	Diversion	Pumpage
434.12	Percha	75,000	30,000	Hatch	0	576.682
	Leasburg	100,000	40,000	Las Cruces	0	11,201.686
	Mesilla (right)	75,000	30,000	Anthony	0	1,153.389
	Mesilla (left)	75,000	30,000			
	El Paso (reach 12)	83,333.333	N/A	El Paso	13,028.762	55,000.000
	El Paso (reach 15)	416,666.667	N/A			

Note: N/A = not applicable.

TABLE 3.8 Breakdown of the Benefits and the Damages for the "Best" Optimal Solution Obtained

Benefit/cost type	EBID	EPCWID No. 1	NM cities/towns*	El Paso
Gross benefit, $	17,290,000	11,319,000	130,815,109	321,790,790
Water cost, $	4,225,000	4,500,000	4,202,821	24,437,587
TDS damage, $	4,120,457	3,416,011	150,000	2,036,815
Net benefit, $	8,944,543	3,402,989	126,462,288	295,316,388
Total net benefit, $			434,126,208	

*These include Hatch, Las Cruces, and Anthony.

water in the river, the TDS mixing dynamics make the total damage higher when the surplus flow in the river is low. This is due to the fact that the drainage water from the agricultural areas always carries higher TDS than the water flow in the river. The marginal increase in the objective function versus the annual available flow in the river, shown roughly on the right half of Fig. 3.13, illustrates the dilution effect of the river water on the drainage water. The other half of the figure shows the dilution effects plus the marginal increase due to agricultural activity as a function of the marginal increase in the acreage.

3.4.2 Seasonal Model

The seasonal model formulation was also prepared as a GAMS/MINOS input code, which was solved using the latest release (version 2.50) of GAMS/MINOS software. The optimal solution obtained is given in Table 3.11. The objective function value obtained for this solution is $435.43 million. The optimal seasonal releases from Caballo Reservoir are obtained as 19,188, 128,648, 555,022, and 87,141 acre·ft for the first, second, third, and fourth seasons, respectively. The TDS concentrations observed in the reaches did not show much disparity among the seasons.

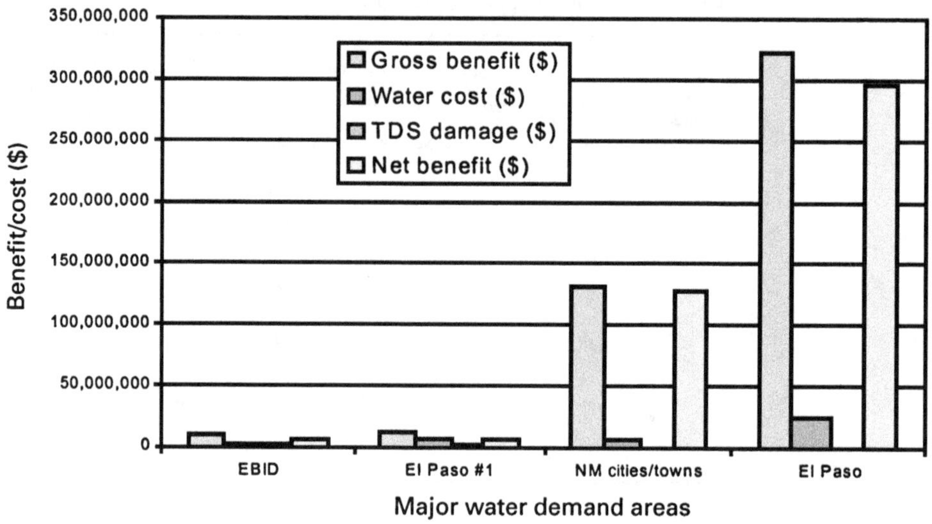

FIGURE 3.11 Comparison of the major benefits and costs for the "best" optimal solution.

TABLE 3.9 Sensitivity Analysis of the Net Benefit Result to Annual Available Flow

Reservoir release, acre·ft/yr	Benefit, million \$	Reservoir release, acre·ft/yr	Benefit, million \$	Reservoir release, acre·ft/yr	Benefit, million \$
400,000	429.78	560,000	431.91	690,000	433.47
410,000	429.91	570,000	432.05	700,000	433.54
430,000	430.17	580,000	432.18	710,000	433.61
450,000	430.44	590,000	432.32	720,000	433.68
460,000	430.57	600,000	432.45	730,000	433.75
470,000	430.71	610,000	432.59	740,000	433.81
480,000	430.84	620,000	432.72	750,000	433.88
490,000	430.97	630,000	432.84	760,000	433.94
500,000	431.11	640,000	432.97	770,000	434.00
510,000	431.24	650,000	433.09	790,000	434.12
520,000	431.38	660,000	433.22	810,000	434.23
540,000	431.64	670,000	433.31	830,000	434.34
550,000	431.78	680,000	433.39	850,000	434.45

3.4.3 Capacity Expansion Model

The seven major water demand areas of El Paso County's water supply system were combined to form a total of three major water demand regions. Herein, the Northwest and Northeast areas are combined to form region 1; the Central, Fort Bliss, and Hueco areas are combined to form region 2; and Lower Valley and the East areas are combined to form region 3. The estimates of the population

TABLE 3.10 Observed 30-Year Average TDS Data and TDS Data Determined Using the Model

Reach	Calculated TDS, ppm	30-year average TDS, ppm
1	450	487
2	450	534
3	494	554
4	544	581
5	544	589
6	636	589
7	631	592
8	631	724
9	787	750
10	787	873
11	809	1037
12	850	1036
13	850	1020

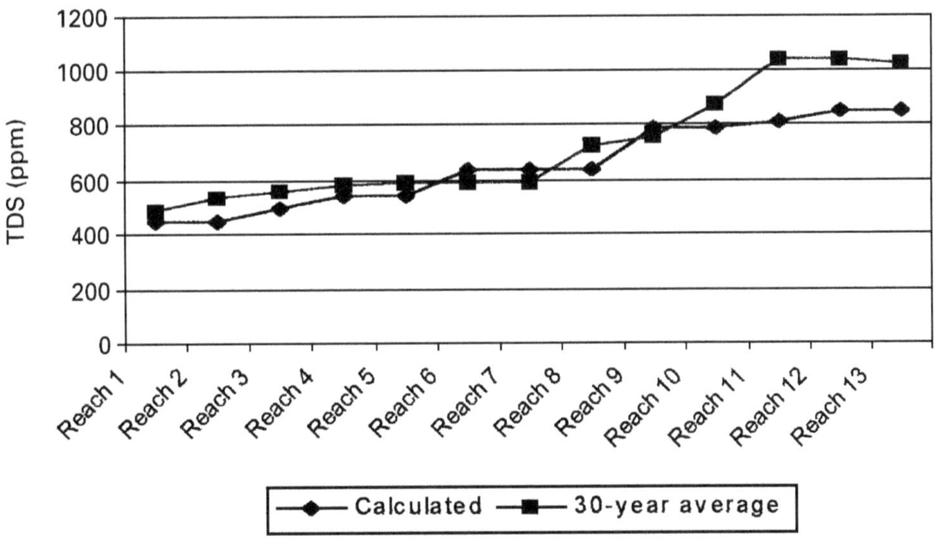

FIGURE 3.12 Graph showing the calculated and 30-year average observed TDS values.

and the households in each region are given in Tables 3.12 and 3.13, respectively. For the purpose of this application, a planning period of 10 years is considered. Therefore, the total number of possible combinations of expansion times equals $(10 - 1)^6 = 531{,}441$. The problem envisages minimizing the total cost over the 10-year planning period.

Using Eqs. (3.51), (3.52), and (3.54) to (3.57) in the problem formulation given by Eq. (3.35), the MINLP problem is solved. The minimum yearly water demands for El Paso for the years 2001 to

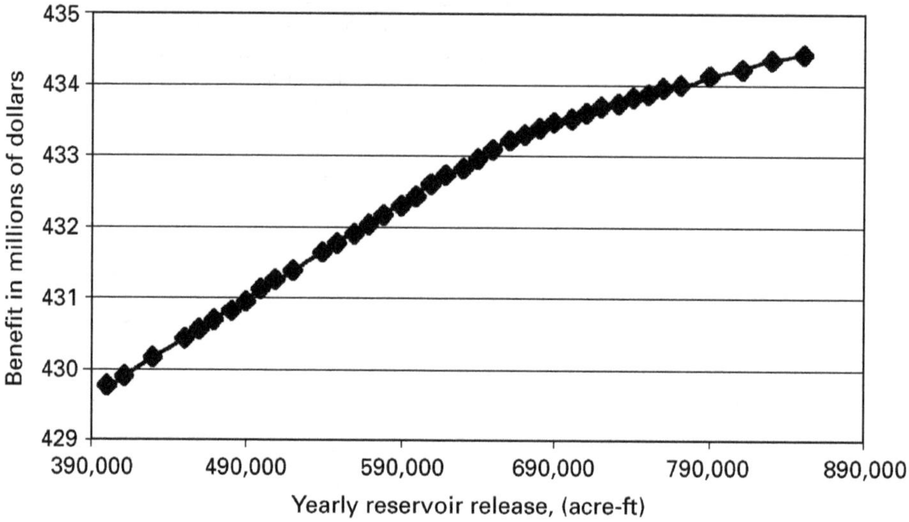

FIGURE 3.13 Net benefit as a function of the annual available flow.

TABLE 3.11 The "Best" Solution Obtained for the Seasonal Model

		Agriculture (left)		Agriculture (right)		Municipal		Industrial	
Reach	Season	Div.	Pump	Div.	Pump	Div.	Pump	Div.	Pump
1	1					0	86		
	2	7,500	22,500			0	143		
	3	54,750	7,500			50	150		
	4	12,750				0	143		
4	1					0	1,680	0	1,625
	2	0	0			0	2,800	0	1,625
	3	0	0			0	3,920	1,045	580
	4	0				0	2,800	0	1,625
7	1								
	2	7,500	22,500	7,500	22,500				
	3	54,750	7,500	54,750	7,500				
	4	12,750		12,750					
9	1					0	172		
	2					0	286		
	3					100	300		
	4					0	286		
12	1					0	10,225	8,513	3,525
	2	8,333				3,292	13,750	12,038	0
	3	60,833				10,109	13,750	12,038	0
	4	14,167				3,292	13,750	12,038	0
15	1								
	2	41,667							
	3	304,167							
	4	70,833							

TABLE 3.12 Estimated Population of Each of the Regions

Year	Region 1	Region 2	Region 3
2001	247,757	177,107	342,685
2002	253,617	177,679	351,613
2003	259,477	178,252	360,542
2004	265,337	178,824	369,470
2005	271,197	179,397	378,398
2006	277,056	179,970	387,326
2007	282,916	180,542	396,254
2008	288,776	181,115	405,183
2009	294,636	181,687	414,111
2010	300,496	182,260	423,039

TABLE 3.13 Estimated Number of Households in Each Region

Year	Region 1	Region 2	Region 3
2001	82,586	59,036	114,228
2002	84,539	59,226	117,204
2003	86,492	59,417	120,181
2004	88,446	59,608	123,157
2005	90,399	59,799	126,133
2006	92,352	59,990	129,109
2007	94,305	60,181	132,085
2008	96,259	60,372	135,061
2009	98,212	60,562	138,037
2010	100,165	60,753	141,013

2010, which are used in the constraints, were determined from a different study by considering supply–benefit analysis. Table 3.14 gives the optimum (minimum) demand by the customers. The maximum yearly groundwater volume that can be pumped per year for each region was limited to 55,000 acre·ft. In the formulation, the length of the pipelines from the diversion points to each of the regions was assumed to be 10 mi each.

The optimal capacity expansion schedule was obtained by solving the above MINLP model. The computed expansion times for the pipelines and the pumping stations to regions 1, 2, and 3 were found to be 9, 8, 9, and 2, 4, 10, respectively. It may be noted that the value 10 for the pumping station to region 3 indicates that there is no expansion necessary and the initial capacity should be built so that it serves during the entire planning period. The total construction, operation, and expansion costs were found to be \$136.20, \$48.04, and \$498.26 million, respectively, with the overall total minimum cost of \$682.50 million. In obtaining the above result, the termination message report by GRG2 was "Termination Criterion Met. Kuhn-Tucker Conditions Satisfied to Within [EPSTOP] at Current Point." This is the best termination message because it ensures that

TABLE 3.14 Minimum Water Supply Requirements for El Paso County

	Region 1		Region 2		Region 3	
Year	Surface water	Groundwater	Surface water	Groundwater	Surface water	Groundwater
2001	84,806	55,000	45,051	55,000	138,677	55,000
2002	88,124	55,000	45,289	55,000	143,222	55,000
2003	91,602	55,000	45,527	55,000	148,969	55,000
2004	94,722	55,000	46,008	55,000	154,017	55,000
2005	97,976	55,000	46,251	55,000	158,240	55,000
2006	101,376	55,000	46,494	55,000	163,764	55,000
2007	104,930	55,000	46,984	55,000	168,394	55,000
2008	108,017	55,000	47,231	55,000	173,224	55,000
2009	111,226	55,000	47,479	55,000	178,268	55,000
2010	114,564	55,000	47,979	55,000	183,540	55,000

the necessary conditions have been satisfied at the current point and thus the current point may be a local optimum point.

REFERENCES

Boyce, G. M. and E. M. Bried (1998). Social Cost Accounting for Trenchless Projects, *Proceedings of the Conference Entitled North American No-Dig' 98,* Albuquerque, N.M.

Boyle Engineering Corp. (1991a). El Paso Water Resources Management Plan, Phase I Completion Report, July.

Boyle Engineering Corp. (1991b). El Paso Water Resources Management Plan, Phase II Completion Report, August.

Boyle Engineering Corp. (1991c). El Paso Water Resources Management Plan, Phase III Completion Report, February 1992.

Boyle Engineering Corp. and Parsons Engineering Science, Inc. (1998). El Paso–Las Cruces Regional Sustainable Water Project, Drain Mitigation Strategies Report, a final report prepared for New Mexico–Texas Water Commission, July.

Brook, A. et al. (1992). *GAMS: A User's Guide,* Release 2.25, Scientific Press, Redwood City, Calif.

Chenery, H. B. (1952). Overcapacity and the Acceleration Principle, *Econometrica,* (20): 1–28.

Clark, R. M. et al. (2002). Cost Models for Water Supply Distribution Systems, *Journal of Water Resources Planning and Management,* 128(5): 312–321.

Engineering-Science, Inc. (1991). Surface Water Supply Alternatives for the City of El Paso and Southern New Mexico Users, Final Report Prepared for the New Mexico Entities, Las Cruces, N.M.

Goldman, F. E. and L. W. Mays (2004). Simulated Annealing Approach for Operation of Water Distribution Systems Considering Water Quality, *Journal of Water Resources Planning and Management* (to be published).

Hinomoto, H. (1974). Determination of Investment Cost Functions of Water Treatment Plants, Report No. 87, University of Illinois at Urbana-Champaign.

Kim, J. H. and L. W. Mays (1994). Optimal Rehabilitation Model for Water Distribution Systems, *Journal of Water Resources Planning and Management,* 120(5): 674–692.

Lall, U. and L. W. Mays (1981). Model for Planning Water-Energy Systems, *Water Resources Research,* 17(4): 853–865.

Lasdon, L. S. and A. D. Waren (1986). *GRG2 User's Guide,* Department of General Business, University of Texas at Austin.

Matsumoto, J. and L. W. Mays (1983). Capacity Expansion Model for Large-Scale Water Energy Systems, *Water Resources Research,* 19(3): 593–607.

Mays, L. W. (1997). *Optimal Control of Hydrosystems,* Marcel Dekker, New York.

Mays, L. W. and Y. K. Tung (1992). *Hydrosystems Engineering and Management,* McGraw-Hill, New York.

Mehrez, A. et al. (1992). Optimal Operation of a Multisource and Multiquality Regional Water System, *Water Resources Research,* 28(5): 1199–1206.

Niemi, E. and T. McGuckin (1997). Water Management Study: Upper Rio Grande Basin, Final Report to the Western Water Policy Review Advisory Commission, July 1997.

Ocanas, G. and L. W. Mays (1980). Models for Water Reuse Planning, Final Report to Texas Department of Water Resources, CRWR-173, Center for Research in Water Resources, University of Texas at Austin.

Ocanas, G. and L. W. Mays (1981a). A Model for Water Reuse Planning, *Water Resources Research,* 17(1): 25–32.

Ocanas, G. and L. W. Mays (1981b). Water Reuse Planning Models: Extensions and Applications, *Water Resources Research,* 17(5): 1311–1327.

Schwartz, M. and L. W. Mays (1983). Models for Water Reuse and Wastewater Planning, *Journal of Environmental Engineering,* 109(5): 1128–1147.

Selvakumar, A. et al. (2002). Costing Models for Water Supply Distribution System Rehabilitation, *Journal of Water Resources Planning and Management,* 128(4): 303–306.

Wolsey, L. A. (1998). *Integer Programming,* John Wiley, New York.

Yang, S. L. et al. (1999). A Multicommodity Formulation for Multisource Water Distribution Network, *Proceedings of the 26th Annual Conference of the Water Resources Planning and Management Division of ASCE,* Tempe, Ariz., June 6–9.

Yeh, W. W-G. et al. (2000). Optimization of Regional Water Distribution System with Blending Requirements, *Journal of Water Resources Planning and Management,* 126(4): 229–235.

CHAPTER 4
RIVER-RESERVOIR SYSTEM OPERATION FOR SEDIMENT CONTROL

John W. Nicklow
Southern Illinois University at Carbondale
Carbondale, Illinois

4.1 PROBLEM STATEMENT

Alluvial rivers and reservoirs naturally undergo continuous morphological changes that are caused by complex interactions between entrained sediment, variable streamflows, and boundaries that are susceptible to erosion. Excess streambed changes induced through sediment degradation, however, threaten stability of underground utilities, bridges, levees, and other hydraulic structures. Similarly, chronic deposition can be problematic in that it reduces capacity of lakes and reservoirs and may cause frequent inundation of floodplain property. For reservoirs, storage depletion limits the useful life of the impoundment and critically reduces the benefits to be earned through hydropower generation, water supply, navigation, or other activities. Although difficult to quantify on a worldwide basis, Mahmood (1987) estimated that reservoirs globally may be losing storage capacity as rapidly as one percent per year and that the associated annual cost of sediment accumulation has approached \$6 billion. More recently, Osterkamp et al. (1998) estimated that annual depleted storage costs in the United States alone are \$2 billion and a Task Committee on Guidelines for Retirement of Dams and Hydroelectric Facilities (1997) concluded that if sedimentation is not given sufficient attention in system design and operation, it may easily control retirement considerations. Finally, in Illinois' rivers and streams, excess sedimentation is recognized as the most critical surface water pollution problem. More than 1700 state drainage districts require periodic maintenance to remove significant amounts of silt, and many of the state's commercially navigable rivers require frequent dredging so that barge traffic remains uninterrupted and profitable (ISWPTF, 1984; Bhomik et al., 1986; IDOT, 1988; IDENR, 1994).

Once sedimentation problems are identified, rehabilitation technologies such as channel dredging or reservoir flushing may be effectively implemented. Unfortunately, such technologies are designed to only temporarily restore a watercourse and are quite costly and difficult to implement on a repeated basis (Fitzpatrick et al., 1987). These considerations have generated a need for control techniques that aid in prevention of the adverse effects of sediment scour and deposition and thus serve as cost-effective means to better balance the problems and benefits that rivers and reservoirs provide. In this respect, a responsible approach would be to manage hydraulic networks in a manner that minimizes future sediment problems. Knowing that streamflow variability plays a dominant role in bed sediment morphology, discharge events, or reservoir releases, could be managed in such a way that changes to stream and reservoir bed elevations are limited. Application of this concept to a multiple-reservoir network is of special interest because any sediment transported from upstream facilities will be deposited downstream and, consequently, affect management strategies of downstream components. The optimization problem associated with the determination of releases could be expressed as

Minimize → Bed elevation changes in rivers and reservoirs of a large-scale network

Subject to → The laws governing network hydraulic and sediment transport dynamics, and operational constraints on reservoir releases and storage levels

In recent years, the development and use of reservoir management models has gained considerable popularity. These models have typically relied on a search algorithm to solve a simplified expression of reservoir dynamics (Yeh, 1985; Wurbs, 1993), whereas others (Unver and Mays, 1990) have gone even further by integrating comprehensive hydraulic simulation modules. The goal in each case, however, is the same: development of a decision support mechanism that identifies *optimal* reservoir operations. The criterion used to define optimal operations in the majority of previously developed models is a prespecified cost-benefit function, or a more explicit objective related to flood control, hydropower generation, or water supply. Carriaga and Mays (1995) extended the underlying concept of these models for sedimentation control by interfacing a differential dynamic programming (DDP) algorithm (Yakowitz and Rutherford, 1984) with a widely used sediment transport simulator, HEC-6 (USACE, 1993). The model created, however, did not consider sedimentation occurring within reservoirs and the authors limit their focus to minimizing adverse effects only in a single downstream river reach. The HEC-6 simulation code handled governing hydraulic and sediment constraints, whereas the search algorithm solved the overall control problem. Nicklow and Mays (2000) subsequently coupled HEC-6 with a gradient-based feedback method known as the Successive Approximation Linear Quadratic Regulator (SALQR) in order to solve the sedimentation problem in large-scale multiple-reservoir river networks. They concluded that although optimal release policies could be achieved, the SALQR still required time-consuming, numerical derivative computations. In addition, resulting policies heavily depended on initial user-specified release policies from which the search begins (Nicklow and Mays, 2000; 2001). Nicklow et al. (2003) further attempted to overcome previous limitations by replacing the SALQR algorithm with a global, heuristic search method known as the genetic algorithm (GA) that does not require derivative computations.

4.2 PROBLEM FORMULATION

The sediment control problem is formulated for a multireservoir river network using J reservoirs and I defined stream and reservoir cross sections, which describe system geometry and boundaries, along T discrete-time intervals. Decision parameters for the system are comprised of stagewise reservoir releases R_j^t, and important state parameters are river and reservoir bed elevations E_i^t and reservoir storage levels S_j^t. The problem is summarized as the determination of a $J \times T$ matrix of reservoir releases that results in a limiting of channel bed morphology.

4.2.1 Mathematical Model

The sediment control problem is mathematically expressed as follows (Nicklow and Mays, 2000):

$$\min Z = \sum_{t=1}^{T-1} \sum_{i=1}^{I} | E_i^{t+1} - E_i^T | \tag{4.1}$$

$$E_i^{t+1} = T_E (E_i^t, S_j^t, R_j^t, t) \qquad \text{for } 1 \le i \le I \tag{4.2}$$

$$S_j^{t+1} = T_s (E_i^t, S_j^t, R_j^t, t) \qquad \text{for } 1 \le j \le J \tag{4.3}$$

$$\underline{R_j^t} \le R_j^t \ \overline{R_j^t} \tag{4.4}$$

$$\underline{S_j^{t+1}} \le S_j^{t+1} \le \overline{S_j^{t+1}} \tag{4.5}$$

$$S_j^T = S_j^{\text{target}} \tag{4.6}$$

where $\underline{Z}$ = objective function value

$\underline{R_j^t}$ and $\overline{R_j^t}$ = lower and upper bounds on reservoir releases, respectively

$\underline{S_j^{t+1}}$ and $\overline{S_j^{t+1}}$ = lower and upper bounds on storage levels, respectively

S_j^{target} = target storage level for terminal storage state S_j^T

Equation (4.1) is the separable objective function to be minimized and represents the *cumulative* bed change at all locations and over the simulation period. The function implicitly depends on a particular release policy through the governing dynamics of hydraulic and sediment transport phenomena. The terms T_E and T_S in Eqs. (4.2) and (4.3) denote that these functions are the transition constraints used to describe the stage-by-stage response of the hydraulic network according to an imposed control policy. The transition equations for the current problem are comprised of relationships for water and sediment continuity, the energy and Manning equations, and a sediment transport function, among others. Equations (4.4) through (4.6) define a feasible range for reservoir release policies and storage levels. These operating constraints, together with the transition constraints, define the feasible solution space for the sedimentation problem.

Note that Eq. (4.1) represents only one possible objective function. For example, Nicklow et al. (2003) provided an alternative formulation that minimizes the maximum incremental bed change in both a spatial and temporal context, or

$$\min Z = \max \mid E_i^{t+1} - E_i^t \mid \tag{4.7}$$

for all i and t. Likewise, other objectives could account for the cost associated with damage caused by excess sedimentation, could focus on sediment control only at critical cross sections in a hydraulic system, or could be aimed at maintenance of a navigation channel, or prespecified design draft, for a river channel.

4.2.2 Constraint Handling

To solve the sediment control problem, it is first necessary to transform the constrained problem into an unconstrained formulation. This can be accomplished through the use of a penalty function, whereby a high cost is assigned to the objective function when one or more bound constraints is violated. Thus, the penalty is designed to prevent the inclusion of reservoir operation policies that are neither desirable nor practical. In the case of an additive penalty function, the augmented objective function for the transformed problem becomes

$$\min Z \sum_{t=1}^{T-1} \sum_{i=1}^{I} \mid E_i^{t+1} - E_i^t \mid + \sum_{t=1}^{T} \sum_{n=1}^{N} R_{P,n} \cdot P_n^t \tag{4.8}$$

where $R_{p,n}$ is a penalty weight and P_n^t is a penalty function term that depends on the magnitude of constraint violation $L_n^t(R_j^t, S_j^{t+1}, t)$ of the nth bound constraint. A similar modification could be made in the case of Eq. (4.7).

Different types of penalty functions are documented; they range in degree of difficulty and the number of parameters required for usage. A simple function called the bracket penalty function (Reklaitis et al., 1983) is incorporated into the current formulation and can be expressed as

$$P_n^t = 0 \qquad \text{if } L_n^t(R_j^t, S_j^{t+1}, t) \le 0 \tag{4.9}$$

$$P_n^t = [L_n^t(R_j^t, S_j^{t+1}, t)]^2 \qquad \text{if } L_n^t(R_j^t, S_j^{t+1}, t) > 0$$

Note that because S_j^{t+1} is explicitly a function of R_j^t and S_j^t, reservoir release policies will be determined such that both the release and storage constraints are satisfied. The bracket penalty function has been used successfully for a soil aquifer treatment problem (Tang et al., 1995) and an estuarine management problem (Li and Mays, 1995). Lootsma (1969) showed that this quadratic function yields better performance than other higher-order power functions. The foremost disadvantage of the bracket penalty function, however, is that with Eq. (4.9), the augmented objective function is not a twice continuously differentiable function (Lin, 1990). The function is then valid only if the iterative solution asymptotically approaches the discontinuity at $L = 0$ from either the feasible side or the infeasible side, exclusively. If this deficiency does not interfere with solution accuracy, however, the simple nature of the bracket penalty function makes it efficient for use.

4.2.3 Transition Constraints

Regardless of the exact nature of the objective function, it is inherently subject to constraining laws that govern open channel hydraulics and sediment transport dynamics, which were expressed generally through Eqs. (4.2) and (4.3). For example, one-dimensional, steady flow is described by the continuity and energy equations, written as

$$\frac{\partial Q_t}{\partial l} - q_l = 0 \tag{4.10}$$

and

$$E_{i+1}{}^t + y_{i+1}{}^t + \frac{\alpha_{i+1}{}^t (V_{i+1}{}^t)^2}{2g} = E_i^t + y_i^t + \frac{\alpha_i^t (V_i^t)^2}{2g} + h_i^t \tag{4.11}$$

where Q = channel discharge

l = distance along the channel
q_l = lateral inflow rate per unit length
y = depth of flow
α = Coriolis coefficient
V = flow velocity
g = gravitational constant
h_i^t = energy loss between sections $i + 1$ and i during time interval t

Energy losses include friction losses and form losses due to expansions and contractions. Friction losses h_f can be evaluated as

$$h_f = \frac{Q^2}{K^2} \tag{4.12}$$

where K denotes the length weighted conveyance of a reach subsection, expressed as

$$K = \sum_{m=1}^{M} K \left[\frac{C_c \dfrac{(A_{i+1} + A_i)}{2} \left(\dfrac{R_{h,\,i+1} + R_{h,\,i}}{2} \right)_m^{2/3}}{n_m l_m^{1/2}} \right] \tag{4.13}$$

where C_c = a constant equal to 1.0, or 1.486 for U.S. customary units
M = total number of longitudinal subsections defined between cross sections i and $i + 1$
A = effective flow area
l_m = the length of the mth strip between sections
n_m = Manning's roughness coefficient
R_h = hydraulic radius

Form losses h_o can be evaluated by

$$h_o = C_o \left| \frac{\alpha_{i+1} (V_{i+1})^2 - \alpha_i (V_i)^2}{2g} \right| \tag{4.14}$$

where C_o is the loss coefficient for channel expansions or contractions.

Sediment transport evaluation is based upon conservation of mass for bed material load, represented mathematically by the Exner equation

$$\frac{\partial G_i^t}{\partial l} + \gamma_m B_o \frac{\partial E_i^t}{\partial t} = 0 \tag{4.15}$$

where G = average sediment discharge during t
γ_m = specific weight of the water–sediment mixture
B_o = movable channel bed width

Computation of sediment discharge focuses on transport capacity, but can be influenced by factors such as sediment inflow, availability of bed material, and armoring opportunities within the channel. Transport capacity can be computed using Toffaleti's (1966) formula, expressed generally as

$$G = (C_u)\,2\,dB_o \tag{4.16}$$

where C and u are the concentration of sediment having a diameter d and flow velocity, respectively, both of which are evaluated at a distance of two grain diameters (i.e., $2d$) above the channel bed. Alternatively, Schoklitsch (1930) proposed

$$G = 7000\,\frac{S_f^{3/2}}{d^{1/2}}\left[Q - \frac{(1.94 \times 10^{25})\,dB_o}{S_f^{4/3}}\right] \tag{4.17}$$

which is written in S.I. units and where S_f is the energy slope. A number of other purported formulae could be used to compute transport capacity as well, including those of Yang (1973), Ackers and White (1973), and Meyer- Peter and Müller (1948). Equation (4.15) represents the means by which any one of these discharge formulas can be interpreted as change in bed elevation (i.e., $\partial E_i^t/\partial t$), or the extent of degradation and/or deposition through time.

4.3 SOLUTION TECHNIQUES

There is no explicit scheme by which a solution to the sediment control problem may be directly ascertained. The problem can be solved, however, by using the discrete-time optimal control approach, whereby a hydraulic and sediment transport simulation model is coupled with a suitable search algorithm. The search algorithm identifies and updates control policies, whereas the simulation model is used to implicitly evaluate the behavioral response of the system by solving governing hydraulic and sediment constraints each time the search mechanism requires their solution. This interfacing strategy has a significant advantage over traditional optimization approaches that may require simplification of the problem due to dimensionality or computational complexities.

4.3.1 Sediment Transport Simulation

The U.S. Army Corps of Engineers HEC-6 computer code (USACE, 1993) is used to solve the simulation constraints that govern open channel hydraulics and sediment transport dynamics. Other, possibly custom-built, sediment transport simulators may be suitable for inclusion in the optimal control methodology. However, the perceived gap between theoretical developments in optimization and systems research and the practical utility of those developments in water resources management are likely to be narrowed by incorporating familiar models, such as HEC-6, and by limiting additional search parameters.

HEC-6 is a one-dimensional, steady, open channel flow simulator designed to evaluate water surface and bed sediment profiles in both rivers and reservoirs by observing the interaction between bed material and the overlying water–sediment mixture. It is capable of simulating uniform scour and deposition over the movable channel width for a wide range of grain sizes in a variety of network configurations and incorporates an evaluation of channel armoring potential. During the simulation process, a series of steady flows is derived from a discretized set of discharges. The energy and continuity equations are solved for flow depths through a standard step backwater computation for each discharge. After evaluating all hydraulic elements, potential sediment transport rates are computed at each section using one of 10 user-selected transport options. For example, one option is a Toffaleti–Schoklitsch combination, in which sediment load is computed using both transport functions. The larger transport capacity obtained from the two formulas is then used in subsequent computations. Sediment routing and determination of new bed elevations are performed using the Exner sediment continuity equation and by evaluating the difference in transport rates at the upstream and downstream sections of each stream reach.

Several modifications to the original HEC-6 computer code are necessary to create the discrete-time optimal control interface. These changes are the addition of a supplementary input file that describes internal boundary conditions in the form of reservoir storage–water surface elevation rating information, operational bounds on releases and storage states, and initial storage level and evaporation loss rates for each reservoir. These modifications allow the integration of bound constraints on storage and releases that are outlined in the current problem formulation.

An additional modification addresses the steady flow characteristics of the simulation model. In its original form, the HEC-6 model requires that reservoir pool elevations be a function of time, as defined by user input. This implies that outflow from any reservoir is equal to inflows to that reservoir during each discretized period. The implication thus neglects the issue of storage accumulation or depletion, which is a key component in any reservoir management model. To overcome this limitation, a storage routing routine is added to force the model to operate under quasi-unsteady, rather than steady, flow conditions during backwater computations. The routine is based on a mass balance of reservoir operation that can be expressed generally as

$$S_j^{t+1} = S_j^t + q_j^t \Delta t + R_j^t \Delta t M - l_j^t \tag{4.18}$$

where q_j^t = vector of inflows to reservoir j
Δt = time step duration
l_j^t = reservoir losses such as net evaporation and infiltration as a function of reservoir storage
M = a $J \times J$ reservoir connectivity matrix with (-1) in diagonal positions, $(+1)$ in position (j, v) if reservoir j contributes to reservoir v's inflow, and zeros elsewhere

This mass balance relationship is used in conjunction with internal boundary conditions to define stagewise pool elevations in each reservoir. Note that the storage rating curve that is used to define water surface elevations is subsequently updated during each duration to reflect the change in storage due to previous scour or deposition in the reservoir.

Other modifications to the sediment transport simulator include the extraction and storing of computed thalweg elevations and reservoir storage levels, or stage-by-stage state transformations, after each time duration; the limiting of output production and file management schemes during operation of the optimal control model; the addition of computational control schemes to obtain derivative information; and the actual simulator–optimizer interface, or conversion of the HEC-6 code to a number of subroutines within the optimal control model.

4.3.2 Search Algorithms

Two alternative algorithms are investigated for solution to the sediment control problem, the SALQR algorithm, and a GA. Although the algorithms are distinctly different, in both cases they use information about current objective function values to provide incrementally improved reservoir operation policies. The SALQR, however, also uses certain derivative information about the objective function and constraints and relies upon a user-specified, initial release policy from which the search begins. As a result, convergence to local optima may, in some cases, occur frequently, and the effects of different starting policies must be evaluated. The method is considered to be a multistage decision technique that is a modification of conventional DDP. It differs from DDP only in that second-order derivatives of the transition constraints are linearized in the optimization step, thus potentially allowing reduced computational times. Figure 4.1 illustrates the SALQR solution approach used to evaluate optimal reservoir release policies that limit bed sediment changes. Further details of a basic SALQR algorithm are provided in App. 4.A and can be found in Culver and Shoemaker (1992) and Chang et al. (1992).

Genetic algorithms represent a robust, heuristic search procedure that relies on stochastic search rules to solve control problems. Developed by Holland (1975), they represent an attempt to adapt the

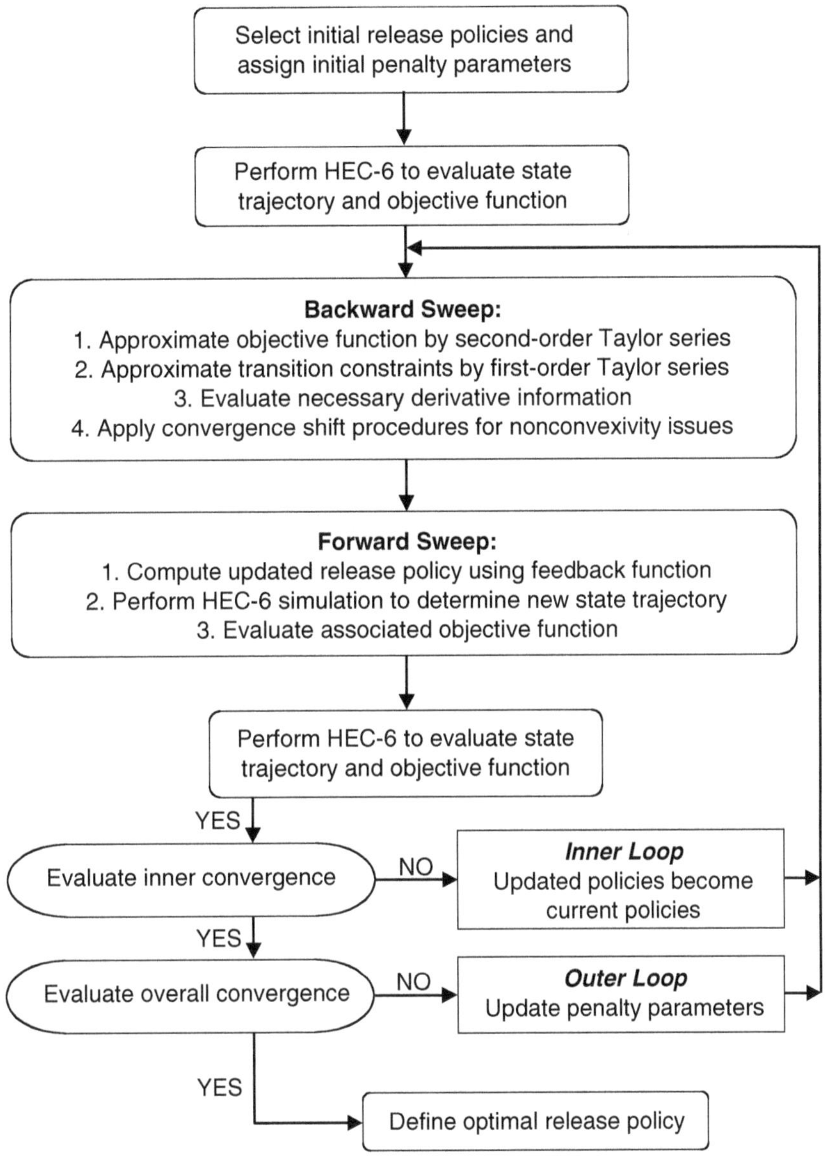

FIGURE 4.1 Constrained SALQR algorithm for the sediment control problem.

evolution observed in nature to problems in which traditional, deterministic search techniques fail. Although there is no universal definition that applies to all GAs, most are at least characterized by the following elements:

1. Generation of an initial population of potential solutions, each identified as a chromosome
2. Computation of the objective function value for each solution and subsequent ranking of chromosomes according to this value

3. Selection of the fittest solutions to undergo crossover and mating of pairs of solutions
4. Completion of a crossover operation in which information describing specific decision variables, or genes, from two parent solutions are combined to create offspring solutions
5. Mutation of a portion of the new offspring to maintain diversity of solutions and discourage convergence to local optima.

These elements are repeated in sequential generations until a final solution is obtained. The general concept behind these components is that solutions having high fitness values contain specific genes and characteristics that are important in optimizing the objective function. By exchanging important genes between two parent alternatives, it is expected that the GA will produce some offspring that contain even more superior characteristics than their parent alternatives. The result is the evolution of policies that are better suited to solve the current problem than the individuals from which they were created.

The major distinction between GAs and their gradient-based counterparts is that they use only the objective function magnitude, rather than derivative information, to display incrementally better solutions, making GAs amenable for application to nonconvex nonlinear problems (Goldberg, 1989). However, they are by nature a heuristic search technique and a globally optimal solution is not guaranteed (Cieniawski et al., 1995). Reliability in locating global optima can be investigated and possibly improved through repeated sensitivity applications of the GA in which the user varies parameters such as population size and mutation frequency. In fact, the majority of GA literature consistently demonstrates an ability to identify global or very near global optima for a range of complicated problems (Nicklow, 2000). Further details of the GA used for the sediment control problem are provided in App. 4.B. For additional discussion of the detailed framework of genetic algorithms, the reader is also referred to Goldberg (1989), Haupt and Haupt (1998), Mitchell (1996), or Belegundu and Chandrupatla (1999).

4.4 EXAMPLE APPLICATIONS

The sediment control models described were applied to several different multiple-reservoir river systems. These include a hypothetical three-reservoir network, the Yazoo River in Mississippi, and the Kaskaskia River in Illinois. The applications were used to assess the capabilities of different solution techniques (i.e., SALQR and GA) and evaluate potential computational issues and concerns.

4.4.1 Hypothetical River-Reservoir Network

Nicklow and Mays (2000) applied the SALQR control model to a hypothetical system containing reservoirs both in series and in parallel, as shown in Fig. 4.2. The system, as well as specified flows, was designed to incur large amounts of geomorphic changes based on governing sediment transport laws. The required information for application of the methodology is listed in Table 4.1. System geometry was defined using 27 rectangular-shaped cross sections that vary in width from 22.9 m to 30.5 m within the river sections and as large as 305 m for some reservoir sections. The total channel length was 31.2 km, and initial bed slopes varied between 0.0012 m/m and 0.008 m/m with an average slope of 0.0034 m/m.

Using Eq. (4.1) as the objective function, the SALQR control model converged to a release policy that yielded 18.69 m of cumulative scour and deposition. Recall that this value represents the sum of the absolute value of all bed changes at all defined river and reservoir cross sections. Details of the solution, including state parameter results, are given in Table 4.2. Nicklow et al. (2003) later applied the GA control model to the same hypothetical problem and demonstrated convergence to an objective function value of 18.24 m. This result indicates that the SALQR methodology is likely to have converged to local, near-global optima and that the GA may, in some cases, be better capable of locating

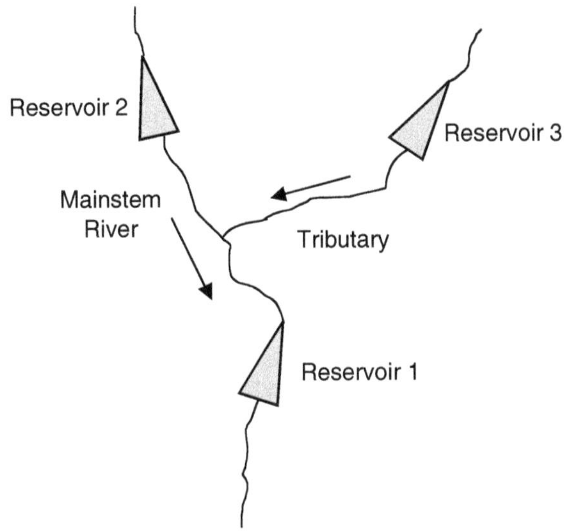

FIGURE 4.2 Schematic of hypothetical network.

TABLE 4.1 Numeric Data for Hypothetical Network

Hydrographic data		
Duration, days	Mainstem inflow, m^3/s	Tributary inflow, m^3/s
0.75	99.1	70.8
0.75	169.8	107.5
0.75	283.0	113.2
0.75	226.4	141.5
0.75	169.8	84.9
0.75	99.1	56.6

Bound constraints	Reservoir 1	Reservoir 2	Reservoir 3
Max. release (m^3/s)	367.9	226.4	226.4
Min. release (m^3/s)	28.3	28.3	28.3
Initial storage (10^6 m^3)	13.6	9.9	11.1
Target storage (10^6 m^3)	9.9	12.3	4.9
Min. storage (10^6 m^3)	2.4	2.4	2.4
Max. storage (10^6 m^3)	14.8	14.8	14.8

Additional Information
Internal boundary conditions—reservoir storage vs. water surface elevation
Downstream boundary condition—stage vs. discharge
Upstream boundary condition—water discharge vs. sediment discharge
Bed material gradation information
Sediment transport function—Toffaleti–Schoklitsch

a global or near-global optimal solution to the sedimentation problem. Table 4.3 provides a convenient comparison of final release policies obtained from the SALQR and GA methodologies.

A general comparison between computational demand of the SALQR model and that of the GA control model can be made by examining the number of calls made to the HEC-6 simulator. The SALQR application required 21,591 calls to HEC-6, 99.6 percent of which were for numerical evaluation of derivative information; whereas the GA required more than 50,000 calls, corresponding to just over three hours of CPU time on a 600-MHz, Pentium III processor. Consider, however, that the convergence of the GA methodology does not depend on user-specified starting points, as with gradient-based techniques. For example, the SALQR method begins its search from, and depends heavily upon, a user-defined operational policy. Thus, the user must repeatedly execute the model with different starting points in efforts to ensure optimality (Lasdon and Waren, 1989). Then again, CPU time for the GA model is heavily dependent on the user-assigned maximum number of generations, which should be large enough to ensure convergence to a global optimum. Note that although computational time was once a significant and constraining issue, it continues to become less of a concern as computing technology (e.g., high-speed and multiprocessor units) advances.

4.4.2 Yazoo River Basin

Nicklow and Mays (2000) applied the SALQR model to the Yazoo River Basin network located in northwest Mississippi. Shown in Fig. 4.3, the system drains an area of approximately 34,706 km^2. Four headwater tributaries, the Coldwater, Little Tallahatchie, Yocono, and Yalobusha Rivers, contribute large volumes of sediment as a result of agricultural practices and other upland

TABLE 4.2 SALQR Solution for the Hypothetical Network

	Release policies, m^3/s			Storage levels, 10^6 m^3			Sediment evaluation, m	
Time interval	Res. 1	Res. 2	Res. 3	Res. 1	Res. 2	Res. 3	Deposition	Scour
1	220.7	158.8	81.0	14.8	6.0	10.4	1.16	2.91
2	236.7	161.2	70.8	14.5	6.6	12.3	0.63	1.93
3	292.5	193.8	96.2	14.3	12.3	13.9	1.58	3.72
4	367.9	191.4	174.9	14.2	14.6	11.8	0.41	1.54
5	324.5	183.2	142.1	14.3	13.7	8.0	0.16	1.83
6	293.5	120.6	104.5	9.9	12.3	4.9	0.67	2.15
						Total*	4.61	14.08

Notes: CPU Requirements—91 Inner Loop Iterations; 245.4 Min. on Sun Ultra 1
*Final Objective Function Value = 18.69m

TABLE 4.3 Comparison of Policies Derived Using SALQR and GAs

	Release, m^3/s computed via GA*			Release, m^3/s computed via SALQR†		
Time period	Res. 1	Res. 2	Res. 3	Res. 1	Res. 2	Res. 3
1	320.1	171.6	130.3	220.7	158.8	81.0
2	292.8	172.9	133.7	236.7	161.2	70.8
3	301.8	177.3	95.2	292.5	193.8	96.2
4	280.8	188.2	108.3	367.9	191.4	174.9
5	280.4	166.2	114.5	324.5	183.2	142.1
6	261.2	133.8	88.0	293.5	120.6	104.5

*Corresponding cumulative bed change = 18.24 m (Reported by Nicklow et al., 2003)
†Corresponding cumulative bed change = 18.69 m (Reported by Nicklow and Mays, 2000)

erosion sources. Flood control reservoirs constructed on each of these tributaries include Arkabutla, Sardis, Enid, and Grenada Lakes. The model application considered the four main tributaries entering the reservoirs, as well as the Skuna river, which flows into Grenada Lake, and one local inflow located upstream of Arkabutla Lake. Additional data associated with the Yazoo network are provided in Table 4.4. In addition, a qualitative evaluation based on the conditions from which each available sediment transport function was derived and recommendations for selection in the absence of adequate historical data revealed that the Toffaleti–Schoklitsch combined transport function may be an appropriate relation for the Yazoo Basin network. Note, however, that Nicklow and Mays (2000) showed that because discharge is either indirectly or directly the dominant term in all transport functions offered by HEC6, function selection has little effect

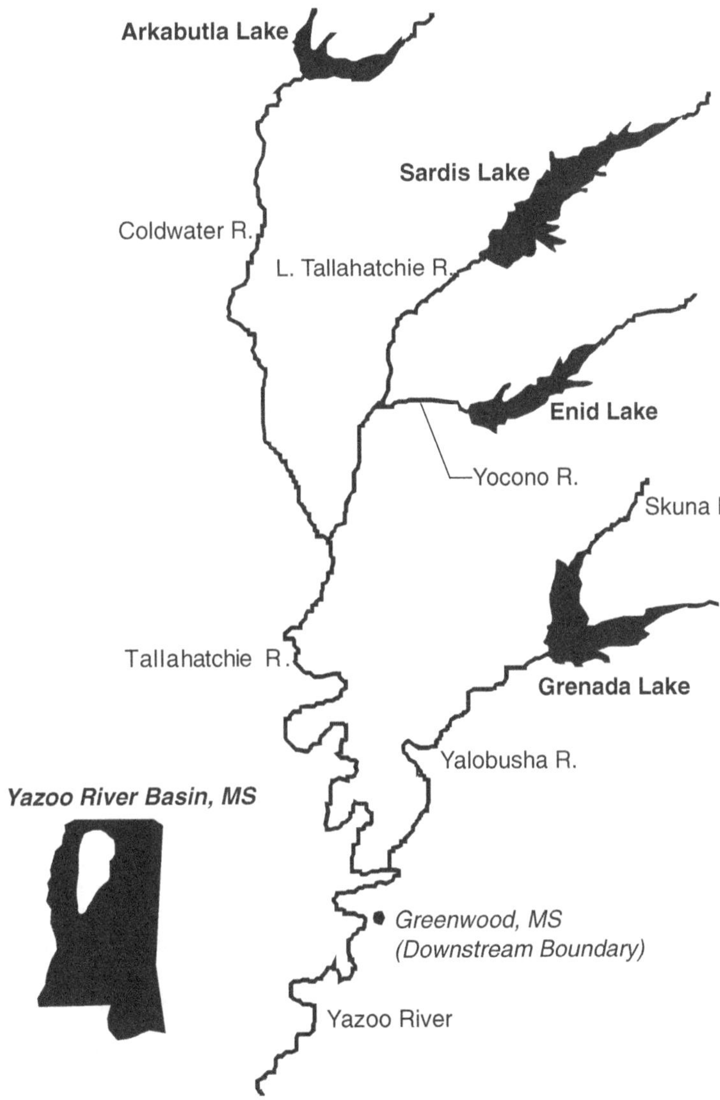

FIGURE 4.3 Yazoo River Basin network.

TABLE 4.4 Numerical Data for the Yazoo River Network

Hydrographic data						
		Discharge, m^3/s				
Hydrograph interval	Interval duration, days	Coldwater River	Yalobusha River	Little Tallahatchie River	Skuna River	Yocono River
1	7.0	58.2	52.4	67.7	17.0	25.9
2	7.0	240.0	162.7	296.9	97.9	171.4
3	7.0	432.8	334.3	703.8	172.6	348.1
4	7.0	261.6	206.6	362.2	83.5	147.2
5	7.0	67.5	52.4	123.8	17.0	25.9

Bound constraints				
	Arkabutla	Grenada	Sardis	Enid
Maximum storage capacity (10^7 m^3)	61.7	37.0	185.2	80.2
Minimum storage capacity (10^7 m^3)	3.7	10.6	13.3	7.1
Initial storage capacity (10^7 m^3)	12.3	18.5	37.0	37.0
Target storage capacity (10^7 m^3)	9.9	12.3	74.1	12.3
Minimum storage capacity (10^7 m^3)	3.7	10.6	13.3	7.1
Maximum storage capacity (10^7 m^3)	61.7	37.0	185.2	80.2
Reservoir loss rate (cm/day)	0.38	0.38	0.38	0.38

Additional information required
Internal boundary conditions—reservoir storage vs. water surface elevation
Downstream boundary condition—stage vs. discharge
Upstream boundary condition—water discharge vs. sediment discharge
Bed material gradation curves
Sediment transport function—Toffaleti–Schoklitsch

on optimal policies, even though the actual objective function values will vary based on resulting bed morphology. Based on the specified data, final release policies that minimize cumulative bed change in the 118 defined river and reservoir cross sections are given in Table 4.5, along with details concerning storage levels and bed sediment evaluations.

4.4.3 Kaskaskia River

Nicklow et al. (2003) applied the GA control model to the Kaskaskia River, shown in Fig. 4.4. Located in central Illinois, this system drains an area of approximately 15,900 km^2. The river and its tributaries originate in flat farmland areas and receive large quantities of sediment as a result of agricultural practices and minimal substantial land cover. The two reservoirs constructed on the Kaskaskia are Lake Shelbyville and Carlyle Lake. Carlyle Lake is located 172 km upstream from the Mississippi River, and Lake Shelbyville is situated approximately 64 km upstream from Carlyle. A portion of the numerical input for the application of the model is listed in Table 4.6. The application used a slightly modified objective function, as described by Eq. (4.7) (i.e., minimizing the maximum incremental bed change). Figure 4.5 displays the specified inflow (i.e., Lake Shelbyville) and optimal releases from each reservoir. Note that Lake Shelbyville releases are tightly constrained by an upper bound imposed through actual flow capacity limits. Convergence of the solution to a final objective function value of 0.41 m is illustrated in Fig. 4.6.

TABLE 4.5 Results for Yazoo River Basin Using SALQR

Final release policies (m^3/s)				
Interval	Arkabutla	Grenada	Sardis	Enid
1	197.1	199.3	283.0	263.6
2	179.7	259.9	283.0	266.0
3	283.0	302.8	222.8	266.0
4	163.0	302.8	84.8	266.0
5	273.9	227.8	60.6	60.0
Storage levels, 10^7 m^3				
Interval	Arkabutla	Grenada	Sardis	Enid
1	3.9	10.5	23.9	22.6
2	7.5	10.6	24.7	16.8
3	16.5	22.9	53.7	21.7
4	22.4	22.0	70.3	14.5
5	9.9	12.3	74.1	12.3
Sediment evaluation, m				
Interval	Deposition	Scour	Cumulative bed change	
1	0.62	3.44	4.06	
2	1.02	1.32	2.34	
3	0.40	0.93	1.33	
4	0.25	0.94	1.19	
5	0.42	0.41	0.83	
Total	2.71	7.04	9.75	

Note: CPU Requirements—144 Inner Loop Iterations; 1,164 Min. on Sun Ultra 1

APPENDIX 4.A: SALQR ALGORITHM

Allowing the state variables S_j^t and E_i^t to be represented by a single term x_t, the SALQR algorithm used to solve the sediment control problem is described as follows.

Step 1. Parameter initialization and computation of initial state trajectory:

a. Initialize penalty weights $R_{P,1}$ for release bound constraint $R_{P,2}$, for the stagewise storage constraint, and $R_{P,3}$, for the terminal storage requirement and initialize the penalty updating parameter $\Delta R_{P,n}$ for $R_{P,n}$.

b. Choose criterion ϵ_n for the maximum permissible violation of the nth constraint.

c. Select a stopping criteria ϵ_1 for the inner optimum, and $\theta_{\min}$ for overall optimum.

d. Select a starting reservoir release policy $R_t^c = (R_j^1, R_j^2, \ldots, R_j^T)$ for $1 \le j \le J$.

e. With initial conditions x_1^c (that is, S_j^1, E_i^1) given, use the sediment transport simulation model to compute the state trajectory, $x_{t+1}^c = G(x_t^c, R_t^c, t)$ for $1 \le t \le T$.

f. Compute the objective function Z for the current release policy.

g. Initialize terms $P_{t+1} = 0$, $Q_{t+1} = 0$ and $\theta_{t+1} = 0$.

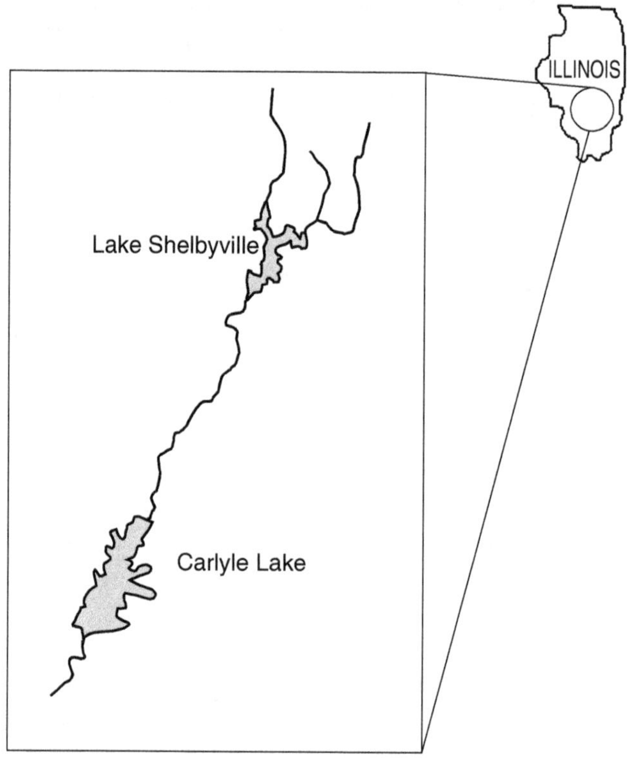

FIGURE 4.4 Kaskaskia River system.

Step 2. Backward recursion for $t = T, T-1,\ldots,1$:

a. Utilize the simulation model to numerically evaluate first- and second-order derivative information to be used in Step 2*b*;

b. Using the current state trajectory and reservoir release policy, compute parameters A_t, B_t^T, C_t, D_t^T, and E_t^T according to

$$A_t = \frac{1}{2}\left(\frac{\partial^2 Z}{\partial x^2}\right)_t + \left(\frac{\partial G}{\partial x}\right)_t^T P_{t+1}\left(\frac{\partial G}{\partial x}\right)_t \tag{4.19}$$

$$B_t^T = \left(\frac{\partial^2 Z}{\partial x \partial R}\right)_t + 2\left(\frac{\partial G}{\partial x}\right)_t^T P_{t+1}\left(\frac{\partial G}{\partial R}\right)_t \tag{4.20}$$

$$C_t = \frac{1}{2}\left(\frac{\partial^2 Z}{\partial R^2}\right)_t + \left(\frac{\partial G}{\partial R}\right)_t^T P_{t+1}\left(\frac{\partial G}{\partial R}\right)_t \tag{4.21}$$

$$D_t^T = \left(\frac{\partial Z}{\partial R}\right)_t + Q_{t+1}\left(\frac{\partial G}{\partial R}\right)_t \tag{4.22}$$

$$E_t^T = \left(\frac{\partial Z}{\partial x}\right)_t + Q_{t+1}\left(\frac{\partial G}{\partial x}\right)_t \tag{4.23}$$

TABLE 4.6 Numerical Data for Kaskaskia River System

Hydrographic Data	
Duration, days	Upstream inflow, m³/s
1.0	37.7
1.0	75.9
1.0	126.9
1.0	260.2
1.0	210.7
1.0	109.9

Bound constraints	Lake Shelbyville	Carlyle Lake
Max. release (m^3/s)	112.0	284.9
Min. release (m^3/s)	0.3	1.4
Initial storage (10^6 m^3)	224.1	348.6
Target storage (10^6 m^3)	237.5	320.7
Min. storage (10^6 m^3)	178.2	211.8
Max. storage (10^6 m^3)	324.7	450.3

Additional information required
Internal boundary conditions—reservoir storage vs. water surface elevation
Downstream boundary condition—stage vs. discharge
Upstream boundary condition—water discharge vs. sediment discharge
Bed material gradation information
Sediment transport function—Toffaleti–Schoklitsch

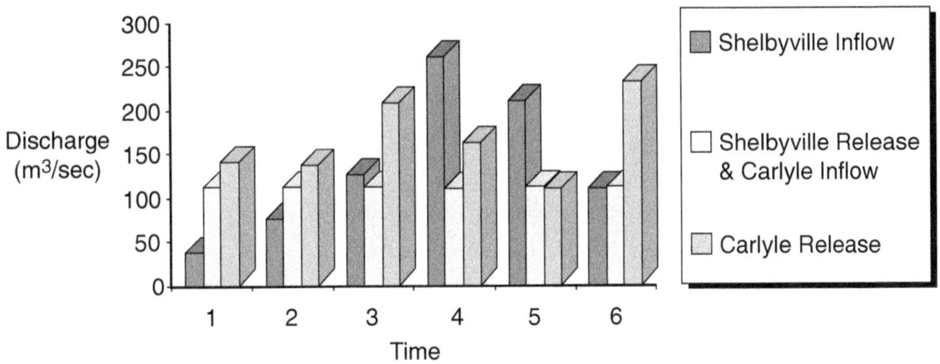

FIGURE 4.5 Comparison of inflows and releases for the Kaskaskia River system.

c. Compute P_t, and Q_t according to

$$P_t = A_t - \frac{1}{4} B_t^T C_t^{-1} B_t \tag{4.24}$$

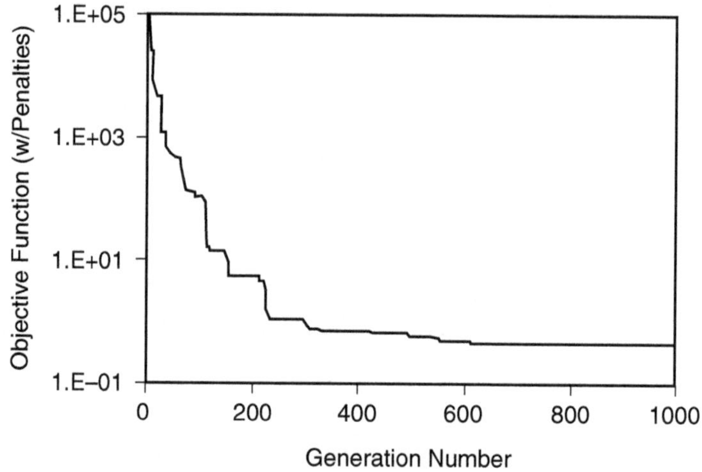

FIGURE 4.6 Convergence for Kaskaskia River system.

and

$$Q_t = -\frac{1}{4} D_t^T C_t^{-1} B_t + E_t^T \tag{4.25}$$

and store P_t and Q_t in memory, replacing P_{t+1}

d. Compute α_t, β_t, and θ_t according to

$$\alpha_t = -\frac{1}{2} C_t^{-1} D_t^T \tag{4.26}$$

$$\beta_t = -\frac{1}{2} C_t^{-1} B_t \tag{4.27}$$

$$\theta_t = -\frac{1}{2} D_t^T C_t^{-1} D_t + \theta_{t+1} \tag{4.28}$$

storing α and β_t, and storing θ_t in place of θ_{t+1}

Step 3. Forward sweep to compute successor policy and corresponding state trajectory;

a. Recursively compute the updated release schedule R_t for $t = 1, \ldots, T$, using a feedback function, expressed as

$$R_t = \alpha_t + \beta_t (x_t - x_t^c) + R_t^c \tag{4.29}$$

and recompute the state trajectory x_{t+1} using the simulation equations.

b. Test inner loop convergence; if

$$\max | R_t - R_t^c | \leq \epsilon_l \tag{4.30}$$

then the inner loop optimum has been reached, so proceed to Step 4; otherwise, set $R_t^c = R_t$ and $x_t^c = x_t$, and return to Step 2 for the next inner loop iteration.

Step 4. Test outer loop convergence; if $\theta_1 \leq \theta_{min}$ and penalty weights $R_{P,n}$ for n = 1, 2, and 3, converge, then the overall optimum has been found and stopping criteria have been satisfied; otherwise, update the penalty weights and return to Step 1 for the next outer loop iteration. The updating scheme for penalty weights is given as

$$R_{P,n}^{k+1} = \Delta R_{P,n} R_{P,n}^{k} \qquad \text{if } |Y_{\max,n}| \leq \epsilon_n \tag{4.31}$$

$$R_{P,n}^{k+1} = R_{P,n}^{k} \qquad \text{if } |Y_{\max,n}| > \epsilon_n$$

where k is the outer loop iteration number, and $Y_{\max,n}$ is the maximum computed bound violation of the nth constraint.

To guarantee convergence of the SALQR algorithm, the stagewise Hessian matrices C_t computed in the backward recursion must be positive definite. For nonlinear transition equations, such as those encountered in the sedimentation problem, the algorithm may not necessarily generate strictly convex matrices even if both the objective function and transition equations are positive definite. Following a preliminary evaluation of different methods, the nonconstant convergence shift (Tang et al. 1995) is suggested for overcoming nonconvexivity concerns. The method is described as follows.

1. Select an initial constant V_c independent of time, such that C_v is positive definite, or strictly convex, and where

$$C_v = C_t + V_c I_m \tag{4.32}$$

and where I_m is the identity matrix.

2. Update V_c at each inner loop iteration according to

$$V_c^{k+1} = \lambda V_c^{k} \tag{4.33}$$

where λ is a specified constant and k represents the iteration number.

Note that the shift procedure can affect the speed of convergence. When the value of the shift parameter V_c is large, the step size taken to reach optimality is reduced and can greatly reduce convergence speed. Likewise, when the parameter is too small, the step size is increased and numerical instability may result. From an examination of available shift techniques, including methods such as the constant shift (Yakowitz and Rutherford, 1984) and the adaptive shift (Liao and Shoemaker, 1991), it is evident that the selection of shift procedures and parameters should be made carefully and may vary significantly according to problem type and size. Note that periodic adjustments to step size may, in some cases, cause forced convergence within the inner convergence loop due to the convergence equation [i.e., Eq. (4.30)]. It is thus imperative that the overall convergence criteria θ_1, which is representative of the objective function gradient, be used as the true indicator of whether an optimal solution has been obtained.

APPENDIX 4.B: GENETIC ALGORITHM

The GA formulated to solve the sediment control problem is described as follows:

Step 1. Parameter initialization:

a. Initialize penalty weights $R_{p,n}$ for the state bound constraints on reservoir storage.

b. Choose a mutation frequency and maximum number of generations.

Step 2. Initial population generation and ranking:

a. Randomly generate an initial population of chromosomes, representing potential reservoir release policies. A default population size of 100 chromosomes is used. Each chromosome has a $J \times T$ length and consists of a number of individual releases R_j^t each referred to as a gene. Note that only feasible policies are retained in the population; thus the constraint on reservoir releases [i.e., Eq. (4.4)] is handled implicitly by the GA. A bracket penalty function is used to handle remaining constraints on state parameters.

b. Use the sediment transport simulation model to compute the state trajectory $x_{t+1}{}^c$ associated with each chromosome.

c. Compute the objective function Z for the current release policy.

d. Sort chromosomes according to their objective function value, with the smallest being assigned the highest rank.

e. For the first generation only, discard the worst 50 percent of chromosomes.

Step 3. Selection, mating, and crossover of chromosomes:

a. Ranked chromosomes are mated; the first is paired with the second, the third with the fourth, and so on.

b. A blending crossover operator is implemented for each pair of chromosomes. The method can be expressed in general terms as

$$p_{\text{new}} = \text{ß}\, p_{mn} + (1 - \text{ß})\, p_{dn} \tag{4.34}$$

where ß is a random number on the interval [0,1], p_{new} is the new offspring parameter, and p_{mn} and p_{dn} is the nth parameter belonging to the designated mother and father chromosomes, respectively.

Step 4: According to the user-assigned mutation frequency, randomly mutate a portion (five to ten percent is recommended for the current problem) of the genes contained in the population. Genes that are selected are assigned a new random value. The resulting chromosomes, those that were mutated and those that were not, collectively represent the new population. Note that an elitist strategy is used, whereby the fittest chromosome is immune from mutation and is guaranteed to remain in the population until another, more fit chromosome replaces it.

Step 5. Check termination criteria; if the maximum number of generations has been reached, the highest ranked chromosome is taken as the final solution; otherwise, return to Step 2b for the next generation.

REFERENCES

Ackers, P. and W. R. White (1973). "Sediment Transport: New Approach and Analysis," *Journal of the Hydraulics Div.*, ASCE, 99(HY11): 2041–2060.

Belegundu, A. D. and T. R. Chandrupatla (1999). *Optimization Concepts and Applications in Engineering,* Prentice Hall, Upper Saddle River, N.J.

Bhomik, N. G., J. R. Adams, A. P. Bonini, A. M. Klock, and M. Demissie (1986). *Sediment Loads of Illinois Streams and Rivers,* Illinois State Water Survey, Report of Investigation 106, Champaign, Ill.

Carriaga, C. C. and L. W. Mays (1995). "Optimal Control Approach for Sedimentation Control in Alluvial Rivers," *Journal of Water Resource Planning and Management,* ASCE, 121(6): 408–417.

Chang, L-C., C. A. Shoemaker and P. L-F. Liu (1992). "Optimal Time-Varying Pumping Rates for Groundwater Remediation: Application of a Constrained Optimal Control Algorithm," *Water Resources Research,* 28(12): 3157–3173.

Cieniawski, S. E., J. W. Eheart, and S. Ranjithan (1995). "Using Genetic Algorithms to Solve a Multiobjective Groundwater Monitoring Problem," *Water Resources Research,* 31(2): 399–409.

Culver, T. B. and C. A. Shoemaker (1992). "Dynamic Optimal Control for Groundwater Remediation with Flexible Management Periods," *Water Resources Research,* 28(3): 629–641.

Fitzpatrick, W. P., W. C. Bogner, and N. G. Bhowmik (1987). *Sedimentation and Hydrologic Processes in Lake Decatur and Its Watershed,* Illinois State Water Survey, Report of Investigation 107, Champaign, Ill.

Goldberg, D. (1989). *Genetic Algorithms in Search Optimization and Machine Learning,* Addison-Wesley, Reading, Mass.

Haupt, R. L. and S. E. Haupt (1988). *Practical Genetic Algorithms,* John Wiley, New York.

Holland, J. H. (1975). *Adaptation in Natural and Artificial Systems,* University of Michigan Press, Ann Arbor, Mich.

IDENR (1994). *The Changing Illinois Environment: Critical Trends, Technical Report of the Critical Trends Assessment Project, Vol. 2: Water Resources,* Illinois Department of Energy and Natural Resources, Springfield, Ill.

IDOT (1988). *Directory of Lake and River Terminals in Illinois,* Illinois Department of Transportation, Division of Water Resources, Springfield, Ill.

ISWPTF (1984). *Illinois State Water Plan: Critical Issues, Cross-Cutting Topics, Operating Issues,* Illinois State Water Plan Task Force, Springfield, Ill.

Lasdon, L. S. and A. D. Waren (1989). *GRG2 User's Guide,* Department of General Business, University of Texas at Austin.

Li, G. and L. W. Mays (1995). "Differential Dynamic Programming for Estuarine Management," *Journal Water Resource Planning and Management,* ASCE, 121(6): 455–462.

Liao, L-Z. and C. A. Shoemaker (1991). "Convergence in Unconstrained Discrete-time Differential Dynamic Programming," *IEEE Transactions on Automatic Control,* 36: 692–706.

Lin, T-W. (1990). "Well Behaved Penalty Functions for Constrained Optimization," *Journal Chinese Institute of Engineering,* 13(2): 157–166.

Lootsma, F. A. (1969). "Hessian Matrices of Penalty Functions for Solving Constrained Optimization Problems," *Philips Res. Reports,* 24: 322–331.

Mahmood, K. (1987). *Reservoir Sedimentation: Impact, Extent, Mitigation,* World Bank Technical Paper No. 71, Washington, D.C.

Meyer-Peter, E. and R. Müller (1948). "Formulas for Bed-Load Transport," *Proceedings of the Second Meeting of the Inter. Assoc. of Hyd. Research,* Stockholm, Sweden.

Mitchell, M. (1996). *An Introduction to Genetic Algorithms,* MIT Press, Cambridge, Mass.

Nicklow, J. W. (2000). "Discrete-Time Optimal Control for Water Resources Engineering and Management," *Water Int.,* 25(1): 89–95.

Nicklow, J. W. and L. W. Mays (2000). "Optimization of Multiple Reservoir Networks for Sedimentation Control," *Journal of Hyd. Engineering,* ASCE, 126(4): 232–242.

Nicklow, J. W. and L. W. Mays (2001). "Optimal Control of Reservoir Releases to Minimize Sedimentation in Rivers and Reservoirs," *Journal of Am. Water Res. Assoc.,* AWRA, 37(1): 197–211.

Nicklow, J. W., O. Ozkurt, and J. A. Bringer (2003). "Control of Channel Bed Morphology in Large-Scale River Networks Using a Genetic Algorithm." *Water Resources Management,* Kluwer, 17(2): 113–132.

Osterkamp, W. R., P. Heilman, and L. J. Lane (1998). "Economic Considerations of a Continental Sediment Monitoring Program," *Int. J. Sediment Res.* 13(4): 12–24.

Reklaitis, G. V., A. Ravindran, and K. M. Ragsdell (1983). *Engineering Optimization: Methods and Applications,* Wiley-Interscience, New York.

Schoklitsch, A. (1930). *Handbuch des Wasserbaues* (2nd ed.), Springer, Vienna, Austria, English translation (1937) by S. Shulits.

Tang, Z., G. Li, L. W. Mays, and P. Fox (1995). "Development of a Methodology for the Optimal Operation of Soil Aquifer Treatment Systems," *Proceedings of the 22nd Annual Conference on Integrated Water Resource Planning for the 21st Century,* Cambridge, Mass.

Task Committee on Guidelines for Retirement of Dams and Hydroelectric Facilities, Guidelines, Hydropower Committee of the Energy Division (1997). ASCE, New York.

Toffaleti, F. B. (1966). "A Procedure for Computation of Total River Sand Discharge and Detailed Distribution, Bed to Surface, Committee on Channel Stabilization," U.S. Army Corps of Engineers.

Unver, O. I. and L. W. Mays (1990). "Model for Real-Time Optimal Flood Control Operation of a Reservoir System," *Water Resources Management,* 4:21–46.

USACE (1993). *HEC-6 Scour and Deposition in Rivers and Reservoirs,* U.S. Army Corps of Engineers, Hydrologic Engineering Center, Davis, Calif.

Wurbs, R. (1993). "Reservoir-System Simulation and Optimization Models," *Journal of Water Resources Planning and Management,* ASCE 119(4): 455–472.

Yakowitz, S. and B. Rutherford (1984)."Computational Aspects of Discrete-Time Optimal Control," Applied Mathematics and Computation, 15:29–45.

Yang, C. T. (1973). "Incipient Motion and Sediment Transport," *Journal of Hyd. Engineering,* ASCE, 99(HY10): 1679–1704.

Yeh, W. W. (1985). "Reservoir Management and Operation Models: A State of the Art Review," *Water Resources Research,* 21(12): 1797–1818.

CHAPTER 5
WATER DISTRIBUTION SYSTEM OPERATION: APPLICATION OF SIMULATED ANNEALING

Fred E. Goldman
Kennedy/Jenks Consultants
Phoenix, Arizona

Larry W. Mays
Department of Civil and Environmental Engineering
Arizona State University,
Tempe, Arizona

5.1 INTRODUCTION

The operation of water distribution systems affects the water quality in these systems. EPA regulations require that water quality be maintained at all points in the system including the point of delivery. Methods to optimize water system operations have been restricted to reducing costs related to pumping and costs related to sizing, construction, and/or maintenance of piping while meeting customer demands, pressure limits, and tank operation restrictions. There have been few attempts to optimize water system operations for both hydraulic and water quality performance and they have been restricted to simplified systems.

A new methodology that formulates the water distribution system problem as a discrete time optimal control problem was developed that linked the method of simulated annealing with EPANET for optimal operation of water distribution systems for both water quality and hydraulic performance. Most optimization techniques require the calculation of derivatives, response functions, or other methods that are limited to specific problems. Simulated annealing allows optimization for a variety of objective functions and can consider many modifications to operational conditions without reprogramming of the optimization procedure. The new methodology was applied to two water systems as examples. The northwest pressure zone in the Austin, Texas application considered pump operation to minimize energy costs, which allowed comparison to published nonlinear optimization methods. The second system was the North Marin Water District, Novato, California system, which considered optimal pump operation for minimizing power costs while meeting hydraulic and water quality constraints. A comparison is made between using a gradient-based method and simulated annealing.

Until 1974, government regulators and water system operators concentrated their efforts on meeting water quality Maximum Contaminant Levels (MCLs) at the water source or water treatment plant. However, with the implementation of the Safe Water Drinking Act in 1974 and amendments in 1986 and 1996, water system operators are being required to meet standards throughout their distribution systems and at the points of delivery. The Surface Water Drinking Rule requires that a detectable level of chlorine be maintained in the system. The Total Coliform Rule requires that total coliform standards be met at the customer's tap. The Lead and Copper rule requires testing at the point of delivery (Clark, 1994b; Clark et al., 1994, 1995; Pontius, 1996).

Water quality can undergo significant changes as it travels through the water distribution system from the point of supply and/or treatment to the point of delivery. For example, chlorine concentration will decrease with time in pipes and tanks through bulk decay and through reaction with the pipe wall. Also, chlorine mass leaves the system at demand nodes or can be added to the system at chlorine booster nodes. Mixing of water from different sources can also affect water quality (Clark et al., 1995).

Pumps are usually operated according to an operating policy, which includes the scheduling of pump operation, which can affect water quality if the system pumps draw water from sources with differing water quality. Pump operation will affect the turnover rate of storage tanks. Chlorine dosage may differ at each pump depending on the water source quality. For example, consider a pump that operates for 6 h divided into 1-h periods. The pump can either be on or off for any period. The number of combinations is $2^6 = 64$. A pump, which operates for 24 h divided into 1-h periods, has $2^{24} = 16{,}777{,}216$ combinations. The 6-h example can be solved by trial and error but the 24-h example is prohibitively large to solve by trial and error. For large combinatorial optimization problems, simulated annealing provides a manageable solution strategy.

These considerations have generated a need for computer models and optimization techniques that consider water quality in the water distribution system as well as system hydraulics.

5.2 GENERAL PROBLEM STATEMENT

Consider a water distribution system with M pipes, K junction nodes, S storage nodes (tanks or reservoirs), and P pumps, which are operated for T time periods. The objective function is the minimization of the total energy cost for pumping:

$$\text{Minimize } Z = \sum_{p=1}^{P} \sum_{t=1}^{T} \frac{\text{UC}_t \, 0.746 \, \text{PP}_{pt}}{\text{EFF}_{pt}} D_{pt} \tag{5.1}$$

where UC_t = unit energy cost of pumping during time period t (\$/kw-hr)
PP_{pt} = power of pump during time period t (hp)
D_{pt} = length of time pump p operates during time period t (hr)
EFF_{pt} = efficiency of pump p during period t.

Conservation of mass and conservation of energy conditions must be met throughout the water distribution system. These conditions make up the system hydraulic constraints. For each link (which can be a pipe or pump) between nodes i and j and node k, the conservation of mass is

$$\sum_i (q_{ik})_t - \sum_j (q_{kj})_t - Q_{kt} = 0 \quad \forall i, j \in M; \qquad k = 1,\ldots,K \quad \text{and} \quad t = 1,\ldots,T \tag{5.2}$$

where q_{ik} = flow in link connecting nodes i and k during time period t and Q_{kt} = flow consumed (+) or supplied (−) at node k during time period t. The conservation of energy for each pipe connecting nodes i and j, in the set of all pipes M is:

$$h_{it} - h_{jt} = f\left(q_{ij}\right)_t \qquad \forall i, j \in M \quad \text{and} \quad t = 1,\ldots,T \tag{5.3}$$

where h_{it} = hydraulic grade line elevation (ground elevation and pressure head) at node i during time t and $f(q_{ij})_t$ = headloss in link connecting nodes i and j during time t. The total number of hydraulic constraints is $(K + M)T$, and the total number of unknowns is also $(K + M)T$, which are the discharges in M pipes and the hydraulic grade line elevations at K nodes.

In addition to the equality hydraulic constraints there are inequality hydraulic bound constraints. The nodal pressure head bounds are

$$\underline{H}_{kt} \le H_{kt} \le \overline{H}_{kt} \qquad k = 1,\ldots,K \quad \text{and} \quad t = 1,\ldots,T \tag{5.4}$$

where H_{kt} and H_{kt} are, respectively, the lower and upper bounds on the pressure head at node k at time t.

The pump operation problem is an extended period simulation. The height of water stored at a tank for the current time period t, y_{st}, is a function of the height of water stored from the previous time period and the flow entering or leaving the tank during the period Δt and can be expressed as

$$y_{st} \le y_{s(t-1)} + \frac{q_{s(t-1)}}{A_s} \Delta t \tag{5.5}$$

where A_s = cross sectional area of tank s and q_s = flow entering or leaving the tank during periods t and $t - 1$. The inequality bound constraints on the height of water in the storage tanks are

$$\underline{y}_{st} \leq y_{st} \leq \overline{y}_{st} \qquad s = 1,\ldots,S \quad \text{and} \quad t = 1,\ldots,T \tag{5.6}$$

where $\underline{y}_{st}$ and $\overline{y}_{st}$ are, respectively, the lower and upper bound tank water levels.

The water quality constraint is the conservation of mass of the chemical in each pipe m connecting nodes i and j in the set of all pipes M and is

$$\frac{\partial (C_{ij})_t}{\partial t} = -\frac{(q_{ij})_t}{A_{ij}} \frac{\partial (C_{ij})_j}{\partial x_{ij}} + \theta(C_{ij})_t \qquad \forall i, j \in M \quad \text{and} \quad t = 1,\ldots,T \tag{5.7}$$

where $(C_{ij})_t$ = concentration of chemical in pipe m connecting nodes i and j as a function of distance and time (mass/ft^3)
x_{ij} = distance along pipe (ft)
A_{ij} = cross-sectional area of pipe connecting nodes i and j
$\theta(C_{ij})_t$ = rate of reaction of chemical within pipe connecting nodes i and j at time t (mass/ft^3/day).

The bounds on chemical concentrations are

$$\underline{C}_{nt} \leq C_{nt} \leq \overline{C}_{nt} \qquad n = 1,\ldots,N \quad \text{and} \quad t = 1,\ldots,T \tag{5.8}$$

where $\underline{C}_{nt}$ and $\overline{C}_{nt}$ are, respectively, the lower and upper bounds of the chemical concentration.

The above formulation results in a large-scale nonlinear programming problem with decision variables $(q_{ij})_t$, H_{kt}, y_{st}, D_{pt}, and $(C_{ij})_t$. In the proposed solution methodology, the decision variables are partitioned into two sets: control (independent) and state (dependent) variables. The operation of a pump during a time period (on or off) is the control variable. The problem is formulated above as a discrete-time optimal control problem.

5.3 SOLUTION METHODOLOGY

5.3.1 Previous Models for Water Distribution System Optimization

Ormsbee and Lansey (1994) reviewed methods used to optimize the operation of water supply pumping systems to minimize operation costs. Ostfeld and Shamir (1993) developed a water quality optimization model that optimized pumping costs and water quality using steady-state and dynamic conditions using GAMS/MINOS [general algebraic modeling system (Brooke et al., 1988); mathematical in-core nonlinear optimization systems (Murtagh and Saunders, 1982)]. Ormsbee (1991) suggested that on–off operation of pumps poses a difficulty for optimization techniques that require continuous functions. Computer-based water quality models exhibit the same difficulties. Most modeling methods include numerical methods rather than continuous functions. Linear superposition was used by Boccelli et al. (1998) for the optimal scheduling of booster disinfection stations by developing system-dependent discretized impulse response coefficients using EPANET and linking with the LP solver MINOS (Murtagh and Saunders, 1987).

Dynamic programming has been used for the optimal scheduling of pumps in a water distribution system by Coulbeck and Orr (1982) and Coulbeck et al. (1987). The method is sensitive to the number of reservoir states and the number of pumps. In general, the increase in number of discretizations and state variables increases the size of the problem dramatically, which is known as the curse of dimensionality (Mays and Tung, 1992). This has restricted the application of this method to small systems.

The nonlinear programming approach by Brion and Mays (1991) evaluates gradients using finite differences or a mathematical approach to provide derivatives required by GRG2, which solves nonlinear optimization problems by using the generalized reduced gradient method (Lasdon et al.,

1978). The method has been used for scheduling pump operations to minimize energy and improve water quality by Sakarya (1998) and Sakarya and Mays (2000), who considered three objective functions: (1) the minimization of the deviations of actual concentrations from a desired concentration, (2) minimization of total pump duration times, and (3) the minimization of total energy while meeting water quality constraints.

5.3.2 The Method of Simulated Annealing

Simulated annealing is a combinatorial optimization method that uses the Metropolis algorithm to evaluate the acceptability of alternate arrangements and slowly converge to an optimum solution. The method does not require derivatives and has the flexibility to consider many different objective functions and constraints. Simulated annealing uses concepts from statistical thermodynamics and applies them to combinatorial optimization problems. Kirkpatrick et al. (1983) explain how the Metropolis algorithm was developed to provide the simulation of a system of atoms at a high temperature that slowly cools to its ground energy state. If an atom is given a small random change there will be a change in system energy ΔE. If $\Delta E \leq 0$, the new configuration is accepted. If $\Delta E > 0$, then the decision to change the system configuration is treated probabilistically. The probability that the new system is accepted is calculated by:

$$P(\Delta E) = \exp(-\Delta E / k_B T) \tag{5.9}$$

A random number evenly distributed between 0 and 1 is chosen. If the number is smaller than $P(\Delta E)$ then the new configuration is accepted; otherwise it is discarded and the old configuration is used to generate the next arrangement. The Metropolis algorithm simulates the random movement of atoms in a water bath at temperature T. By using Eq. (5.9), the system becomes a Boltzmann distribution.

Entropy is a measure of the variation of energy at a given temperature. At higher temperatures there is significant energy variation, which reduces dramatically as the temperatures are lowered. This implies that the annealing process should not get "stuck" because transitions out of an energy state are always possible. Also, the process is a form of *adaptive divide-and-conquer.* Gross features of the eventual solution appear at the higher temperatures with fine details developing at low temperatures (Kirkpatrick et al., 1983).

The requirements for applying simulated annealing to an engineering problem are: (1) a concise representation of the configuration of the decision variables, (2) a scalar cost function, (3) a procedure for generating rearrangements of the system, (4) a control parameter (T) and an annealing schedule, and (5) a criterion for termination. (Dougherty and Marryott, 1991).

5.3.3 Configuration of Decision Variables

The decision variables consist of the operational schedule of each pump during discrete time periods. For this application, the 24-h day was divided into 1-h periods. The program finds the optimum pump schedule where each of the pumps is either on or off during each of the time periods. A configuration is a given schedule of pump operation, which determines in which 1-h period each pump is operating. EPANET is then run using that pump configuration. The program results are used to rate the performance of the simulation. The decision on whether to retain or change the pump operation configuration is based on the Metropolis algorithm as shown in Fig. 5.1, a pseudocode description of the adaptation of the simulated annealing algorithm to pump operation optimization.

5.3.4 Development of Cost Function

The cost function $g(C_i)$ for a given configuration C_i is used in place of energy of a system of atoms. The temperature T can be considered a control parameter, which has the same units as the cost function. The cost function should include the cost of pumping and penalties for violations of the storage tank

```
INITIALIZE;
T = T0;
C = Ck0;
g = g(Ck0);
DO WHILE (stopping criterion not satisfied);
 DO WHILE (equilibrium not satisfied);
   PERTURB:
   Ck+1 = rearrangement of Ck;
   Δg = g(Ck+1)−g(Ck);
   IF Δg ≤ 0 THEN
     ACCEPT
     C = Ck|1;
   ELSE
     IF random(0,1) < exp (−Δg/T)THEN
       ACCEPT:
       C = Ck+1;
     ENDIF
   ENDIF
 ENDDO
   T = α T
 ENDDO
```

T = pseudotemperature

C_k = pump operation schedule configuration

g = cost of pump operation schedule configuration C_k including pseudocosts for constraint violations

FIGURE 5.1 Pseudocode description of the simulated annealing algorithm. [*After Dougherty and Marryott (1991)*).]

level, pressure, and water quality bounds. The pumping cost is calculated for each period for each pump running and is a function of the pump flow rate, head, efficiency, and electricity rate during that period. The cost of violations of the pressure and water quality bounds is calculated using penalty functions. By adjusting the penalty functions, the optimization problem can be adjusted to bias one constraint over another constraint.

The cost of pumping is added to the pseudocost penalty functions:

$$\text{COST}_{\text{pumps}} = K \sum_{ij} \frac{Q_{ij}\,\text{TDH}_{ij}\,P_j\,t_j}{\eta_{ij}} \tag{5.10}$$

where K = unit conversion factor

Q_{ij} = flow from pump i during period j

TDH_{ij} = the operating point total dynamic head for pump I during period j

P_j = power rate per kilowatt hour during period j

t_j = length of time that pump operates during period j

η_{ij} = wire to water efficiency of pump i during period j.

Storage tanks have a minimum level to provide emergency fire flow storage. If the tanks are depleted below this level then fire protection is compromised. A penalty term τ_1 has been developed to account for this constraint which is based on the constraint penalty terms developed by Brion (1990) given by:

$$\tau_1 = \sum_j \beta_s (\min [0, E_{sj} - E_{\min,s}])^2 \tag{5.11}$$

where τ_1 = penalty for violations of the tank low water level bound
E_{sj} = water level for tank s during period j
$E_{\min,s}$ = lower bound or the minimum level in tank s
β_s = penalty term for tank low-level constraint for tanks

Cohen (1982) has stated "optimization of a network over a limited horizon of, say, 24 hours has no meaning without the requirement of some periodicity in operation; a simple way to do that is to constrain the final states to be the same as the initial ones." A constraint τ_2 is developed to generate a cost if the tank levels do not return to their starting elevation.

$$\tau_2 = \sum_s \beta_{s2} (\min [0, 1 - / E_{s,1} - E_{sj} /])^2 \tag{5.12}$$

where $E_{s,1}$ = water level for tank s at beginning of the simulation, during period 1
$E_{s,j}$ = water level for tank s at the end of the simulation, during the final period j
β_{s2} = penalty term for beginning and ending tank level constraint for tank s

For a 24-h simulation the concept of returning tanks to their starting level is somewhat addressed by using a starting configuration where the pumps supply a volume of water equal to the sum of the nodal demands. Providing a total volume of pumped water equal to the total volume of the system demands will return the tanks exactly to their original level if there is only one tank. Even if pumping equals demand exactly, the tanks may not return to their original level if there are several tanks unless the tanks all start full; because one tank may supply water to another tank during the simulation, the total volume stored will be the same but shifted from one tank to another.

The Cohen condition may be unnecessarily restrictive. The second example involved modifying pump operation for a 24-h period that was repeated for 12 days to allow the water quality variations to overcome the initial conditions. It was observed that the tanks exhibited periodic behavior and adjusted themselves until the pumps were supplying a quantity equal to the demands (see Fig. 5.2). If the pumps operated for longer periods, the tanks remained closer to full and the pump heads moved to the left of the system curve reducing the flow. If the pumps operated for shorter periods, the tank levels lowered until the flows increased to meet the demands. By running the simulations over several days, the pump operation can be scheduled to optimize efficiency and perhaps reduce cost.

A water distribution system needs to deliver water to its customers at sufficient pressure to service the water system customers but at a pressure that will not damage water systems or customer's facilities. The Uniform Plumbing Code sets the normal pressure range as 15 psi to 80 psi (IAPMO, 1994), although a city may have a range in a pressure zone from 40 to 80 psi with a 20 psi residual during a fire flow (Malcolm Pirnie, 1996). The system operation needs to operate between two extreme pressures, $p_{\min}$ and $p_{\max}$.

The constraints and penalty functions for pressure bounds at each node are:

$$\tau_3 = \sum_{kj} \gamma_1 (\min [0, P_{kj} - P_{\min}]^2 \tag{5.13}$$

$$\tau_4 = \sum_{kj} \gamma_2 (\min [0, P_{\max} - P_{kj}]^2 \tag{5.14}$$

where $p_{\min}$ = minimum system pressure bound
$p_{\max}$ = maximum system pressure bound
p_{kj} = pressure at node k during time period j
γ_1, γ_2 = penalty terms for minimum and maximum pressure violations

The EPANET (Rossman, 1993) program solves distribution systems hydraulically and then routes chemicals or contaminants through the water system during the time period. EPANET can also consider chemical reactions and calculate the decay of chlorine with time. The combination of hydraulic solver and water quality calculations allows the program to predict the chlorine residual at

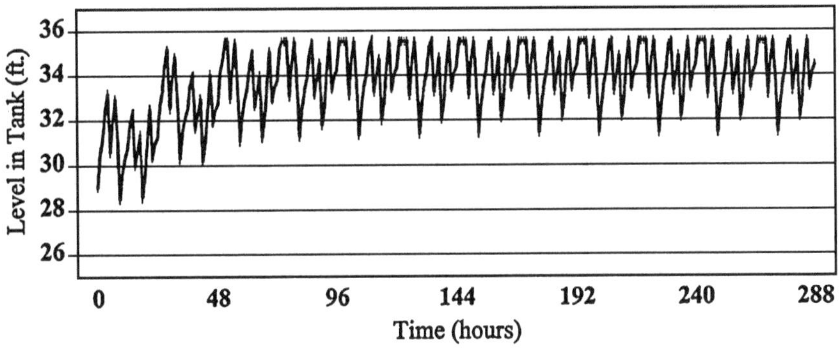

FIGURE 5.2 Periodic behavior.

any node in the system at any time period. A penalty term will be used to consider upper and lower limit bounds on free chlorine to adjust the cost function for chlorine violations.

The penalty terms for minimum and maximum free chlorine concentration are:

$$\tau_5 = \sum_{kj} \gamma_3 (\min [0, c_{kj} - c_{\min}])^n \tag{5.15}$$

$$\tau_6 = \sum_{kj} \gamma_4 (\min [0, c_{\max} - c_{kj}])^n \tag{5.16}$$

where $c_{\min}$ = minimum free chlorine concentration
$c_{\max}$ = maximum free chlorine concentration
c_{kj} = chlorine residual concentration at node k during time period j
γ_3 = penalty term for minimum chlorine residual pressure bound violations
γ_4 = penalty term for maximum chlorine residual pressure bound violations

The value of n will usually be 2. In cases where the lower concentration bound is more important, a value of $n < 1$ will place a higher penalty on minimum free chlorine violations. There will also be a term for the amount of chlorine used. The goal will be to meet the chlorine residual bounds while using the smallest amount of chlorine. Not only will the operational cost be decreased but also the creation of Total Trihalomethane (TTHM) will also be reduced.

5.4 DEVELOPMENT OF SOFTWARE

Version 1.1e of EPANET (5/1/96) authored by Dr. Lewis A. Rossman, Risk Reduction Engineering Laboratory of the Office of Research and Development, U.S.E.P.A., Cincinnati, Ohio, was used as the basis of the annealing computer program. The program source files, in the C language, were provided by Dr. Rossman. The basic strategy was to use EPANET as the simulator and to develop around EPANET new functions that calculate penalties or costs, generate pump operation configurations, and evaluate the acceptance of potential configurations using the Metropolis algorithm. The EPANET program was developed for a single simulation as described by the flowchart in Fig. 5.3. In order to carry out the numerous simulations required for annealing, several modifications were required. New functions were written in the C programming language to carry out penalty calculations, generate pump operation configurations, modify the temperature according to the annealing schedule, and analyze each result using the Metropolis algorithm. Pump scheduling is determined by dividing the 24-h day into 1- or 2-h periods. The pump schedule is repeated if the simulation is carried out for several days. The schedule specifies the on–off condition for each pump for each period. The starting pump operation configuration is completely arbitrary. Each pump operation schedule after the beginning schedule is determined using a random number generator that chooses the pump and period for change. The change is to turn the chosen pump on during the chosen period if it was off or vice versa. This new pump schedule becomes the new data set for EPANET and the program output is checked for energy costs and penalty functions to determine the value of the objective function. The code also detected special "flags" generated by the EPANET program to identify nonfeasible data sets caused by violations of conservation of mass and energy [Eqs. (5.2) and (5.3)]. Special routines were developed to collect information useful in understanding the progress of the annealing method and to develop the current value of the objective function.

Figure 5.4 is a flowchart that shows the operation of the annealing program. The chart shows that the number of loops and iterations per loop are set before the process begins. Each loop carries out a number of iterations at a given temperature. If a pump configuration is held for 50 times it is considered the optimal configuration. If the program finishes normally, the final pump configuration is considered optimal. Frustration between constraints in a water distribution system provides many good configurations to recommend to the system operators. There may be pump configurations that provide better solutions than the final solution. A routine was added, to save the 20 best configurations for examination by the user.

5.5 APPLICATIONS

5.5.1 The Austin Data Set

Brion (1990) and Brion and Mays (1991) applied their model to the City of Austin Northwest B Pressure Zone which consisted of three pumps, 126 pipes of approximately 38.5 miles in total length, 98 junction nodes of which 5 were pressure "watch points," and one storage tank (Fig. 5.5). The 24-h simulation was broken down into twelve 2-h time periods. The system served about 31,000 residents located in about 32,000 acres.

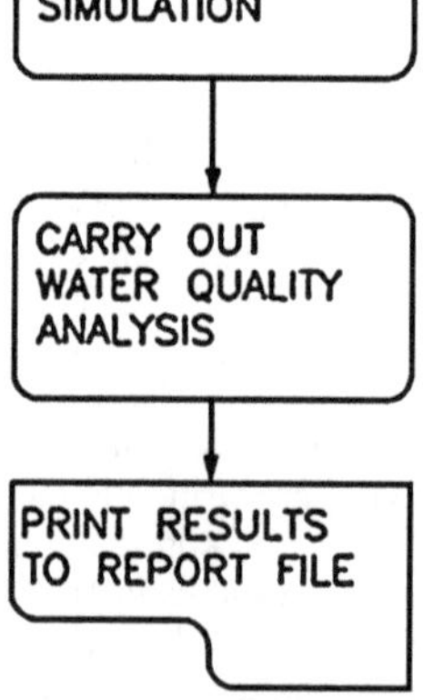

FIGURE 5.3 EPANET flowchart.

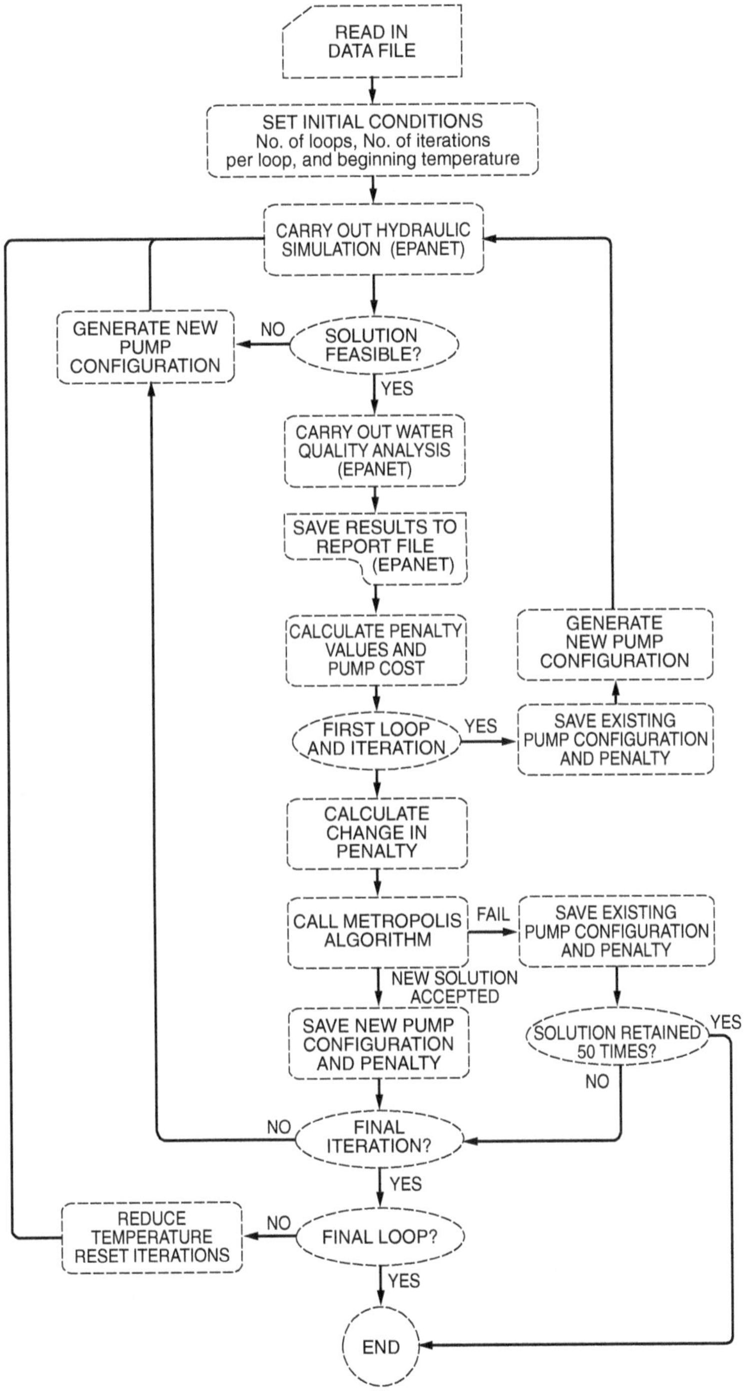

FIGURE 5.4 Simulated annealing flowchart.

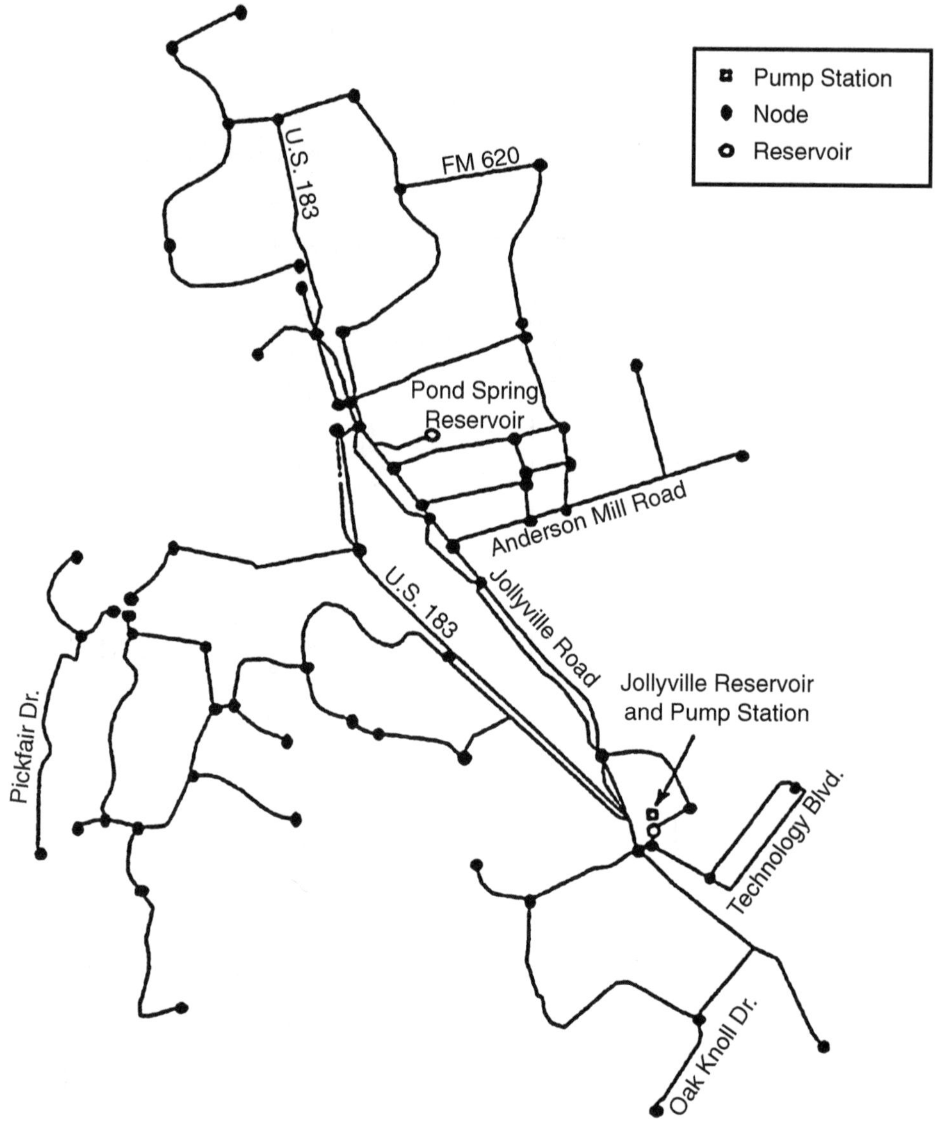

FIGURE 5.5 Water distribution system for City of Austin, Northwest B Pressure Zone. (*Brion and Mays, 1991.*)

The pipes, junctions, pumps, tanks, and diurnal flow distribution can be found in Goldman (1998) or Brion (1990). Brion and Mays (1991) collected information on the actual pumping times used on September 29, 1988 which resulted in a pumping cost of $231 during the day. Using NLP with an augmented Lagrangian description of the pressure and tank constraints and linking the GRG2 program with KYPIPE the cost was reduced to $219, a savings of 5.2 percent. The resulting pump operation schedule is shown in Fig. 5.6.

Simulated annealing was applied to the same data set. Four loops of 200 iterations were performed. The computer time on the Arizona State University Research UNIX cluster was 19 min 38 s. The cost function was the sum of the cost of pumping and constraint violation penalties. The

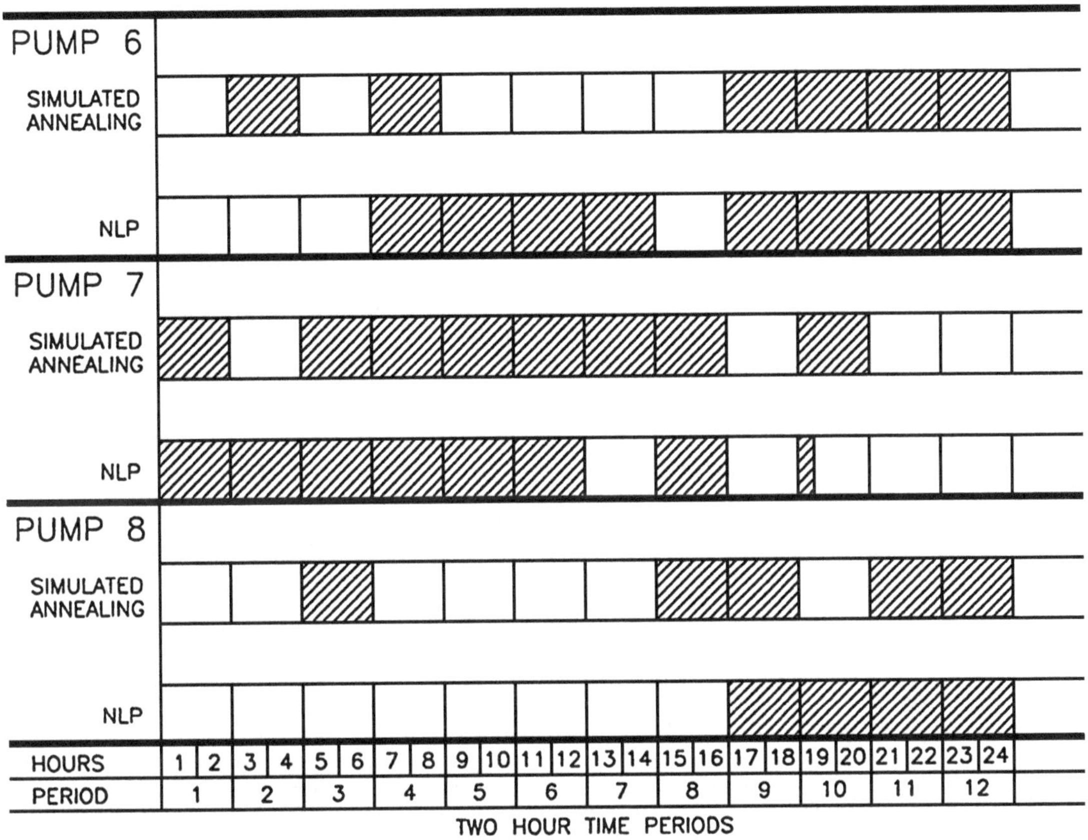

FIGURE 5.6 Austin example—simulated annealing and NLP pump operation schedules.

beginning temperature was 160, which was reduced using the Huang Method (Dougherty and Marryott, 1991).

The series of trial pump configurations was generated starting with the last trial configuration, using a random number generator to choose a pump and a period, and switching the operation. If the chosen pump was operating during the chosen period in the previous trial it is turned off. If it was off during the chosen period it is changed to operating. The stopping condition was having a pump schedule withstand 20 challenges from new pump schedules. Because of the high level of frustration with many pump schedules having very similar cost, the ending condition was not met.

The best result was $221.38, which occurred in trial 67 of loop 3. This resulted in a savings of 4.1 percent. The annealing method did not isolate one solution because of the high level of frustration. The program found 173 solutions, which met the pressure and tank constraints and resulted in a cost less than $235; 50 pump schedules had a cost less than $230. The feasible solutions were reviewed to find the least costly pump operation schedule that met the pressure and tank constraints. The optimum pumping schedule found using simulated annealing and NLP is shown in Fig. 5.6.

A review of Fig. 5.6 shows the NLP solution had a 15-min sliver of operation occurring for pump 7 during the period between hours 18 and 20, which is undesirable for operating pumps and most likely would need to be adjusted by the operator. Short period results can be expected from the NLP method, which is optimizing the duration of pump operation for a 2-h time period. Simulated annealing resulted in a minimum 2-h pumping period because it is a combinatorial optimization method where the pump has to be on or off during the entire period; thus the minimum operating time is 2 h.

The tank trajectory for the simulated annealing optimization pump schedule is shown in Fig. 5.7. The minimum height in the tank was 1 ft. A penalty term was used to return the tank level to the original tank level, within a tolerance of 1 ft.

5.5.2 North Marin Example

This example is a water quality optimization problem that was solved simultaneously by simulated annealing as part of this research project and by the mathematical programming approach using the GRG2 optimizing code linked with EPANET as part of the research project carried out by Sakarya (1998). The water distribution system is taken from example 3 of the EPANET users' manual (Rossman, 1993). It consists of the piping system for the North Marin Water District—Navato, California (Fig. 5.8), which was the location of the sampling and modeling study carried out as part of a recent AWWA Research Project (Vasconcelos et al., 1996).

The goal of the optimization is to find the optimal operation of two pumps with differing capacity characteristics pumping from two different sources of water. Dr. Sakarya's original code minimized the operating time of the two pumps considering pressure and water quality constraints without considering pumping cost. Her code was modified to consider pumping cost by assuming a constant efficiency.

Two systems were studied. In the first system, both sources had free chlorine concentrations of 0.5 mg/L and the minimum free chlorine concentration at any demand node was taken as 0.05 mg/L. Pressures at demand nodes were to be kept between 20 and 100 psi. Global bulk decay was taken as 0.1/day and global wall decay was taken as 1.0 ft/day. The second system had two different chlorine residuals. The free chlorine for the larger pump was 0.4 mg/L and the smaller pump was 0.2 mg/L. The minimum free chlorine concentration was taken as 0.05 mg/L.

It takes many days for initial chemical concentrations to be overcome and repeatable steady-state behavior to be observed. This periodic behavior was discussed previously and was reported for water quality by Boccelli et al. (1998). Figure 5.9 shows the periodic behavior of the chlorine concentration in Tank 1 of this example. To reach steady-state conditions, each iteration was run for 288 h. This resulted in an average of 10 s for each iteration. The results for the last 24 h were used to calculate pumping cost and constraint violation penalties.

The piping system consisted of 125 pipes, 102 nodes, three tanks, and two pumps as described in Goldman (1998) and Rossman (1993). Twenty-four 1-h periods were considered with the pumping schedule repeated every 24 h. The two optimization techniques, simulated annealing and NLP using reduced gradients with Lagrangian multipliers for constraints and GRG2, obtained similar results. For the case where the source chlorine concentration for both sources was 0.5 mg/L, there were no pressure or chlorine violations and the minimum cost found by NLP was \$399.95 per day versus \$429.53 for simulated annealing. For the case where water qualities for the two sources were 0.2 and

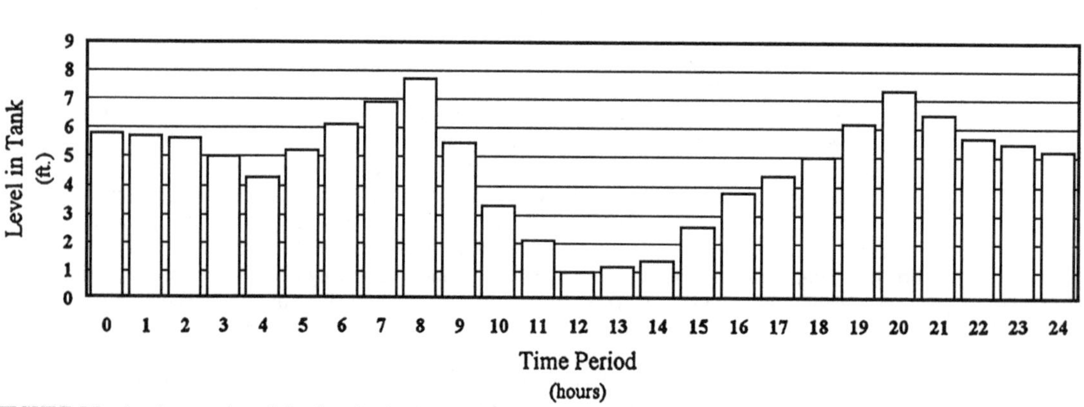

FIGURE 5.7 Austin example tank levels—simulated annealing.

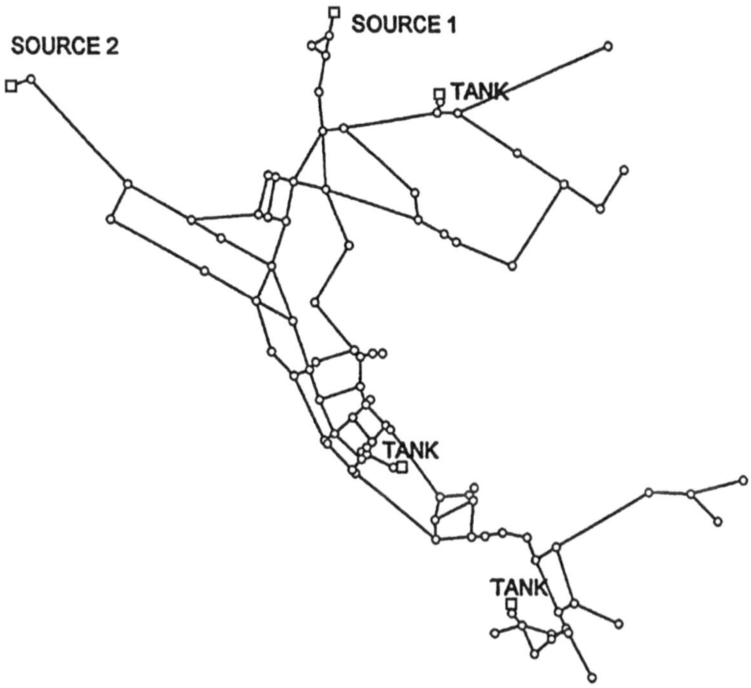

FIGURE 5.8 Water distribution system for North Marin Water District—Navato, California. (*Rossman, 1993 and Vasconcelos et al., 1996.*)

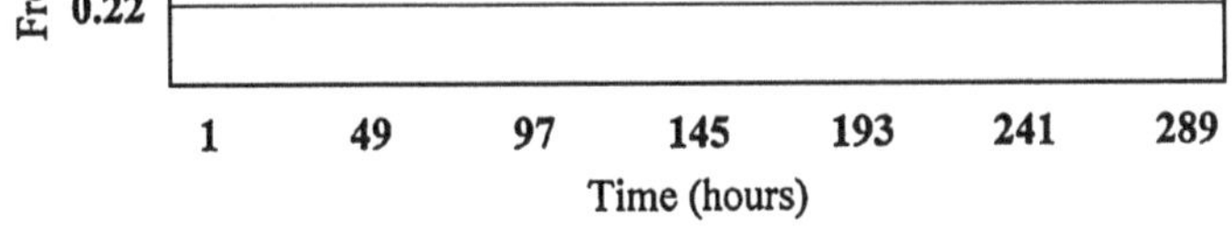

FIGURE 5.9 North Marin example, chlorine concentration: Tank 1.

0.4 mg/L the NLP cost was \$408.39 versus \$407.66 for simulated annealing. There were no pressure violations but the total chlorine violations measured in total chlorine excursions for all nodes for 24 h in μg/L, was 69.76 for NLP versus 5.89 for simulated annealing.

There were distinct differences in time to arrive at a solution. Dr. Sakarya used a PC with a Pentium processor and the simulated annealing was run on the ASU UNIX computer cluster which consists of five IBM computers including two RS/6000 Model 390 interactive servers and three RS/6000 Model 590 mathematical computational processing. The ASU machines run the IBM AIX operating system, which is IBM's version of UNIX. In both instances the time to run a 12-day long

EPANET simulation was about 10 s. The NLP optimization required about one-third the iterations of the simulated annealing optimization although there was no attempt to streamline the simulated annealing and it turned out that the best solutions occurred before the final loop started. It is believed that even with some improvement in efficiency the simulated annealing would require at least twice the iterations that NLP requires.

Simulated annealing has been shown to be more flexible and adaptable than NLP optimization. The requirement that many parts of the distribution system such as node quality, pressures, tank levels, and pump operation require derivatives with respect to pump operation time during each period restricts the flexibility of the method to accept changes to the system or consider a variable pump efficiency. On the other hand, simulated annealing can accept changes to the distribution system because it is only concerned with cost calculation and deriving a Markov chain of operation schedules.

The NLP method's efficiency in finding optimum solutions appears to be very sensitive to the Lagrangian coefficients used in the outer loop of the optimization. Also, the significant differences in the pump capacities result in much more rapid changes in energy cost, pressures, and water quality to changes in the operation of the larger pump. This resulted in the preferred changes in the larger pump, which reduced its operating time until any more restrictions would have resulted in an unbalanced solution. If a smaller pump was included in the pump operation schedule, it was rarely changed by NLP because such changes were likely to result in violations of conservation of mass and energy and infeasible solutions from the EPANET simulator.

A fundamental tenet in using NLP optimization linked to simulators such as EPANET is the implicit function theorem (Brion, 1991) which states that if D_t is the set of pump durations during periods t (the control variable) then $H(D_t)$ the node pressure matrix, and $E(D_t)$ the tank level matrix (the dependant or state values) exist in the neighborhood of D^*, the set of optimum pump operations. This implies that the solutions exist in the neighborhood. However, in running the simulated annealing routines many unbalanced infeasible solutions existed in the vicinity of the optimum solution. This becomes important if continuity is assumed in the vicinity of the NLP optimums. This problem does not reveal itself in NLP that includes water quality because the water quality portion of EPANET does not include the solving of equations that could be unbalanced. Dr. Sakarya was aware of the continuity problem and her optimum solutions are continuous. However, optimizations that use NLP and do not include water quality could be affected by this problem.

Another important issue is global versus local optimum and the concept of frustration. A total of 36 pump operation configurations were found by simulated annealing that have total water quality violations less than 5.89 μg/L and energy costs less than \$420 per day, where the water quality violations are the sum of the violations at all demand nodes for the last day of the simulation. There are many "nearly equal" optimal pump operation schedules that have nearly the same penalty values. This is most likely the reason that simulated annealing found optimum solutions in the next to the last loop.

The actual pump operation schedules developed by the two optimization techniques are shown in Fig. 5.10 for the case where both sources had 0.5 mg/L chlorine concentration.

As previously mentioned, the NLP solution where the sources had identical 0.5 mg/L concentrations resulted in a cost of \$399.95 versus \$429.53 which appears to be a preferred solution. However, a review of Fig. 5.10 reveals a problem with the NLP optimization. NLP is optimizing the amount of pumping time for a given period. This can result in several periods where the pumps operate only part of the time. This occurs during periods 2,3,8,10,12, and 14. In order to operate pumps more continuously there would need to be some adjustment of the pumping schedule shown in Fig. 5.10, which could affect the ultimate cost of pumping.

5.6 SUMMARY AND CONCLUSIONS

Simulated annealing has been successfully linked with EPANET to optimize the operation of a water distribution system for hydraulic behavior and water quality.

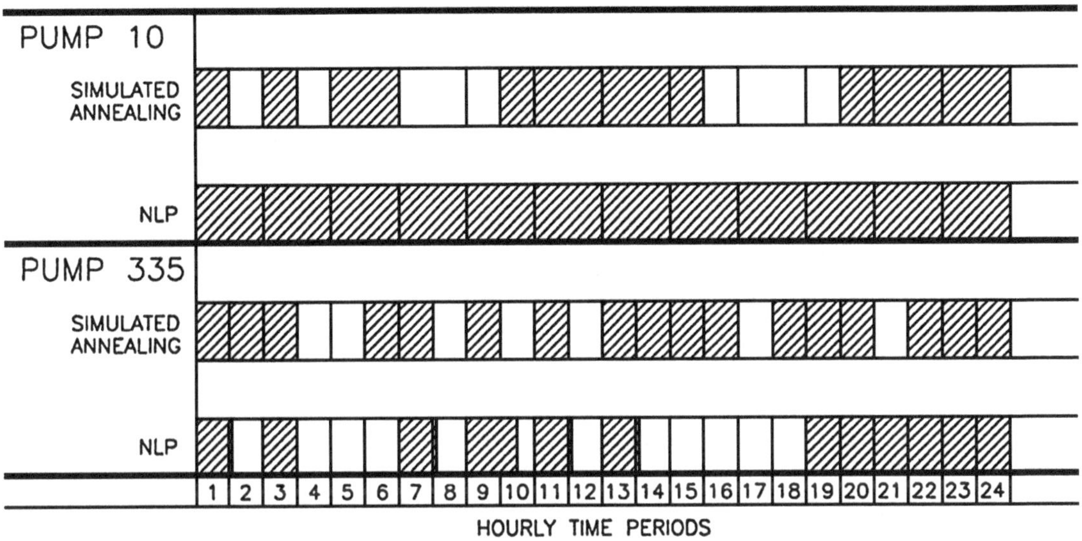

FIGURE 5.10 Simulated annealing and NLP pump operation schedules for system 1 with 0.5 mg/L free chlorine at both sources and minimum free chlorine at nodes as 0.05 mg/L.

REFERENCES

Boccelli, D. L., M. E. Tryby, J. G. Uber, L. A. Rossman, M. L. Zierolf, and M. M. Polycarpou (1998). "Optimal Scheduling of Booster Disinfection in Water Distribution Systems," *Journal of Water Resources Planning and Management,* ASCE, 124(2): 99–111.

Boulos, P. F., T. Altman, P. Jarrige, and F. Collevati (1995). "Discrete Simulation Approach for Network-Water-Quality Models," *Journal of Water Resources Planning and Management,* ASCE, 121(1): 49–60.

Brooke, A., D. Kendrick, and A. Meeraus (1988). *GAMS: A User's Guide.* Scientific Press, Redwood City, Calif.

Brion, L. M. (1990). "Methodology for Optimal Operation of Pumping Stations in Water Distribution Systems," Ph.D. Thesis, University of Texas, Austin, Texas.

Brion, L. M. and L. W. Mays (1991). "Methodology for Optimal Operation of Pumping Stations in Water Distribution Systems," *Journal of Hydraulic Engineering,* ASCE, 117(11): 1551–1569.

Clark, R. M. (1994a). "Development of Water Quality Models," *Computer Modeling of Free-Surface and Pressurized Flows,* M. H. Chandhry and L. W. Mays, eds., Kluwer Academic, Dordrecht, The Netherlands.

Clark, R. M. (1994b). "Applying Water Quality Models," *Computer Modeling of Free-Surface and Pressurized Flows,* M. H. Chandhry and L. W. Mays, eds., Kluwer Academic, Dordrecht, The Netherlands, 581–612.

Clark, R. M., W. M. Grayman, J. A. Goodrich, R. A. Deininger, and K. Skov (1994). "Measuring and Modeling Chlorine Propagation in Water Distribution Systems," *Journal of Water Resources Planning and Management,* ASCE, 120(6): 871–887.

Clark, R. M., L. A. Rossman, and L. J. Wymer (1995). "Modeling Distribution System Water Quality: Regulatory Implications," *Journal of Water Resources Planning and Management,* 121(6): 423–428.

Cohen, G. (1982). "Optimal Control of Water Supply Networks," Chapter 8 in *Optimization and Control of Dynamic Operational Research Models,* vol. 4, S. G. Tzafestas, ed., North-Holland, Amsterdam, pp. 251–276.

Coulbeck, B. and C. H. Orr (1982). "Optimized Pump Scheduling for Water Supply Systems," *IFAC Third Symposium, Control of Distributed Parameter Systems,* Toulouse, France.

Coulbeck, B., M. Brdy, and C. H. Orr (1987). "Exact Incorporation of Maximum Demand Charges in Optimal Scheduling of Water Distribution Systems," *IFAC Tenth Triennial World Congress,* Munich, Federal Republic of Germany, pp. 359–364.

Dougherty, D. E. and R. A. Marryott (1991). "Optimal Groundwater Management—1. Simulated Annealing," *Water Resources Research,* 27(10): 2493–2508.

Elton, A., L. F. Brammer, and N. S. Tansley (1995). "Water Quality Modeling in Distribution Networks," *Journal AWWA,* 87(7): 44–52.

Goldman, F. E. (1998). "The Application of Simulated Annealing for Optimal Operation of Water Distribution Systems," dissertation presented to Arizona State University, at Tempe, Arizona, in partial fulfillment of the requirements for the degree of Doctor of Philosophy.

Grayman, W., R. A. Dieninger, A. Green, P. Boulos, R. W. Bowcock, and C. Godwin (1996). "Water Quality and Mixing Models for Tanks and Reservoirs," *Journal AWWA,* 88(7): 60–73.

Huang, M.D., F. Romeo, and A. Sangiovanni-Vincentelli (1986). "An Efficient General Cooling Schedule for Simulated Annealing," *IEEE Transactions on Computer Aided Design,* CAD-5(1): 381–384.

IAPMO (1994). *Uniform Plumbing Code,* International Association of Plumbing and Mechanical Officials.

Kirkpatrick, S. (1984). "Optimization by Simulated Annealing: Quantitative Studies," *Journal of Statistical Physics,* 34(5/6): 975–986.

Kirkpatrick, S., C. D. Gelatt, and M. P. Vecchi (1983). "Optimization by Simulated Annealing," *Science,* 220(4598): 671–680.

Lasdon, L. S., A. D. Waren, A. Jain, and M. Ratner (1978). "Design and Testing of a Generalized Reduced Gradient Code for Nonlinear Programming," *ACM Transactions on Mathematical Software,* 4(1): 34–50.

Li, Guihua and L. W. Mays (1995). "Differential Dynamic Programming for Estuariane Management," *Journal of Water Resources Planning and Management,* ASCE, (121)6: 455–462.

Malcolm Pirnie (1996). *Peoria Water Master Plan—Technical Memorandum No. 3—Distribution and Flow Model (Draft).*

Nitivattananon, V., E. C. Sandowski, and R. G. Quimpo (1996). "Optimization of Water Supply System Operation." *Journal of Water Resources Planning and Management,* ASCE, 122(5): 374–384.

Mays, L. W. and Y. K. Tung (1992). *Hydrosystems Engineering and Management.* McGraw-Hill, New York, pp. 140–157.

Miller, L. H. and A. E. Quilici (1997). *The Joy of C.* John Wiley, New York.

Metropolis, N., A. Rosenbluth, M. Rosenbluth, A. Teller, and E. Teller (1953). "Equation of State Calculations by Fast Computing Machines," *Journal of Chemical Physics,* 21: 1087–1092.

Mulligan, A. E. and L. C. Brown (1998). "Genetic Algorithms for Calibrating Water Quality Models," *Journal of Environmental Engineering,* 124(3): 202–211.

Murtagh, B. A. and M. A. Saunders (1982). "A Projected Lagrangian Algorithm and Its Implementation for Sparse Nonlinear Constraints." *Mathematical Programming Study,* 16: 84–117.

Murtagh, B. A. and M. A. Saunders (1987). *Minus 5.1 User's Guide.* Stanford Optimization Lab., Stanford University, Stanford, Calif.

Okun, D. A. (1996). "From Cholera to Cancer to Cryptosporidiosis," *Journal of Environmental Engineering,* 122(6): 453–458.

Ormsbee, L. E., editor (1991). "Energy Efficient Operation of Water Distribution Systems," A Report by the ASCE Task Committee on the Optimal Operation of Water Distribution Systems, Univ. of Kentucky, Lexington, Ky.

Ormsbee, L. E. and K. E. Lansey (1994). "Optimal Control of Water Supply Pumping Systems," *Journal of Water Resources Planning and Management,* 120(2): 237–252.

Ormsbee, L. E. and S. L. Reddy (1995). "Nonlinear Heuristic for Pump Operations," *Journal of Water Resources Planning and Management,* 121(4): 302–309.

Ormsbee, L. E., T. M. Walski, D. V. Chase, and W. W. Sharp (1989). "Methodology for Improving Pump Operation Efficiency," *Journal of Water Resources Planning and Management,* 115(2): 148–164.

Ostfeld, A. and U. Shamir (1993). "Optimal Operation of Multiquality Networks. I: Steady State-Conditions," *Journal of Water Resources Planning and Management,* 119(6): 645–662.

Ostfeld, A. and U. Shamir (1993). "Optimal Operation of Multiquality Networks. II: Unsteady Conditions," *Journal of Water Resources Planning and Management,* 119(6): 663–684.

Otten, R. H. J. M. and L. P. P. P. van Ginnekin (1989). *The Annealing Algorithm,* Kluwer Academic, Boston.

Pontius, F. W. (1996). "Overview of the Safe Drinking Water Act Amendments of 1996," *Journal AWWA,* 88(10): 22–33.

Rossman, L. A. (1993). *EPANET Users Manual,* Risk Reduction Engineering Laboratory, Office of Research and Development, U.S. Environmental Protection Agency, Cincinnati, Ohio.

Rossman, L. A. and P. F. Boulos (1996). "Numerical Methods for Modeling Water Quality in Distribution Systems: A Comparison," *Journal of Water Resources Planning and Management,* 122(2): 137–146.

Rossman, L. A., P. F. Boulos, and T. Altman (1993). "The Discrete Volume Element Method for Network Water Quality Models," *Journal of Water Resources Planning and Management,* 119(5): 505–517.

Rossman, L. A., R. W. Clark, and W. M. Grayman. (1994). "Modeling Chlorine Residuals in Drinking-Water Distribution Systems," *Journal of Environmental Engineering,* 120(4): 803–820.

Sadowski, E., V. Nitivattananon, and R. E. Quimpo (1995). "Computer-Generated Optimal Pumping Schedule," *Journal AWWA,* 87(7): 53–63.

Sakarya, A. B. (1998). "Optimal Operation of Water Distribution Systems for Water Quality Purposes," dissertation presented to Arizona State University, at Tempe, Arizona, in partial fulfillment of the requirements for the degree of Doctor of Philosophy.

Sakarya, A. B. and L. W. Mays (2000). "Optimal Operation of Water Distribution Pumps Considering Water Quality," *Journal of Water Resources Planning and Management,* 126(4): 210–220.

van Laarhoven, P. J. M. (1980). *Theoretical and Computational Aspects of Simulated Annealing,* Center for Mathematics and Computer Science, Amsterdam, The Netherlands.

Vasconcelos, J. J., P. F. Boulos, W. M. Grayman, L. Kiene, O. Wable, P. Biswas, A. Bhari, L. A. Rossman, R. M. Clark, and J. A. Goodrich (1996). *Characterization and Modeling of Chlorine Decay in Distribution Systems,* AWWA Research Foundation and American Water Works Association, Denver, Colo. pp. 258–271.

Vidal, René V. V., ed. (1993). *Applied Simulated Annealing,* Springer-Verlag, New York.

Wanakule, N., L. W. Mays, and L. S. Lasdon (1986). "Optimal Management of Large-Scale Aquifers: Methodology and Applications." *Water Resources Research,* 22(4): 447–465.

Wiffen, G. J. and C. A. Shoemaker (1993). "Nonlinear Weighted Feedback Control of Groundwater Remediation under Uncertainty," *Water Resources Research,* 29(9): 3277–3289.

Wood, D. J. (1980). *User's Manual—Computer Analysis of Flow in Pipe Networks Including Extended Period Simulations,* University of Kentucky.

Zessler, U. and U. Shamir (1989). "Optimal Operation of Water Distribution Systems," Journal of Water Resources Planning and Management, 115(6): 735–752.

CHAPTER 6
OPTIMAL FEEDBACK CONTROL OF IRRIGATION WATER DELIVERY SYSTEMS

B. T. Wahlin
WEST Consultants, Inc.
Tempe, Arizona

6.1 INTRODUCTION

Water is becoming an increasingly scarce worldwide commodity. In many parts of the world, water consumption is nearing the limits of the natural system and in some areas these limits have already been surpassed (Postel, 1992). The three major areas of water use are: (1) agricultural, (2) industrial, and (3) municipal. Agriculture is the largest consumer of water. Postel (1992) estimated that 65 percent of all water taken from rivers, lakes, and aquifers is used for agricultural purposes. Over the last century, the world's population has increased from around 1.6 billion to 5.4 billion, and to supply enough food for this surge in population, agricultural water demands have increased fivefold (Postel, 1992).

To overcome the problem of water scarcity, many engineering projects have been developed with the goal of giving people more control over nature's supply of water. The operation of these water projects has become more and more difficult because of the increasing number of political, financial, and ecological constraints. For example, the U.S. Bureau of Reclamation (USBR) has developed criteria for evaluating water management plans in response to the Central Valley Project Improvement Act of 1992 and the Reclamation Reform Act of 1982. These criteria require urban and agricultural water users to promote the highest level of water use efficiency reasonably achievable using the best available, cost-effective technology as well as the best management practices.

Postel (1992) estimated that agricultural users can cut their water use by 10 to 50 percent simply by using water more efficiently. However, problems with reliability and flexibility of the water supplies can hinder an agricultural user's efforts to conserve water. The primary goal of automatic control of irrigation water delivery systems is to provide a more reliable and flexible water supply for the user. With the restrictions on reliability and flexibility removed, agricultural water users are then better equipped to implement water conservation measures.

This chapter concentrates on the automatic downstream feedback control of irrigation water delivery systems. Two different types of automatic controllers are examined: linear quadratic regulator (LQR) control and model predictive control (MPC). Anticipatory feedforward control and local flow control are also extremely important aspects of the overall automatic control of irrigation water delivery systems; however, these topics are beyond the scope of this chapter. Any attempt to automatically control an irrigation water delivery system should include three components:

1. Feedback control to reject unknown disturbances and correct for errors
2. Feedforward control to route known flow changes through the system
3. Flow control to maintain the flow rate at each check structure in the system

6.2 IRRIGATION WATER DELIVERY SYSTEMS

6.2.1 Basic Layout

The main purpose of an irrigation water delivery system is to deliver a controlled amount of water to an agricultural water user. In many parts of the world, users have little or no choice with regard to the actual amount, rate, or timing of the water delivery. In other areas, especially in the arid southwestern United States, the desired amount of water is delivered to the users at a prearranged rate and time. This work deals strictly with the latter type of arrangement.

The type of irrigation water delivery system considered in this chapter consists of a system of open-channel canals. Water is diverted from a reservoir into the main, or primary, canal. Along this main canal are several check structures that control the water levels. The section of the canal between two check structures is referred to as a pool. Just upstream from these check structures are gravity-based turnouts, or offtakes. These turnouts are structures within the canal system that are used to deliver water into either a smaller secondary canal or directly into a user's field. The flow through the gravity-based turnout is a function of the head acting on the turnout. Thus, the flow through these turnouts is more easily regulated if the water level around the turnout is kept constant. Check, or control, structures are typically located just downstream from the turnouts and are used to maintain the water levels near the turnouts at constant, preset values called the water-level setpoints. These setpoints are maintained regardless of the flow rate in the canal. There is a variety of check structures that can be used in an irrigation canal system. Check structures can be either orifice-based, such as sluice and radial gates, or weir-type gates, such as overshot gates. The secondary canals also have various check structures that are used to control the water levels near gravity-based turnouts. The discharge through these turnouts flows into smaller tertiary canals or directly to a user's farm canal or field. A typical irrigation canal consisting of three pools is illustrated in Fig. 6.1.

6.2.2 Reliability and Flexibility

The reliability and flexibility of the water supply are two important aspects of irrigation water delivery systems. For this chapter, both reliability and flexibility are defined in qualitative terms only. *Reliability* is defined as the ability of an irrigation water delivery system to follow through on the promised water deliveries. Secondary aspects of reliability include: (1) the flows through the offtakes remain constant if they have been set to remain constant, and (2) the flows and water levels in the canals are controlled well enough that the canal banks and structures are not damaged (Burt and Plusquellec, 1990). The flexibility of an irrigation water delivery system is the degree to which users have control of the rate, frequency, and duration of the water deliveries.

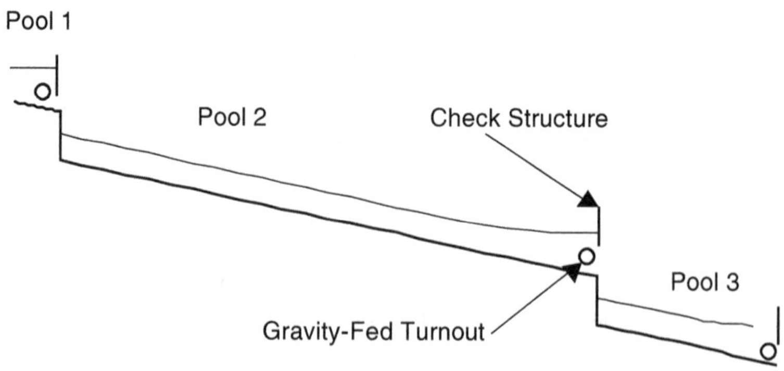

FIGURE 6.1 A typical three-pool irrigation canal.

6.2.3 Water Delivery Scheduling

To deliver the appropriate amount of water to a user at the appropriate time, a scheduling technique must be utilized. Demand, rotation, and arranged schedules are the typical scheduling methodologies used. Ideally, farmers would like a system that allows them to turn the water on and off at any time, in accordance with the crops' needs and farm management constraints. This highly flexible type of system is called a demand schedule because users are free to take any amount of water at any time. To build a canal system that could handle all of the users' demands simultaneously would be very costly because it requires large in-line storage and frequent changes at the control structures. Thus, a pure demand schedule is typically not used in practice (Johnston and Robertson, 1990).

To make it easier to build and operate a canal system, some operators fix the supply and the demand. In this type of system, called a rotation schedule, canal operators fix the rate, frequency, and duration of a delivery for the entire irrigation season (Johnston and Robertson, 1990). The drawback of the rotation schedule is that there are frequent over/underapplications and operational spills, which means that the users' demands rarely match the supplies.

If water demands are known or can be estimated, then operators can manipulate the system to match the supply with the demand. This system is known as an arranged schedule; it is more flexible than a rotation schedule but less flexible than a demand schedule. Under an arranged schedule, the user schedules the rate, frequency, and duration of a water delivery with the canal operators in advance (Johnston and Robertson, 1990). There are variations of the rotation and arranged schedule. For example, users in some irrigation districts are able to schedule the rate and frequency of the water delivery but not the duration; the water is only delivered for a certain block of time (e.g., 12-h blocks).

The automatic control systems discussed in this chapter are most applicable to arranged deliveries. Although automatic control would still be useful for rotation and demand systems, these types of systems were not considered.

6.2.4 Flow Measurement

With accurate flow measurements at each check structure, the canal operators are able to route the scheduled flow changes through the canal system efficiently. Typically, the check gate itself is used as a flow measurement device. In this case, steady-state flow passing through the gate can be calculated using measurements of the upstream water level, the downstream water level (if the gate is submerged), and the gate opening. Other more accurate flow measurement devices, such as broad-crested weirs, may be placed in the canal just downstream from the check structures to measure the flow rate. Many irrigation water delivery systems are operated without flow measurement at the control gates or with only roughly accurate measurements. In these situations, the canal operators rely primarily on the water levels to control the canal. Canal operators assume that if the water levels are at their desired setpoints and all the demands are met, then the flow rates in the canal must be correct.

6.3 CONTROL OF IRRIGATION WATER DELIVERY SYSTEMS

6.3.1 Manual Control

Most irrigation districts use manual control to operate their irrigation water delivery system. Under this type of control, canal operators manually adjust each gate in the canal system in an attempt to keep the water levels at the desired setpoints while routing the scheduled deliveries through the canal. Irrigation water delivery systems operating under manual control are able to deliver a stable and reliable water supply only if the users are allowed very little flexibility (i.e., a relatively rigid arranged schedule). However, in order to utilize water more efficiently, agricultural water users must have more flexibility over their water deliveries. Increasing the flexibility of the water deliveries increases the unsteadiness of the flow in the canal system and can lead to poor delivery performance (Clemmens et al., 1998a). Thus, under manual control it is difficult to deliver water both flexibly and reliably.

The basic problem with manual control is that upstream flow changes arrive gradually at the downstream end and the flow rates are not measured adequately, making it difficult to determine the correct settings for the check structures. If the check gates operate under submerged conditions, it is even more difficult to determine the correct settings because flow measurements are even less accurate under these conditions. Furthermore, operators in manually controlled systems adjust the check structure settings based on current conditions. However, the flow in open channels is unsteady, which implies that the operators are using incomplete information when they adjust the check structures. Unsteady flows are difficult to predict because of the complications due to delay time, wave dispersion, and changes in pool storage volume. These factors can make the task of routing water deliveries both difficult and time consuming.

Local Manual Control. Under local manual control, canal operators physically travel from gate to gate and cannot see all of the control structures at once. The adjustments to a check gate are based only on the canal operators' judgment and observations made from traveling up and down the canal (Clemmens, 1998). It is difficult for a single operator to control a canal where changes are taking place at multiple locations on the canal simultaneously. As an example, assume that a delivery change is required at the third check gate structure in Fig. 6.1. Experienced operators have an idea of how long it takes the scheduled change to move through pools 1, 2, and 3. If the operators estimate that it takes one hour for the water to travel that distance, then one hour before the delivery, the operators open the head gate and send the water into pool 1. The operators travel to the first check gate and wait until a significant portion of the wave has arrived. The operators then open the check gate and send the delivery into pool 2. This process is repeated for the next two pools. Finally, when most of the wave arrives at the end of pool 3, the operators open the turnout to the field. However, the initial check gate settings are never correct. The canal operators must then return to the head gate and make further adjustments. If the initial gate settings are poor, there will be large errors in the system and two passes may not be enough to get the canal in balance (Clemmens, 1998).

Supervisory Manual Control. To overcome the problems with local manual control, some irrigation districts have replaced local manual control with supervisory manual control. In a supervisory control system information from every site is gathered in a centralized location using *Supervisory Control And Data Acquisition* (SCADA) systems. Canal operators can view what is happening in the entire canal system at once using SCADA systems. This information allows operators to route the scheduled deliveries through the canal system more quickly because they do not have to physically travel up and down the canal. As an example of how supervisory manual control can improve operations, one irrigation system that switched from local manual control to supervisory manual control was able to reduce their operating staff by seven people on about 80 km of main canals (Clemmens et al., 1994).

6.3.2 Automatic Control

In automatic control, flows are routed through an irrigation water delivery system and the water levels are maintained at their setpoint without human intervention. Many efforts to develop automatic control of irrigation canals have taken place over the last 40 years. Initially, the controllers were single-variable, local, electromechanical devices that controlled the water level just upstream from a check gate or just downstream from an offtake gate (e.g., the Little-Man controllers). Now SCADA systems can be used for the automatic supervisory control of irrigation water delivery systems from a central site. This type of supervisory control can either be single-variable or multivariable.

Upstream Control. Upstream control is the control of a check gate's movement based on the water level immediately upstream from the gate (Burt, 1987). With upstream control, a flow change at the upstream end of a canal cascades to the end of the canal unless it is removed through an offtake change. All the delivery errors move to the downstream end of the canal, resulting in an excess or deficit water delivery at that location. In Fig. 6.2, gate 1 is under upstream control.

Downstream Control. Downstream control is the control of a check gate's movement based on the water level at some location downstream from the gate. Typically, the water level used is located just downstream from the gate (called near downstream control) or at the downstream end of the pool immediately downstream (called remote downstream control). In Fig. 6.2, gate 2 is under remote downstream control (i.e., the water level at the downstream end of pool 3 determines the control actions at gate 2). A change in water level at the downstream end of pool 3 causes a change in the position of gate 2. If the entire canal is under downstream control, then the delivery errors cascade all the way to the top of the canal (Burt, 1987). Thus, downstream control is more user-oriented than upstream control because it can respond better to user demands. All of the feedback controllers in this chapter were designed based on remote downstream control.

Local Automatic Control. Local control refers to the situation where there are separate independent controllers on each pool in the canal system (Burt, 1987). A local controller is simply a single-variable controller; all the control actions are computed independently at each site. Local control of a single pool is fairly easy. However, local control of a series of pools is difficult because a series of local automatic controllers placed on a canal system interferes with each other. In this situation, the controllers are usually de-tuned. That is, the control algorithm is made less aggressive so the effects of the hydraulic interactions between the pools are dampened.

Supervisory Automatic Control. With supervisory automatic control, all the gates are automatically controlled from a single remote location (Burt, 1987). All the measured variables are gathered and sent to a single location. The control algorithm uses all of the centrally located data to determine the appropriate changes in the gate positions. Supervisory control alone does not overcome the interfering control loops experienced in local automatic control; supervisory control does not imply multivariable control. For example, a SCADA system can be used to collect all the measured parameters from the canal system and assemble them in a single location. However, if the appropriate gate positions are still determined independently for each pool, then the control loops still interact with each other. One benefit of having supervisory control is that all of the measured variables are gathered in a single location and this information can be used to design hydraulic decouplers that eliminate or reduce the control loop interactions. Decoupling the system is a more efficient way of handling the interacting control loops than de-tuning the controllers.

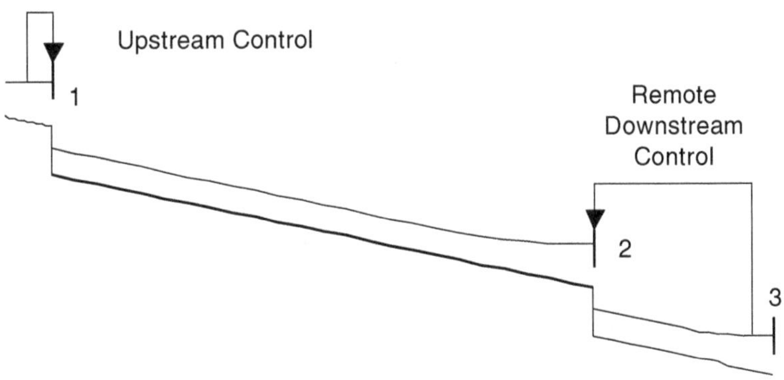

FIGURE 6.2 Upstream vs. downstream control. (*Note:* As shown, the combination of control schemes is not viable.)

6.3.3 Justifications for Automatic Control of Irrigation Water Delivery Systems

The main justification for automatic control of irrigation water delivery systems is that it improves the flexibility of the water deliveries while also maintaining the reliability. Another justification for canal automation is that it reduces the social conflict between water users that is always present when the water deliveries are unreliable (Burt and Plusquellec, 1990). In addition to these primary justifications for canal automation, Burt and Plusquellec (1990) give several secondary justifications.

1. There is an improvement in the safety of the operation of irrigation water delivery systems through reduced spillage, decreased water-level fluctuations, and less damage to the structures and the canal lining
2. The irrigation water delivery system becomes easier to operate
3. There is a reduction in the operating costs of the irrigation water delivery system
4. The capacity of the system is increased

6.3.4 Difficulties in the Automatic Control of Irrigation Water Delivery Systems

Common Difficulties. The basic problem inherent in an irrigation water delivery system is determining the appropriate schedule of check gate openings that keeps the water levels at their desired setpoints while routing the scheduled water deliveries through the system. Initially, control of an irrigation system may appear to be fairly straightforward because, for any pool in the system, the inflow into the pool equals the outflow from the pool plus the change in storage volume over time (Clemmens, 1998). However, there are many aspects of open-channel flow that make the control of water levels much more difficult than it initially appears. First, canal operators must be able to handle unscheduled changes in demand, which can arise from check structure failures, vandalism, or other emergency situations. The canal operators must also deal with safety concerns. For example, operators do not want to overtop the canal because it can cause serious structural damage to the canal system. Also, the water levels in the canal cannot be lowered too quickly without causing problems in the system. Although rapid increases in water levels are not a problem unless the canal overtops, rapid decreases can damage the canal even when the depths remain at an acceptable level (Buyalski et al. 1991).

Another difficulty arises because canal operators must account for uncertainty in various aspects of the canal system. These uncertainties can manifest themselves in inaccurate flow measurements or in unaccounted-for losses or gains in the system. As an example of how uncertainties make canal control difficult, consider the uncertainties in flow measurements. All flow measurements have some uncertainty associated with them, which may be large or small, depending on the equipment used and the skill of the operators. The hydraulics of a flow measurement device may also change slowly over time due to algae growth, rounding of the gate blade, and other factors. All of these flow measurement errors accumulate in the system, and if they are not accounted for, the water deliveries to the users are never as intended (Clemmens, 1998).

Difficulties Due to Unsteady Flow. The transient effects that are characterized by unsteady open-channel flow also make it difficult to control the water levels in a canal. These transient effects can be characterized by (1) delay times, (2) dispersion effects, (3) changes in storage volumes, and (4) hydraulic interactions between the pools. Open-channel flow is characterized by delay times. If a disturbance is introduced in the upstream end of the pool, it takes some time before the disturbance is felt at the downstream end of the pool. This disturbance also does not arrive at the downstream end all at once. Instead, it arrives gradually over a period of time because of wave dispersion (Clemmens, 1998). Figure 6.3 demonstrates a specific example of this behavior for a single pool. In this example, there are initially 0.6 m^3/s flowing into and out of the canal pool. Two hours into the simulation, there is a step change in inflow from 0.6 m^3/s to 0.9 m^3/s. It takes approximately 15 min for this step change to be noticed at the downstream end of the pool because of the time it takes for the disturbance to travel down the canal. It also takes several hours for the entire step change to be felt at the downstream end of the canal due to wave dispersion and changes in canal pool volume.

Another difficulty in the control of open-channel systems is associated with the storage volume in a pool. To go from one steady-state flow to another, a certain volume of water must be added to or taken out of the pool. The volume of water stored in a pool is a function of both the flow rate and the resistance to flow, expressed in terms of Manning *n,* as illustrated in Fig. 6.4 (Clemmens, 1998). Also, if the water level at the downstream end of a pool changes, then the backwater effects as well as the volume stored in the pool changes. In many cases, canal operators change flow rates or water levels in the canal without considering the required change in pool storage volume. This practice makes it more difficult for the canal operators to effectively control the canal.

Yet another difficulty arises from the fact that a series of canal pools does not act independently of each other because the outflow from one pool is also the inflow into the next pool. Thus, if a check gate is adjusted on one pool, the outflow from that pool as well as the inflow to the next pool is changed. Because the inflow to the second pool is changed, the water levels in that pool are also changed. If the backwater effects extend across the entire pool and the gate at the upstream end becomes submerged, then adjusting the gate at the downstream end of the pool also affects the inflow into that pool. The delay times, dispersion characteristics, and storage volume of the individual pools change as a result of these hydraulic interactions. If individual control loops are designed for each pool in a canal system without considering these interactions, then the control loops do not perform

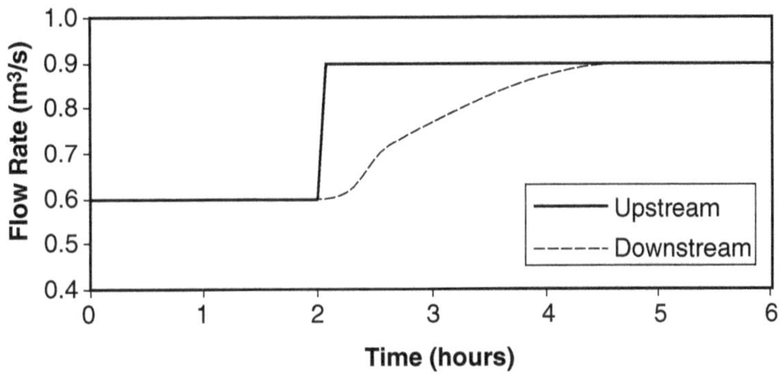

FIGURE 6.3 Outflow from a canal pool for a step change in inflow.

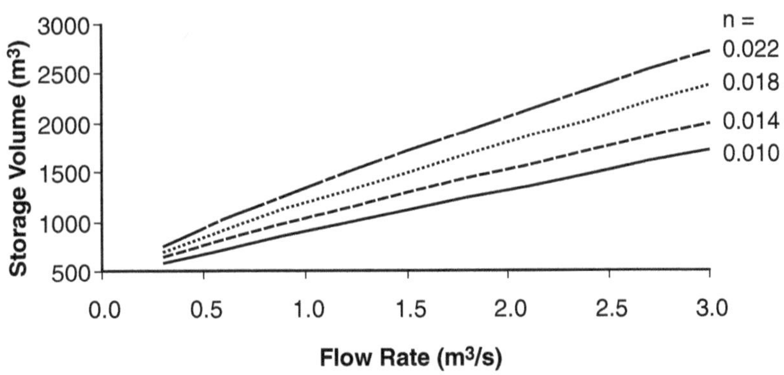

FIGURE 6.4 Storage volume for a pool as a function of flow rate and flow resistance.

as expected because the underlying characteristics of the pools are different from what was assumed when the controllers were designed. Also, control loops designed without considering the interactions between the pools interfere with each other and introduce hidden feedback loops that can destabilize the system (Shinskey, 1996). Although the individual controllers can be de-tuned to restore stability, the overall control system does not perform as well as if controllers were acting on individual pools alone.

Because formal hydraulic procedures for handling all of these situations may be difficult to apply or may not be available, canal operators must rely on experience to effectively operate an irrigation water delivery system. In time, the canal operators learn the "feel" of the irrigation water delivery system and can heuristically control it to a reasonable degree of accuracy using manual operations. An automatic control system for an irrigation water delivery system is designed to more efficiently route scheduled flow changes through the system as well as account for the various uncertainties and control difficulties associated with open-channel flow.

Nontechnical Difficulties. Although there have been many advances in hardware and software, canal automation has not become routine. One reason is that canal automation is not as lucrative as other technology industries so it has not experienced the explosive growth seen in the electronics and telecommunications industries (Rogers, 1999). Another reason is that each irrigation system control project must be uniquely designed which prevents the equipment from being mass-produced (Rogers, 1999).

There are many obstacles when developing an automatic control system for an irrigation project. Some of these obstacles are technical, such as hardware, software, or controller problems. However, the most difficult problems to overcome are usually nontechnical, such as resistance to change. It is more difficult for engineers to overcome the nontechnical obstacles than the technical ones because the nontechnical obstacles deal more with sociology rather than with engineering principles. Rogers (1999) gives a detailed summary of some of the technical and nontechnical obstacles experienced in canal automation projects.

6.4 BASIC CONTROL DEFINITIONS

6.4.1 Process Model

The goal of any automatic control system is to control some sort of process. Thus, a model is needed to mathematically describe the process to be controlled. This response model, or process model, is the underlying core of any automatic controller, whether the controller is a feedback controller or a feedforward controller. For flow in open channels, the response model is the Saint Venant equations, which are a set of hyperbolic, nonlinear, partial differential equations distributed in time and space. However, it would be incredibly difficult to design and implement a controller using the Saint Venant equations as the response model. Thus, the overall control problem would be simplified if a linear response model could be used to describe the flow in open channels. Over the years, investigators have used a wide variety of linear process models for automatic control of irrigation water delivery systems. Some of these linear process models include linearized versions of the Saint Venant equations, lumped-parameter transfer function models, and models developed using system identification. This work uses a lumped-parameter model called the integrator-delay (ID) model as the process model (Schuurmans et al., 1995).

6.4.2 Control Variables

Control variables are the target values or setpoints that indicate the success of the control algorithm (Malaterre et al., 1998). In irrigation water delivery systems, the control variables are usually the water levels at some location in the pools (e.g., at the downstream end of the pool, immediately downstream from the check structure, or some place in the middle of the pool).

6.4.3 Measured Variables

The measured variables are the inputs to the control algorithm (Malaterre et al., 1998). For irrigation water delivery systems, typical measured variables are water levels and discharges.

6.4.4 Control Action Variables

The control action variables are the outputs from the control algorithm (Malaterre et al., 1998). The control action variables can either be gate positions or discharges. A control action variable is sent to a gate's actuator and the gate is moved accordingly. If the control action variable is gate position, then the control algorithm must take into account the complex hydraulics that exist among the gate position, the flow through the gate, and the upstream and downstream water levels (Malaterre and Baume, 1999). If the control action variable is discharge, then the discharge must be converted to a gate position by inverting the gate's discharge equation. One benefit of using discharge as the control action variable is that it allows for some decoupling of the hydraulic interactions between the pools (Malaterre and Baume, 1999). However, the main benefit of using discharge as the control action variable is that the complex hydraulics among gate position, flow through the gate, and upstream and downstream water levels are placed outside the control algorithm. Removing the gate hydraulics from the control algorithm greatly simplifies the controller design process. However, these dynamics must be accounted for in some manner or the overall performance of the control algorithm suffers. A local flow controller, as explained later, can be used to account for the hydraulics of the gate.

If the control action variables are expressed in absolute terms, then a problem can occur. For a control algorithm to work with absolute gate position, the gate position must be accurately measured because the control algorithm instructs the gate to move to a specific location. If the gate position indicators are inaccurate or if they are not zeroed properly, then the gate is not moved to the correct location and control performance degrades (Burt et al., 1998). To overcome this problem, most control algorithms are expressed in incremental form; the control action variables are expressed as changes in gate position or discharge instead of absolute values. The control action variables used in this work are changes in flow rate at each check structure.

6.4.5 Performance Indicators

There are many different ways to measure the performance of an automatic control system. Clemmens et al. (1998b) present many performance indicators that can be used to judge the overall effectiveness of automatic control systems for open-channel water level control. These performance indicators are geared toward canal systems and four of them are presented here.

Maximum Absolute Error. The maximum absolute error (MAE) is a measure of the maximum deviation in water level from the desired setpoint:

$$\text{MAE} = \frac{\max\left(|y_t - y^{\text{stp}}|\right)}{y^{\text{stp}}} \tag{6.1}$$

where y_t is the computed water level at time t and y^{stp} is the water level setpoint. The MAE does not give an indication of how long the error persists.

Integrated Absolute Error. The integrated absolute error (IAE) indicates the magnitude of the water level errors and how long they persist:

$$\text{IAE} = \frac{\frac{\Delta t}{T}\sum_{t=0}^{T} |y_t - y^{\text{stp}}|}{y^{\text{stp}}} \tag{6.2}$$

where Δt is the regulation time step and T is the length of the test. The IAE does not give an indication about the relative importance of the error occurring late in the response versus the error occurring early in the response.

Steady-State Error. The steady-state error (StE) is the maximum of the average error over the end portion of a simulation. The StE is defined as the maximum of the average error over the last two hours of the test:

$$\mathrm{StE} = \frac{\Delta t}{t_0 + \Delta t} \frac{\left| \sum_{t=t_2-t_0}^{t_2} (y_t - y^{\mathrm{stp}}) \right|}{y^{\mathrm{stp}}} \tag{6.3}$$

where t_0 equals two hours and t_2 is the end time of the test. The StE is an indication of how well the controller was able to return the control variables to their setpoints.

Integrated Absolute Discharge Change. Another performance indicator is the integrated absolute discharge change (IAQ), which is an indication of the number of gate movements required to achieve control:

$$\mathrm{IAQ} = \sum_{t=t_1}^{t_2} (|Q_t - Q_{t-1}|) - |Q_{t1} - Q_{t2}| \tag{6.4}$$

where Q_t is the check gate discharge at time t. The second term in Eq. (6.4) is simply the difference between the initial and final flow rate and allows the minimum IAQ to be zero.

6.5 TYPES OF AUTOMATIC CONTROL

6.5.1 Feedback Control

In a feedback control algorithm, deviations of the control variables (e.g., water levels) from the setpoints produce a set of corrective control action variables (e.g., changes in check gate opening) that move the control variables back to their setpoints (Malaterre et al., 1998). For a typical irrigation system under downstream control, water levels at the downstream ends of the pools are the control variables. If the water levels start to deviate from the setpoints, then corrective actions are determined from the feedback control algorithm. These corrective actions are then applied to the gates at the upstream ends of the pools. A feedback control system has three main objectives: (1) stability, (2) robustness, and (3) performance. A feedback system also accounts for unknown disturbances and flow measurement inaccuracies. The feedback controller operates on a simplified process, or response, model of the system. As explained earlier, it does not operate on the "true" system.

Single-Variable Feedback Control. Single-variable feedback control uses one measured variable to determine the appropriate corrective control action. Because there is one measured variable and one control variable, this situation is called a Single-Input-Single-Output (SISO) system. If there are multiple control variables (e.g., a canal system with several pools), then several single-variable controllers can be used in series. As mentioned earlier, the drawback to using multiple single-variable controllers on a coupled system is that these controllers can interfere with each other and destabilize the system.

Multivariable Feedback Control. In multivariable control, there are multiple measured variables and multiple control actions. This Multiple-Input-Multiple-Output (MIMO) feedback control system uses all of the control variables simultaneously to determine the appropriate set of corrective control actions. Multivariable control is a more efficient way of handling the interactions that exist between multiple single-variable controllers. However, the drawback is that a multivariable controller is much harder to tune and implement.

6.5.2 Feedforward Control

A feedforward, or open-loop, controller applies corrective efforts to the system with the assumption that these corrective efforts will achieve the desired results. The main purpose of a feedforward controller is to eliminate the effects of prescheduled disturbances (i.e., known flow changes). The basic idea of a feedforward controller is to use the prescheduled disturbances to anticipate the influence of these disturbances on the system and to make suitable corrective control actions beforehand (Åström and Wittenmark, 1997). Unlike a feedback system, the feedforward corrective control actions take place before the measured disturbances are introduced into the system. The corrective feedforward control actions are designed to attenuate unwanted water level fluctuations and to stabilize the flow more quickly (Clemmens et al., 1998a). However, determining the appropriate feedforward corrective actions is not a trivial problem because of the nonlinear and dispersive nature of open-channel flows and the constraints that need to be imposed on the changes to ensure reliable deliveries and safe operation of the delivery system (Clemmens et al., 1998a).

Systems that have predictable processes with long delay times, such as irrigation water delivery systems, are particularly suited for feedforward control. The heuristic manual control employed by canal operators can be considered a form of open-loop control, which explains why canal operators can effectively control an irrigation water delivery system using manual control once they obtain a "feel" for the system. If there are no uncertainties or unscheduled flow changes in the system, then the canal can be adequately controlled using a feedforward routine alone. However, in the presence of uncertainties and disturbances, there is no guarantee that the feedforward controller will have the desired effect on the control variables. Thus, it is good practice to use feedback control in combination with feedforward control. Feedforward control is used to route scheduled flow changes through the canal system whereas feedback control is used to account for uncertainties and unscheduled flow changes. Both feedforward and feedback control can be manual or automatic.

Two different feedforward controllers are briefly explained next. These feedforward controllers are the gate-stroking method and the volume compensation method.

Gate-Stroking. A feedforward algorithm based on hydraulic principles is the gate-stroking method (Wylie, 1969). The gate-stroking method is based on the inverse solution of the Saint Venant equations. Although this method is theoretically appealing, it has not been successfully applied to real canal operations. Gate-stroking attempts to produce a step change in flow at the downstream end of the pools. In reality, this is impossible due to wave dispersion. Thus, the inverse solution of the Saint Venant equations can yield extreme flow changes that are difficult or impossible (e.g., negative flow rates) to implement, which is one reason gate-stroking has not had any real-world success (Bautista and Clemmens, 1999).

Volume Compensation Method. The volume compensation method (Bautista et al., 2002) assumes a succession of steady-states, each resulting from a change in demand (offtake) flow or a change in a pool's setpoint. The transition from an initial to a final steady-state requires a change in volume ΔV for the pool requesting the change, pool P_r, and for each pool upstream from it, pools 1 through $(P_r - 1)$:

$$\Delta V_i = V_i^1 (Q_i^1, y_i^{\text{stp1}}, n_i) - V_i^0 (Q_i^0, y_i^{\text{stp0}}, n_i) \tag{6.5}$$

where

$$Q_i^1 = Q_i^0 + \Delta q \tag{6.6}$$

$$y^{stp1} = y^{stp0} + \Delta y^{stp} \tag{6.7}$$

In the above expressions, V_i^0 and V_i^1 represent the volume of the ith pool for successive steady-states, Q_i^0 and Q_i^1 the corresponding pool inflow for successive steady-states, Δq an offtake flow rate change, y^{stp0} and y^{stp1} the initial and final pool setpoint, Δy^{stp} the setpoint change, and n_i the pool's Manning roughness coefficient. With setpoint changes, only pool P_r's volume is affected.

A schedule of check structure flow changes needed to deliver a requested Δq is computed by taking into account ΔV_i and the delay of each affected pool $\Delta\tau_i$. Rather than conducting repeated steady flow calculations to determine ΔV_i, a simpler approach is to tabulate the relationship between pool inflow rate and volume at selected values of y^{stp} and n, and to interpolate from those tables. Various approaches can be used to estimate the delay for varying flow conditions, including canal operator experience. In cases where the change in steady-state is due to an offtake flow change, a pool's delay estimate $\Delta\tau_i$ can be obtained as follows (Bautista et al., 2003):

$$\Delta\tau_i = \frac{\Delta V_i}{\Delta q} \tag{6.8}$$

Such a delay can be interpreted as a delay based on kinematic shock velocity. It generates a flow schedule at the top of pool i consisting of a single flow change of magnitude $\Delta Q_i^1 = \Delta q$. The timing of this change $t_i^{\Delta Q1}$, depends on t_d, the time of the demand change and the accumulated delay of all pools between pools i and P_r:

$$t_i^{\Delta Q1} = t_d - \sum_{K=i}^{P_r} \Delta\,\tau_K \tag{6.9}$$

In Eq. (6.9), $i \le P_r$ and K is a pool index.

If the new steady-state is the result of a setpoint change, then the delay needs to be estimated by other means (e.g., kinematic wave theory). The resulting schedule then consists of two flow changes, the second of which is needed to restore the original pool inflow rate:

$$\Delta Q_i^1 = \frac{\Delta V_{P_r}}{\Delta\tau_i} \qquad \Delta Q_i^2 = -\Delta Q_i^1 \tag{6.10}$$

As indicated by Eq. (6.10), in this case the inflow schedule for each upstream pool depends exclusively on pool P_r's volume change.

A final schedule is obtained by superimposing the individual schedules computed for each Δq and Δy^{stp}. Care must be taken in determining the correct initial conditions when processing each individual demand change. More details on how to apply this approach to complex scheduling situations are provided in Bautista et al. (2002).

6.5.3 Flow Control

Using discharge as the control action variable removes the complex gate hydraulics from the feedback controller. Although this removal simplifies the controller design, the gate hydraulics still must be taken into account or the overall controller performance degrades. One way to take the gate hydraulics into account is to introduce a secondary, or slave, controller on each gate called a flow controller. The purpose of the flow controller is to vary the gate opening so that the gate delivers the target flow (Clemmens et al., 1998a). Thus, the flow controller takes into account the uncertainties of the gate hydraulics by tweaking the gate opening and forcing the gate to pass the target flow. If the flow controller's corrective actions are quick enough (e.g., at least as fast as the feedback controller's corrective actions), then the performance of the overall controller (feedback, feedforward, and flow controllers) can be as efficient as expected (Malaterre and Baume, 1999).

6.6 INTEGRATOR-DELAY MODEL (PROCESS MODEL)

A feedback or feedforward control can be designed to control a process, or response, model of the real-world situation. Schuurmans et al. (1995) developed a linear, lumped-parameter response model for flow in open-channel systems that can handle backwater effects as well as normal flow conditions. This model, called the integrator-delay (ID) model, is a good choice for a process model for the feedback controller because it is a lumped parameter, linear model. These characteristics of the ID model make the design of the feedback controller easier than it would be if the process model were a nonlinear or distributed model. The ID model assumes the flow in a canal pool has two parts: (1) the normal depth flow section, and (2) the backwater section (see Fig. 6.5).

The ID model is developed by first linearizing the Saint Venant equations and then taking the Laplace transform of these linearized equations. For each pool, the boundary conditions are specified in terms of the flow rates. Because the dynamics of the flow in the backwater portion are different from those in the normal depth portion, a separate approximation model is derived for each part. The model for the overall system is obtained by connecting the backwater model to the normal depth model (Schuurmans et al. 1995). In the backwater part, the waves do not deform very much and the dynamics are complex due to wave reflections (Schuurmans et al., 1995). Schuurmans et al. (1995) propose the following linear approximation for the backwater portion:

$$\frac{de}{dt} = \frac{1}{A_s} \left(q_{\text{in}} (t - \tau) - q_{\text{out}} (t)\right) \tag{6.11}$$

where e = deviation of the downstream water level from its desired level
q_{in} = deviation of the inflow from its initial value
q_{out} = deviation of the outflow from its initial value
A_s = surface area of the backwater part
τ = delay time of the normal depth portion of the pool

As shown in Fig. 6.5, q_{out} is applied to the downstream gate and q_{in} is applied to the upstream gate.

For the normal depth part, waves traveling in the upstream direction dampen out quickly, and it was assumed that waves travel only in the downstream direction (Schuurmans et al., 1995). Schuurmans et al. (1995) recommend using the kinematic wave model in the normal depth part of the canal. In the kinematic wave model, the important parameter to determine is the delay time (i.e., the time it takes for an upstream disturbance to travel to the downstream end of the pool). The ID model assumes that there is no delay time for a disturbance traveling through the backwater part. Thus, only the delay time through the normal depth portion of the pool needs to be determined. In the kinematic wave model, the speed of the wave is a function of the water surface elevation h only,

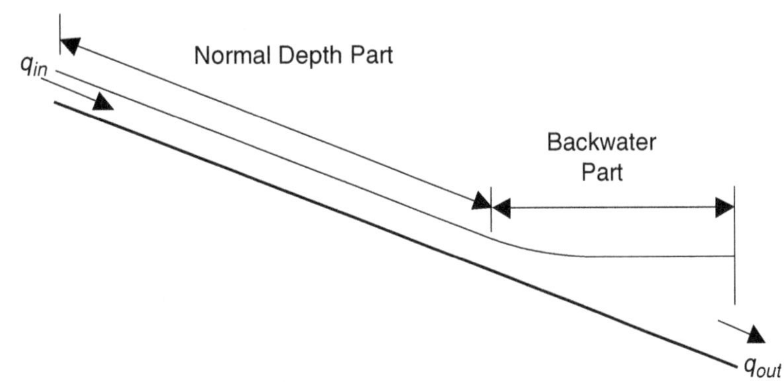

FIGURE 6.5 Approximate model of flow in an irrigation pool used for the ID model.

and the speed of the wave (dx/dt) can be derived from Eq. (6.11) by making the substitution $dA = Bdh$, where B is the top width of the canal, and rearranging the terms.

$$\frac{dx}{dt} = \frac{1}{B}\frac{dQ}{dh} \tag{6.12}$$

The delay time τ can then be computed knowing the length of the normal depth portion of the pool L_n.

$$\tau = \frac{L_n}{\dfrac{1}{B}\dfrac{dQ}{dh}} \tag{6.13}$$

If the Manning equation is used to estimate the dQ/dh term, then the delay time can be expressed as (Schuurmans et al., 1995)

$$\tau = \frac{L_n}{\left[1 + \left(\dfrac{4}{3}\dfrac{P_0}{B_0}\dfrac{dR}{dh}\Big|_0\right)\right]v_0} \tag{6.14}$$

where P_0 = wetted perimeter
R = hydraulic radius
v_0 = average flow velocity

and the subscript zero indicates that parameters are evaluated under the initial flow conditions.

The ID model depends on only two hydraulic parameters per pool: the delay time τ, and the surface area of the backwater part A_s. Once these two properties are determined, the hydraulic characteristics for the pool are defined. However, the ID model is still a linearized approximation of the true system, which implies that a feedback or feedforward controller based on the ID model may not function well over a wide range of flow conditions.

6.6.1 Validation of the Integrator-Delay Model

To validate the ID model, Ellerbeck (1995) performed open-loop tests on the WM canal in the Maricopa-Stanfield Irrigation and Drainage District (MSIDD), which is an irrigation district located in central Arizona. The WM canal is a very steep canal with very little storage consisting of eight pools of various lengths. Ellerbeck (1995) performed various open-loop tests on this canal and the results were compared to the ID model predictions. These tests showed that the ID model accurately predicts the water levels in the canal (Ellerbeck, 1995). As an example, the results for one of Ellerbeck's (1995) open-loop tests are shown in Figs. 6.6 and 6.7. For this test, the gate at the downstream end of pool 2 (gate 2) was moved as shown in Figs. 6.6 and 6.7. The water levels at the downstream end of pools 2 and 3 were then measured in the field. MATLAB was used to simulate the response of the ID model. The actual water levels agree very well with the ID model even though a turnout was opened by mistake during this test. The other open-loop tests gave similar results as shown above. Ellerbeck (1995) concluded that the ID model accurately predicts the water levels in the WM canal.

6.6.2 Determining the Hydraulic Properties of the Integrator-Delay Model

The delay time τ and the backwater surface area A_s can be determined either through numerical calculations or unsteady flow simulation. The ID model assumes that there is no delay across the backwater section of a pool. If a pool is completely under backwater, then the delay time is zero. If a pool is partially under backwater, the delay time is the time it takes for the disturbance wave to travel

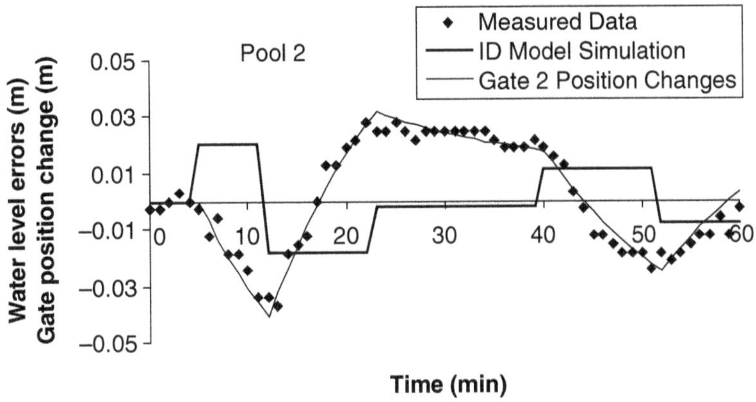

FIGURE 6.6 Actual water levels vs. ID model predicted water levels in pool 2 of the WM canal. (*From Ellerbeck, 1995.*)

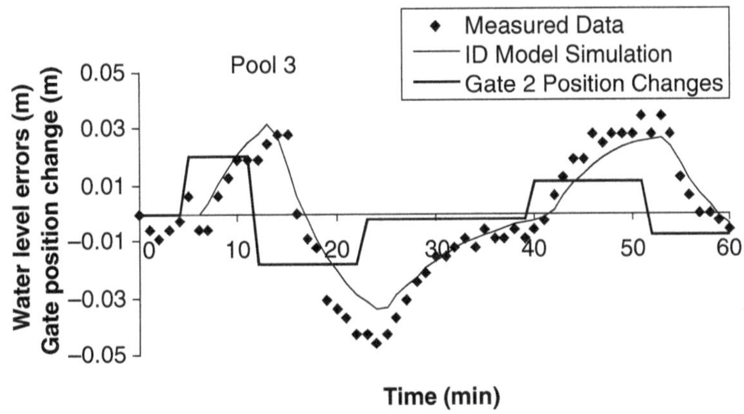

FIGURE 6.7 Actual water levels vs. ID model predicted water levels in pool 3 of the WM canal. (*From Ellerbeck, 1995.*)

through the normal depth part. This delay time can be calculated using kinematic wave theory [see Eq. (6.14)] or by using unsteady flow simulation.

To determine the backwater surface area, an estimate of the backwater length is needed. Numerically solving the Saint Venant equations and determining the flow profile is one way to estimate the backwater length. This procedure requires two different flow profiles to be calculated. One profile is of the pool with the gate in place at the downstream end. The second profile is of the pool with a normal depth control at the downstream end. These two profiles are then superimposed, and the backwater length is the length from the downstream end of the pool up to where the difference between the two profiles is not significant. The backwater surface area can then be calculated from the backwater length and the geometric properties of the pool. Under many circumstances, it is difficult to determine backwater length and some engineering judgment must be used.

If there are culverts or supercritical flow sections within the pool, kinematic wave theory cannot be used to determine the delay time. In this situation, unsteady flow simulation must be used. Clemmens et al. (1997) present a method for determining the delay time and backwater surface area through unsteady flow simulation. In this method, each pool in the canal system is analyzed individually. First, a hydraulic model of the pool is built within a simulation program. Next, the simulation program is run until the system reaches steady-state at the design flow rate and the downstream water level is at the desired setpoint. Then, the inflow is increased over one time step by 20 percent from the steady-state value while the outflow is kept constant. The water level at the downstream end of the pool rises when the wave reaches the end of the pool. Finally, the ID model parameters are graphically fit to the downstream water level response.

Figure 6.8 shows a typical example of how the downstream water level varies as a function of time. In this example, the initial steady-state flow rate is 0.1 m^3/s and the initial downstream water level is 0.8 meters. Two hours into the simulation, the inflow is increased from 0.1 m^3/s to 0.12 m^3/s (a 20 percent increase). The delay time τ can be determined graphically as the amount of time between the step change in inflow and the time it takes the downstream water level to begin to rise. A straight line can be fitted through the water level response, and the backwater surface area is proportional to the slope of this straight line dy/dt as well as the change in flow rate ΔQ.

$$A_s = \Delta Q \frac{dt}{dy} \tag{6.15}$$

The initial water level rise should be used to fit the straight line. Later portions of the water level response curve may be linear for a longer time period than the initial rise, but the initial water level rise should be used to determine the delay time and backwater surface area. The initial linear water rise also is an indication of the robustness of the linear ID approximation of the Saint Venant equations. When the water level response deviates from the straight line, the nonlinear characteristics of open-channel flow become significant and the ID model is no longer applicable. If the initial linear water surface rise is very long, then the ID model is valid over a wide range of flow depths. If the initial linear water surface rise is short, then the ID model is valid over only a very narrow range of flow depths.

If the pool is completely under backwater, the water level response may oscillate around some straight line. In this case, the straight line should not be fit through the initial water level rise. Instead, the straight line should be fit through the middle of the oscillations as shown in Fig. 6.9. For the example shown in Fig. 6.9, the step change in inflow occurs two hours into the simulation. There is no delay time for the pool in this example as would be expected for a pool completely under backwater. Note that pools completely under backwater may or may not exhibit the oscillations shown in Fig. 6.9.

The delay time and the backwater surface area are functions of the flow through the pool. Thus, the estimated parameters are valid only for a limited range of flow conditions. Because these values are used in the design of the controller, the controller may not function properly if the actual flow rates are significantly different from the design flow rates.

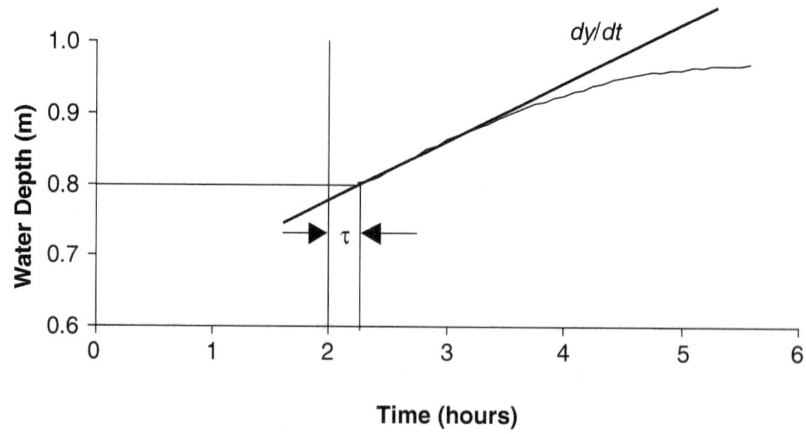

FIGURE 6.8 Response in water level at the downstream end of a pool for a step change in inflow (pool not completely under backwater).

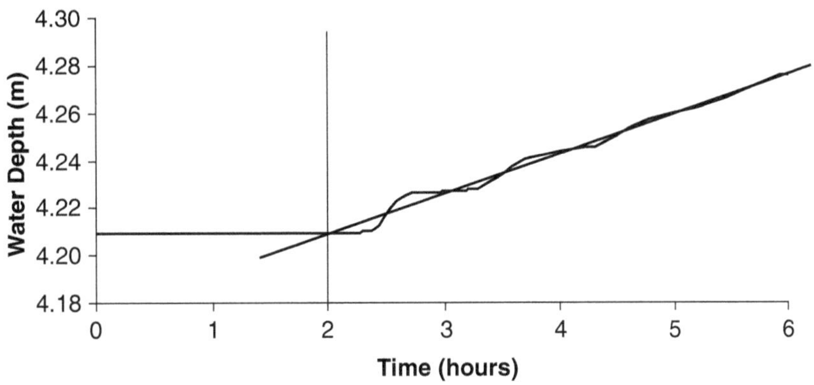

FIGURE 6.9 Response in water level at the downstream end of a pool for a step change in inflow (pool completely under backwater).

6.7 FUNDAMENTALS OF AUTOMATIC CONTROL SYSTEMS

6.7.1 Tuning

Tuning is the process of determining the various constants that are used in an automatic control algorithm. For a one-pool system the controller constants can be determined using the Ziegler–Nichols method (Franklin et al., 1994). The Ziegler–Nichols method was developed for SISO control systems. If this type of tuning procedure is applied to a multiple-pool system in which water levels at multiple locations are controlled through flow changes at multiple locations, then the Ziegler–Nichols tuned control loops will interact with each and degrade or destabilize the control system. Thus, a different tuning method, such as optimization or trial-and-error, must be used to determine the controller constants for a MIMO system. The tuning procedure is further complicated by the fact that controller constants that give good results at one flow rate may give poor results at another flow rate. In many situations, a different set of controller constants must be determined for different operating conditions of the canal.

6.7.2 Decoupling

Hydraulic interactions among the pools complicate automatic control of irrigation water delivery systems. A mathematical device, known as a decoupler, can be designed to cancel these interactions. In a system that is completely decoupled, each individual control loop has the ability to manipulate all the inputs in the system to control the outputs without upsetting the other system outputs. Theoretically this is very appealing, but tuning the decoupler is difficult. Open-channel flow is nonlinear and time-variant; thus, if the decoupler constants are chosen to be linear and constant, then perfect decoupling does not occur and, in some cases, this could lead to instabilities (Shinskey, 1996).

Schuurmans (1992) suggests two different types of decouplers for irrigation water delivery systems, which he named Decoupler I and Decoupler II. Decoupler I is a partial decoupler that attempts to prevent the control loop disturbances from moving in the upstream direction by passing a fraction of a gate's movement to all of the upstream gates. A block diagram of this decoupler for a three-pool canal under remote downstream feedback control is shown in Fig. 6.10. The blocks labeled "PI" represent a Proportional-Integral (PI) feedback controller, which is explained in detail later. The blocks labeled "K_{DI}" represent the transfer functions for Decoupler I. Schuurmans (1992) recommends using a static version of Decoupler I (i.e., all of the transfer functions for Decoupler I are constants).

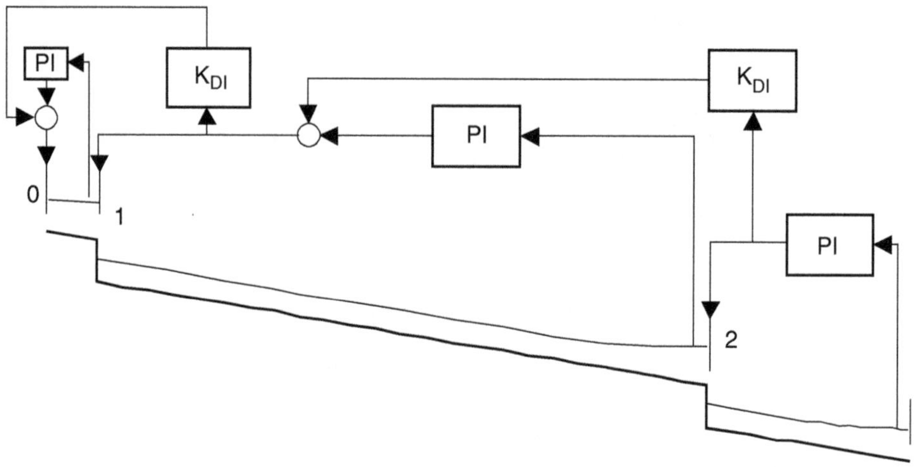

FIGURE 6.10 Block diagram for a remote downstream feedback controller with Decoupler I on a three-pool canal.

Decoupler II is also a partial decoupler and it attempts to prevent the control loop disturbances from moving in the downstream direction by passing a fraction of a gate's movements to the next gate downstream. The block diagram for a remote downstream feedback controller with Decoupler II for a three-pool canal is shown in Fig. 6.11. The blocks labeled "K_{DII}" represent the transfer functions for Decoupler II. Decoupler II can be represented both statically and dynamically (Schuurmans, 1992). In the static version, the transfer functions are constants; in the dynamic version, the transfer functions take into account the dynamics and the delay times of a given pool and the pool immediately downstream. Schuurmans (1992) recommends using the dynamic version of Decoupler II; however, this version makes the tuning process even more cumbersome.

Using Decouplers I and II together should theoretically prevent disturbances from moving into either the upstream or downstream pools (i.e., they form a complete decoupler). Although Schuurmans (1992) mentions this configuration, he did not study it.

6.7.3 State-Space Formulation

The state-space formulation is a very convenient method of expressing the dynamics of a linear system because it is much easier for computers to handle matrices than a large set of individual equations. It is also ideal for MIMO systems, and the ID model can be expressed very conveniently using the state-space formulation. Another advantage of the state-space formulation is that the internal dynamics of the system are still accessible. If transfer function methods are used instead of state-space theory, then the internal dynamics of the system become a black box that the user cannot analyze.

Continuous-Time Systems. Any linear response model can be expressed in continuous state-space form as

$$\dot{\mathbf{x}}(t) = \mathbf{A}\mathbf{x}(t) + \mathbf{B}\mathbf{u}(t) \qquad \mathbf{y}(t) = \mathbf{C}\mathbf{x}(t) + \mathbf{D}\mathbf{u}(t) \tag{6.16}$$

where $\mathbf{x}$ = vector of state variables
$\mathbf{u}$ = vector of inputs
$\mathbf{y}$ = vector of outputs
$\mathbf{A}$ = system matrix

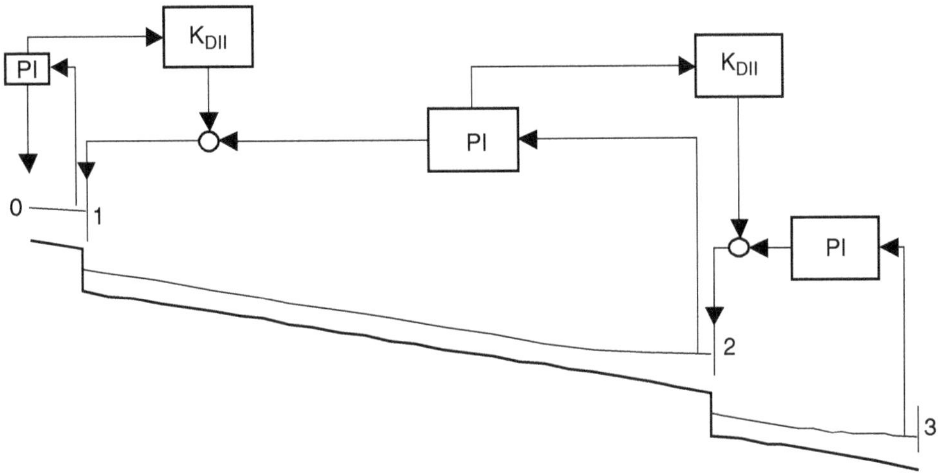

FIGURE 6.11 Block diagram for a remote downstream feedback controller with Decoupler II on a three-pool canal.

B = control input matrix
C = output matrix
D = direct feed matrix

The state variables of the system are the variables that are used to define the system. For example, in analyzing a spacecraft, the state variables may be the pitch, roll, and yaw of the spacecraft. For an irrigation water delivery system, one possible set of state variables consists of the water level errors and the previous control actions. The water level errors and the previous control actions give an overall picture (or describe the state) of the canal system. The set of parameters used as the state variables is not unique; the state vector can be defined as needed by the user.

Discrete-Time Systems. Control of an irrigation water delivery system as proposed here is a discrete-time problem. That is, control actions only take place at a given time interval (e.g., the gates are moved every 10 min). Thus, the continuous-time state-space model must be discretized. To do this, a sampling time Δt must be chosen. Typically, the sampling time is chosen to be the same as the time interval between control actions. It is common to convert a continuous-time system to a discrete-time system using zero-order-hold sampling. Zero-order-hold sampling assumes that the analog signals remain constant during the time between sample periods (Åström and Wittenmark, 1997). The state-space equations for a zero-order-hold sampled system are given by

$$\mathbf{x}(k+1) = \mathbf{\Phi}\mathbf{x}(k) + \mathbf{\Gamma}\mathbf{u}(k) \qquad \mathbf{y}(k) = \mathbf{C}\mathbf{x}(k) + \mathbf{D}\mathbf{u}(k) \tag{6.17}$$

where k is the index for the regulation time step (i.e., the time between k and $k + 1$ is Δt). Typically, the matrix $\mathbf{D} = \mathbf{0}$ so it will be dropped. The matrices $\mathbf{\Phi}$ and $\mathbf{\Gamma}$ can be determined from matrices **A** and **B** with

$$\mathbf{\Phi} = e^{\mathbf{A}\Delta t} \tag{6.18}$$

$$\mathbf{\Gamma} = \int_0^{\Delta t} e^{\mathbf{A}s}\, ds\mathbf{B} \tag{6.19}$$

Discrete-Time Systems with Delay Time. The discrete-time state-space model shown in Eq. (6.17) cannot be used as currently presented because of the delay times in the inputs. Continuous systems with delays are difficult to handle because the state vector becomes infinite in size (Åström and Wittenmark, 1997). Other methods, such as the Smith Predictor, must be incorporated to handle the delays. The infinite-dimensional problem does not exist for discrete-time systems. Because the control signal is constant between samplings, the state vector for a discrete-time system with delays becomes finite in size (Åström and Wittenmark, 1997). Assume that each pool in the system has the same delay time τ and sample time Δt. Then, the number of delay periods for each pool d is the integer rounded up from

$$d = \text{int}\left(\frac{\tau}{\Delta t}\right) \tag{6.20}$$

Using this definition, the discretized state-space equations for a system with delays can be expressed as

$$\mathbf{x}(k+1) = \mathbf{\Phi x}(k) + \mathbf{\Gamma}_0\mathbf{u}(k-(d-1)) + \mathbf{\Gamma}_1\mathbf{u}(k-d) \tag{6.21}$$

where $\Phi = e^{\mathbf{A}\Delta t}$

$$\mathbf{\Gamma}_0 = \int_0^{\Delta t - \tau_1} e^{\mathbf{A}s}\, ds\mathbf{B}$$

$$\mathbf{\Gamma}_1 = e^{\mathbf{A}(\Delta t - \tau_1)} \int_0^{\tau_1} \mathrm{e}^{\mathbf{A}s}\, ds\mathbf{B}$$

$$\tau_1 = \text{mod}(\tau/\Delta t)$$

The corresponding state-space description is (Åström and Wittenmark, 1997)

$$\begin{bmatrix} \mathbf{x}(k+1) \\ \mathbf{u}(k-(d-1)) \\ \vdots \\ \mathbf{u}(k-1) \\ \mathbf{u}(k) \end{bmatrix} = \begin{bmatrix} \mathbf{\Phi} & \mathbf{\Gamma}_1 & \mathbf{\Gamma}_0 & \cdots & \mathbf{0} \\ \mathbf{0} & \mathbf{0} & \mathbf{I} & \cdots & \mathbf{0} \\ \vdots & \vdots & \vdots & \ddots & \vdots \\ \mathbf{0} & \mathbf{0} & \mathbf{0} & \cdots & \mathbf{I} \\ \mathbf{0} & \mathbf{0} & \mathbf{0} & \cdots & \mathbf{0} \end{bmatrix} \begin{bmatrix} \mathbf{x}(k) \\ \mathbf{u}(k-d) \\ \vdots \\ \mathbf{u}(k-2) \\ \mathbf{u}(k-1) \end{bmatrix} + \begin{bmatrix} \mathbf{0} \\ \mathbf{0} \\ \vdots \\ \mathbf{0} \\ \mathbf{I} \end{bmatrix} \mathbf{u}(k) \tag{6.22}$$

$$\mathbf{y}(k) = \mathbf{Cx}(k)$$

Notice that there are now additional state variables that represent the previous control actions. There are many ways to calculate the matrices $\mathbf{\Phi}$, $\mathbf{\Gamma}_0$, and $\mathbf{\Gamma}_1$ based on the continuous-time model and the system matrices **A, B,** and **C.** The easiest way is to use the built-in numerical methods found in programs such as MATLAB. Typically, all the pools in the system do not have the same delay time. When this situation occurs, the delay times and the number of delay periods for each pool are expressed as a vector of values in Eq. (6.22).

Incremental Form. It is desirable to express the system in terms of changes in the state variables and inputs rather than in terms of their absolute value. Converting a state-space model to incremental form introduces additional state variables that represent the system's output. The incremental state-space model is given by (Prett and Garcia, 1988)

$$\begin{bmatrix} \Delta\mathbf{x}(k+1) \\ \mathbf{y}(k+1) \end{bmatrix} = \begin{bmatrix} \mathbf{\Phi} & \mathbf{0} \\ \mathbf{C\Phi} & \mathbf{I} \end{bmatrix} \begin{bmatrix} \Delta\mathbf{x}(k) \\ \mathbf{y}(k) \end{bmatrix} + \begin{bmatrix} \mathbf{\Gamma} \\ \mathbf{C\Gamma} \end{bmatrix} \Delta\mathbf{u}(k)$$

$$\mathbf{y}(k) = [\mathbf{0} \quad \mathbf{I}] \begin{bmatrix} \Delta\mathbf{x}(k) \\ \mathbf{y}(k) \end{bmatrix} \tag{6.23}$$

6.7.4 Classical Feedback Control Theory

Before delving into the details of optimal control, it is helpful to review the fundamentals of classical feedback control. Classical feedback control theory refers to the application of some form of a

Proportional-Integral-Derivative (PID) controller that keeps the control variables as close as possible to their setpoints despite the variety of inputs and disturbances that enter the system. The proportional (P) gain of these controllers is used to calculate the magnitude of the control actions that must be taken to drive the control variables back to their setpoints. It is called a proportional controller because its output is proportional to the measured error. Typically, integral (I) control is required to eliminate the steady-state offset error that is encountered by using a proportional controller alone. The derivative (D) term tells the controller when the steady-state offset is approaching zero, and it can be used to reduce overshoot and dampen out oscillations. Typically, derivative control is not used in the automatic feedback control of irrigation water delivery systems because derivative control does not perform well on systems with delay times (Shinskey, 1996).

Proportional-Integral Control. The proportional-integral (PI) feedback controller has been used extensively in process industries. It has also been applied to the downstream water level control of irrigation canals for which reasonable control has been obtained under only a limited range of operating conditions [see Pongput and Merkley (1997) and Rogers and Goussard (1998)]. In continuous form, the PI feedback controller can be expressed as (Åström and Wittenmark, 1997)

$$u(t) = K_p e(t) + \frac{K_p}{K_i} \int_0^t e(s)\, ds \tag{6.24}$$

where u = control action variable
K_p = proportional gain of the controller
K_i = integration time
e = difference between setpoint and measured output

In this case, u is either a gate position or a flow rate and e is the water level error. Trapezoidal integration is typically used to approximate the integral in Eq. (6.24). The discretized form of the PI controller can be expressed as

$$u(k) = K_p e(k) + \frac{K_p \Delta t}{K_i} \sum_{i=1}^{k} \frac{e(i) + e(i-1)}{2} \tag{6.25}$$

where k is the time index and Δt is the sampling time. To express the PI controller in incremental form, decrement k by 1 in Eq. (6.25) and subtract the result from Eq. (6.25).

$$\Delta u(k) = u(k) - u(k-1) = \left(K_p + \frac{K_p \Delta t}{2K_i}\right) e(k) + \left(-K_p + \frac{K_p \Delta t}{2K_i}\right) e(k-1) \tag{6.26}$$

Next, define K_{p1} as

$$K_{p1} \equiv K_p + \frac{K_p \Delta t}{2K_i} \tag{6.27}$$

and substitute that expression into Eq. (6.26).

$$\Delta u(k) = K_{p1}\, e(k) + \left(-K_{p1} + \frac{K_p \Delta t}{K_i}\right) e(k-1) \tag{6.28}$$

Now, define K_{i1} as

$$K_{i1} \equiv \frac{K_p \Delta t}{K_i} \tag{6.29}$$

and substitute that expression into Eq. (6.28). After some algebraic manipulation, Eq. (6.28) becomes

$$\Delta u(k) = K_{p1}\Delta e(k) + K_{i1}e(k-1) \tag{6.30}$$

where $\Delta e(k) = e(k) - e(k-1)$. In incremental form, the change in the control action variable (either a change in flow or gate position) is related to the change in water level error $\Delta e(k)$ and the previous water level error $e(k-1)$ through the constants K_{p1} and K_{i1}. Determining the appropriate values for the controller constants is done through the tuning process.

Control Law. In state-space controller design, the control law is defined as the feedback of a linear combination of the state variables (Franklin et al., 1994)

$$\mathbf{u}(k) = -\mathbf{K}\mathbf{x}(k) \tag{6.31}$$

where **u** is the control action vector, **x** is the state vector, and **K** is the gain matrix. A block diagram of this control law is shown in Fig. 6.12. This setup assumes there is no natural process disturbance or measurement noise in the system.

State-Space Representation of Classical Control Theory. The classical PI controller from Eq. (6.30) can be expressed in the control law form:

$$\mathbf{x}(k) = \begin{bmatrix} \Delta e(k) \\ e(k-1) \end{bmatrix}$$

and

$$-\mathbf{K} = \begin{bmatrix} K_{p1} & 0 \\ 0 & K_{i1} \end{bmatrix}$$

resulting in

$$\Delta u(k) = -\mathbf{K}\mathbf{x}(k) = K_{p1}\Delta e(k) + K_{i1}e(k-1)$$

A benefit of using the control law is that a wide variety of controllers can be designed by changing the form of the gain matrix. Consider a four-pool in-line canal system under remote downstream control with no delay in the first pool, two delay periods in the second pool, and one delay period in both the third and fourth pools. In this case, state and control action vectors are given by:

$$\mathbf{x}(k) = \begin{bmatrix} \Delta e_1(k) \\ \Delta e_2(k) \\ \Delta u_1(k-2) \\ \Delta u_1(k-1) \\ \Delta e_3(k) \\ \Delta u_2(k-1) \\ \Delta e_4(k) \\ \Delta u_3(k-1) \\ e_1(k-1) \\ e_2(k-1) \\ e_3(k-1) \\ e_4(k-1) \end{bmatrix}$$

and

$$\mathbf{u}(k) = \begin{bmatrix} \Delta u_0(k) \\ \Delta u_1(k) \\ \Delta u_2(k) \\ \Delta u_3(k) \end{bmatrix}$$

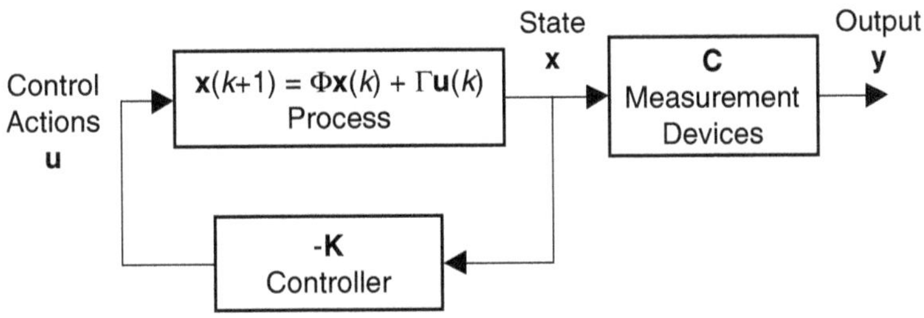

FIGURE 6.12 Block diagram for control-law design. (*Franklin et al., 1994.*)

The gain matrix derived for a simple PI controller in incremental state-space form would look like

$$
-\mathbf{K} = \begin{bmatrix} K_{P1} & 0 & 0 & 0 & 0 & 0 & 0 & 0 & K_{I1} & 0 & 0 & 0 \\ 0 & K_{P2} & 0 & 0 & 0 & 0 & 0 & 0 & 0 & K_{I2} & 0 & 0 \\ 0 & 0 & 0 & 0 & K_{P3} & 0 & 0 & 0 & 0 & 0 & K_{I3} & 0 \\ 0 & 0 & 0 & 0 & 0 & K_{P4} & 0 & 0 & 0 & 0 & 0 & K_{I4} \end{bmatrix} \tag{6.32}
$$

If a PI controller is desired that takes into account the delay time, then the control law for the second pool, for example, can be expressed as

$$
\Delta u_2(k) = K_{P2}\Delta e_2(k) + K_{-2L2}\Delta u_1(k-2) + K_{-1L2}\Delta u_1(k-1) + K_{I2}e_2(k-1)
$$

where K_{-2L2} and K_{-1L2} are the controller constants that take into the account the delay, or lag, from two and one previous time steps, respectively. This type of controller is referred to as a PIL (i.e., Proportional-Integral-Lag) controller. For a PIL controller, the gain matrix for the four-pool system would be

$$
-\mathbf{K} = \begin{bmatrix} K_{P1} & 0 & 0 & 0 & 0 & 0 & 0 & 0 & K_{I1} & 0 & 0 & 0 \\ 0 & K_{P2} & K_{-2L2} & K_{-1L2} & 0 & 0 & 0 & 0 & 0 & K_{I2} & 0 & 0 \\ 0 & 0 & 0 & 0 & K_{P3} & K_{-1L3} & 0 & 0 & 0 & 0 & K_{I3} & 0 \\ 0 & 0 & 0 & 0 & 0 & 0 & K_{P4} & K_{-1L4} & 0 & 0 & 0 & K_{I4} \end{bmatrix}
$$

If a PIL controller with Decoupler I is desired, then a portion of the control actions in a given pool is passed to all upstream pools. The notation for this type of controller is PIL^+. The + in this notation indicates that a portion of the control actions is passed to all upstream gates. Filling in the appropriate values in the gain matrix can accomplish this decoupling. For the four-pool system, the gain matrix for a PIL^+ controller would look like

$$
-\mathbf{K} = \begin{bmatrix}
K_{P1} & K_{DP21} & K_{-2DL21} & K_{-1DL21} & K_{DP31} & K_{-1DL31} & K_{DP41} & K_{-1DL41} & K_{I1} & K_{DI21} & K_{DI31} & K_{DI41} \\
0 & K_{P2} & K_{-2L2} & K_{-1L2} & K_{DP32} & K_{-1DL32} & K_{DP42} & K_{-1DL42} & 0 & K_{I2} & K_{DI32} & K_{DI42} \\
0 & 0 & 0 & 0 & K_{P3} & K_{-1L3} & K_{DP43} & K_{-1DL43} & 0 & 0 & K_{I3} & K_{DI43} \\
0 & 0 & 0 & 0 & 0 & 0 & K_{P4} & K_{-1L4} & 0 & 0 & 0 & K_{I4}
\end{bmatrix}
$$

where the additional coefficients are the Decoupler I coefficients. If Decouplers I and II are incorporated, the control actions are passed to one gate downstream as well as all the gates upstream. This type of controller is referred to as PIL_{-1}^{+} (the -1 indicates that the control actions are passed to one gate downstream) and the gain matrix looks like

$$
-\mathbf{K} = \begin{bmatrix}
K_{P1} & K_{DP21} & K_{-2DL21} & K_{-1DL21} & K_{DP31} & K_{-1DL31} & K_{DP41} & K_{-1DL41} & K_{I1} & K_{DI21} & K_{DI31} & K_{DI41} \\
K_{DP12} & K_{P2} & K_{-2L2} & K_{-1L2} & K_{DP32} & K_{-1DL32} & K_{DP42} & K_{-1DL42} & K_{DI12} & K_{I2} & K_{DI32} & K_{DI42} \\
0 & K_{DP23} & K_{-2DL23} & K_{-1DL23} & K_{P3} & K_{-1L3} & K_{DP43} & K_{-1DL43} & 0 & K_{DI23} & K_{I3} & K_{DI43} \\
0 & 0 & 0 & 0 & K_{DP34} & K_{-1DL34} & K_{P4} & K_{-1L4} & 0 & 0 & K_{DI34} & K_{I4}
\end{bmatrix}
$$

where the additional elements in the gain matrix are the Decoupler II coefficients. Other controller configurations can be obtained by manipulating the constants in the gain matrix. If the entire gain matrix is full of nonzero elements, then every control action is passed to every upstream and downstream gate. This controller is referred to as a fully centralized PI controller, or PIL_{-}^{+}. For more details on this terminology, refer to Clemmens and Schuurmans (2004).

6.8 OPTIMAL FEEDBACK CONTROL OF IRRIGATION WATER DELIVERY SYSTEMS

Two different feedback algorithms are discussed in this section. The first feedback controller uses Linear Quadratic Regulator (LQR) theory to tune the controller. The second feedback controller is Model Predictive Control (MPC). MPC is a multivariable control algorithm that was developed in the late 1970s and has been used extensively in the process control industry. Qin and Badgwell (1997) report over 2200 applications of MPC in a wide variety of areas such as refining, petrochemicals, mining, food processing, and aerospace.

To implement an overall control scheme for an irrigation water delivery system, these optimal feedback controllers should be used in tandem with anticipatory feedforward controllers and flow controllers.

6.8.1 Linear Quadratic Regulator

Optimization techniques have been applied to control problems in an attempt to simplify the tuning process. Assume that a linear response model, such as the ID model, is used as the underlying process model, and the control law in Eq. (6.31) is used to define the controller. This controller can be tuned by minimizing a quadratic performance indicator or objective function. This type of problem is referred to as a linear quadratic regulator problem. The term *linear* refers to the fact that a linear underlying process model is used, and the term *quadratic* refers to the fact that the objective function is a quadratic function.

Standard Formulation. Linear quadratic regulator theory can be used to determine the appropriate values for the nonzero coefficients of **K**. In LQR theory, the gain matrix **K** is chosen so that the control sequence *u* moves the state-space system Eq. (6.17) along its trajectory in such a way that the value of the objective function *J* is minimized (Lewis and Syrmos, 1995). The objective function *J* is a performance index of the controller and is a quadratic function of the state vector and the control actions. A typical objective function is given by

$$J = \sum_{k=0}^{\infty} \mathbf{x}(k)^T \mathbf{Q}\mathbf{x}(k) + \mathbf{u}(k)^T \mathbf{R}\mathbf{u}(k) \tag{6.33}$$

where **Q** and **R** are cost-weighting matrices on the state vector and control actions, respectively. Note that Eq. (6.33) is only one possible form of the objective function. For canal control problems, it makes sense to use only the current water level errors in the objective function. The water level errors are the output from the state-space model ($\mathbf{y}(k) = \mathbf{C}\mathbf{x}(k)$). Thus, the objective function can be modified as follows:

$$J = \sum_{k=0}^{\infty} \mathbf{y}(k)^T \mathbf{Q}\mathbf{y}(k) + \mathbf{u}(k)^T \mathbf{R}\mathbf{u}(k) \tag{6.34}$$

The problem with the objective function in Eq. (6.34) is that the summation goes to infinity. To circumvent this problem, a Lyapunov function can be used. A Lyapunov function is a function that decreases along the trajectories of the system. If the objective function can be expressed in terms of a Lyapunov function, then the value of the Lyapunov function decreases after each summation step. This implies that the values of **x** and **u** go to zero as *k* approaches infinity. Thus, the infinite sum in Eq. (6.34) converges to some finite value. If *J* can be minimized, the system is asymptotically stable because the values of **x** go to zero as *k* approaches infinity. To express the objective junction in terms of a Lyapunov function, first substitute Eq. (6.31) into Eq. (6.34). This yields the equation:

$$J = \sum_{k=0}^{\infty} \mathbf{y}(k)^T \mathbf{S}\mathbf{y}(k) \tag{6.35}$$

where

$$\mathbf{S} = \mathbf{Q} + \mathbf{K}^T\mathbf{R}\mathbf{K} \tag{6.36}$$

Equation (6.36) is a Lyapunov function. Using this Lyapunov function and assuming that the initial water level errors are given by **y**(0), Eq. (6.35) can be written as

$$J = \sum_{k=0}^{\infty} \mathbf{y}(k)^T \mathbf{S}\mathbf{y}(k) = \mathbf{y}(0)^T \mathbf{P}_L\mathbf{y}(0) \tag{6.37}$$

where $\mathbf{P}_L$ is the stabilizing matrix satisfying the Lyapunov function

$$\mathbf{\Phi}_c{}^T\mathbf{P}_L + \mathbf{P}_L\mathbf{\Phi}_c = -\mathbf{S} \tag{6.38}$$

where $\mathbf{\Phi}_c$ is the closed-loop system matrix given by

$$\mathbf{\Phi}_c = \mathbf{\Phi} - \mathbf{\Gamma}\mathbf{K} \tag{6.39}$$

Here it is assumed that $\mathbf{P}_L$ and **S** are symmetric, positive semidefinite matrices and all the eigenvalues of $\mathbf{\Phi}_c$ are in the open left-hand plane. The objective function can be further simplified to

$$J = \mathrm{tr}(\mathbf{P}_L\mathbf{Y}) \tag{6.40}$$

where tr is the trace function (the sum of the diagonal elements of a matrix) and

$$\mathbf{Y} = \mathbf{y}(0)\mathbf{y}(0)^T \tag{6.41}$$

The objective function is quadratic, so large water level errors and large inputs are penalized more than small ones. Also, the summation accumulates the water level errors over time; long-term errors are penalized more than short-term errors. Although the water level errors can be kept to a minimum if there are large input changes, large input changes are not desirable on a canal system. Thus, the input is included in the cost function to find a good balance between the conflicting goals of small water level errors and reasonably sized gate movements.

Initial Condition Problem. One of the problems with this formulation of the LQR problem is that the solution is dependent on the initial value of the state vector and different gain matrices are obtained for different initial conditions. In some cases, the initial state may be unknown. This problem can be avoided by minimizing the expected value of J, $E\{J\}$, instead of J (Lewis and Syrmos, 1995). When this is done, Eq. (6.41) is replaced by

$$\mathbf{Y} = E\{\mathbf{y}(0)\mathbf{y}(0)^T\} \tag{6.42}$$

It is usual to assume that nothing is known of $\mathbf{y}(0)$ except that it is uniformly distributed on a surface described by $\mathbf{Y}$ (Lewis and Syrmos, 1995). When trying to drive the arbitrarily nonzero states to zero, Lewis and Syrmos (1995) recommend that the initial states be uniformly distributed on a unit sphere (i.e., all the water level errors in $\mathbf{y}(0)$ are set equal to one).

Assumptions. There are several assumptions that are made in the formulation of the LQR problem. First, the cost-weighting matrix for the control actions $\mathbf{R}$ must be a positive definite matrix. Second, the cost-weighting matrix for the water level errors $\mathbf{Q}$ must be a positive semidefinite matrix. Also, it is assumed that $\mathbf{Q}$ can be factored as

$$\mathbf{Q} = \mathbf{H}^T\mathbf{H}$$

and the pair of matrices $(\mathbf{\Gamma}, \mathbf{H})$ is observable. If this assumption is not valid, there are unobservable modes that do not contribute anything to the objective function. These unobservable modes are not penalized and the optimal controller does not affect them. Finally, it is assumed that the pair of matrices $(\mathbf{\Gamma}, \mathbf{\Phi})$ is controllable.

Unconstrained Linear Quadratic Regulator Solution. If there is no restriction on the final value of $\mathbf{x}$ and on the elements in $\mathbf{K}$, the optimal gain matrix can be determined using the discrete-time algebraic Riccati equation (Åström and Wittenmark, 1997).

$$\mathbf{P} = \mathbf{\Phi}^T\mathbf{P}\mathbf{\Phi} - \mathbf{\Phi}^T\mathbf{P}\mathbf{\Gamma}(\mathbf{\Gamma}^T\mathbf{P}\mathbf{\Gamma} + \mathbf{R})^{-1}\mathbf{\Gamma}^T\mathbf{P}\mathbf{\Phi} + \mathbf{Q} \tag{6.43}$$

where $\mathbf{P}$ is a symmetric matrix, called the stabilizing matrix, that satisfies the Riccati equation. The optimal gain, called the Kalman gain $\mathbf{K}_k$, is then determined from (Lewis and Syrmos, 1995).

$$\mathbf{K}_k = (\mathbf{\Gamma}^T\mathbf{P}\mathbf{\Gamma} + \mathbf{R})^{-1}\mathbf{\Gamma}^T\mathbf{P}\mathbf{\Phi} \tag{6.44}$$

The value of $\mathbf{P}$ and $\mathbf{K}_k$ can be determined numerically using standard programs such as MATLAB. Note that the Kalman gain depends only on the stabilizing, system, and weighting matrices; it does not depend on the state vector. Thus, it can be computed before the control is ever applied to the system (Lewis and Syrmos, 1995).

Constrained Linear Quadratic Regulator Solution. If there are constraints imposed on the gain matrix $\mathbf{K}$ (i.e., some of the elements in $\mathbf{K}$ are set to zero), then the Riccati equation cannot be used to determine the elements of the gain matrix. Instead, a constrained nonlinear optimization technique must be used. There is a wide variety of techniques that can be used to solve a constrained nonlinear optimization problem. For example, the function FMINCON in MATLAB can be used. FMINCON uses a Sequential Quadratic Programming (SQP) method to determine the value of $\mathbf{K}$ that minimizes the objective function subject to the constraint that certain elements in $\mathbf{K}$ must remain zero. By forcing some of the elements in $\mathbf{K}$ to be zero, other controllers such as the PIL, PIL$^+$, or PIL$^+_{-1}$ controllers mentioned earlier can be designed [see Clemmens and Schuurmans (2004) for more details]. Any controller tuned using nonlinear optimization techniques is referred to as centrally tuned. Thus, a centrally tuned PI controller refers to a classical PI controller that was tuned using optimization techniques instead of a series of locally tuned PI controllers.

Tuning. The tuning procedure is the process of determining the gain matrix **K**. LQR theory can be used to automatically determine these elements; however, this does not imply that there is no additional tuning to be done. The cost-weighting matrices **Q** and **R** must still be found through trial-and-error. Values within **Q** and **R** provide a tradeoff between minimizing water level errors and minimizing check flow changes. Because only the relative size of **Q** and **R** are important, the **Q** matrix can be set equal to **I** and the **R** matrix can be adjusted to express the relative tradeoff between water level accuracy and amount of gate movements. Using **I** as the **Q** matrix implies that water level deviations are given the same penalty weight in all the pools. The cost weighting matrix **R** was weighted as a function of the relative capacity of each pool because it was assumed that control actions in pools with low capacity have more of an effect on the system than control actions in pools with high capacity. For example, a 1 m^3/s flow change for a gate with a capacity of 2 m^3/s is much more extreme than for a 10 m^3/s capacity gate. The first element of **R**, R_1, represents the control action penalty for the first gate, and this element is set as part of the trial-and-error tuning process. The remaining values within each **R** matrix are scaled according to the relative gate capacity. For example, if R_1 equals 3 and check 2 has 80 percent of the capacity of check 1, then the second element in **R**, R_2, is equal to $3/(0.8)^2$ or 4.7.

6.8.2 Model Predictive Control

The term MPC does not refer to a specific control strategy. Instead, it refers to a broad range of control methods that explicitly use a simplified process model of the real system to obtain the control actions by minimizing an objective function (Camacho and Bordons, 1999). MPC has three basic components, which can be summarized as follows (Camacho and Bordons, 1999):

1. A process model is used to predict the system output for some time into the future $\hat{\mathbf{y}}(k + i \mid k)$. The distance that is predicted into the future is called the prediction horizon p. The output predictions have both a free response and a forced response (Clarke, 1994). The free response is the expected behavior of the system assuming no future control actions. The forced response is the additional component of the process output that corresponds to the unknown future control actions $\Delta\mathbf{u}(k + i \mid k)$. The forced response is considered over a time horizon called the control horizon m, and the free response is considered over the entire prediction horizon p. The control horizon is less than or equal to the prediction horizon. While in the control horizon, the process model is used to obtain

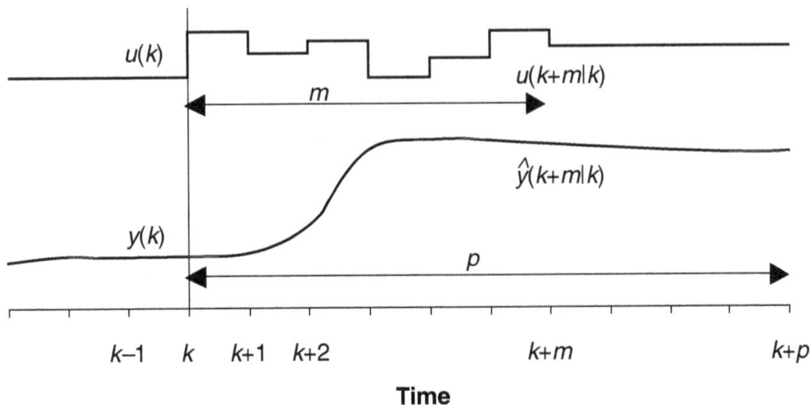

FIGURE 6.13 MPC prediction strategy.

output predictions based on both the free and forced responses. After the control horizon is passed, the remaining output predictions are based only on the free response of the system. This prediction strategy is shown in Fig. 6.13.

2. An objective function, which is typically the sum of some combination of the water level errors and control actions over the prediction horizon, is minimized by adjusting the future unknown control actions. This optimization problem is subject to the many constraints that are imposed on the system.
3. Once the sequence of future control actions that minimizes the desired objective function is determined, only the first set of control actions is implemented on the system. The system is then updated and the process is repeated. This overall strategy is known as the receding horizon strategy. A flowchart of the basic MPC structure can be seen in Fig. 6.14.

Camacho and Bordons (1999) compare the MPC control algorithm to the strategy used to drive a car. The driver looks down the road and determines the desired trajectory for the vehicle. The driver determines the desired route and the series of control actions (i.e., steering, accelerator, brakes, etc.) needed to move the vehicle along the desired route. The driver implements the first of these control actions on the vehicle. Next, the driver reassesses the situation and determines a whole new sequence of control actions based on what has happened since the last control action was implemented. This process is repeated as the driver moves along the desired path. Thus, the driver is constantly predicting into the future and calculating a new series of control actions as the vehicle moves down the road.

Benefits of Model Predictive Control. The main benefits to using MPC are that it is a multivariable controller, it can explicitly take into account constraints that are imposed on the system, and it has the ability to route known disturbances through the system (i.e., an anticipatory feedforward routine can be built directly into MPC).

Multivariable Aspects. Traditional local control techniques do not work very well on multivariable systems if the variables are coupled (i.e., changing one process variable affects other process variables in the system). Controller performance can improve if a decoupler is implemented [see Schuurmans (1992) for more information on decouplers]. An irrigation water delivery system must be considered a coupled system because changing the gate position in one pool affects the water levels in the other pools. With MPC, the hydraulic coupling between pools is automatically taken into account in the output predictions. These predictions, which are based on the process model used to describe the real system, include the effect that moving a gate in one pool has on the water levels in the other pools.

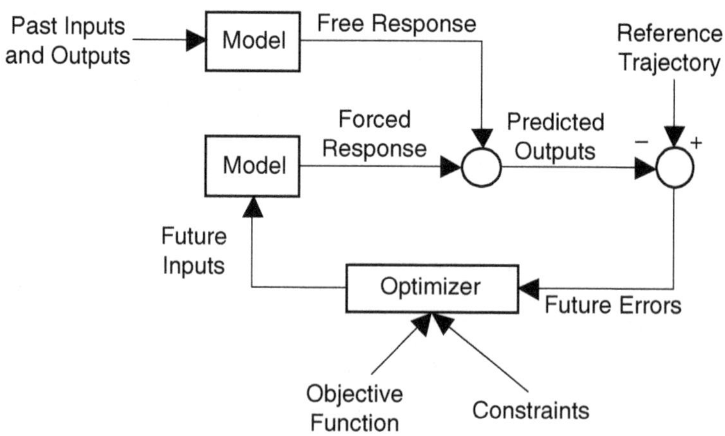

FIGURE 6.14 Basic structure of MPC. [*Adapted from Clarke (1994).*]

Explicit Constraint Handling. MPC has the ability to explicitly account for constraints that are placed on the system. Constraints can be in the form of either input constraints or output constraints. Input constraints are constraints that are imposed on the actuators of the system (i.e., the check gates). Output constraints are imposed on the output of the underlying process model (i.e., the water level errors). For an irrigation water delivery system, typical input constraints are limits on the magnitude of the check gate openings or limits on the rate at which the check gates are moved. Typical output constraints are limits on the magnitude and rate of change of the water level errors in the pools. These constraints can also be changed on the fly as the flow situations in the irrigation water delivery system change.

Feedforward Routine. Because MPC predicts the system response a certain distance into the future, known disturbances can be placed at the appropriate places in time along the prediction horizon. As MPC predicts into the future, it "sees" these known disturbances coming and begins correcting for them before they are actually encountered.

Drawbacks of Model Predictive Control. MPC is not a panacea and, as such, there are also many drawbacks to this control algorithm.

On-Line Optimization. MPC must solve an optimization problem at every controller time step. Solving an optimization problem on-line may present difficulties if the optimization problem is complex because it requires a significant amount of computing power. In these situations, the computational power needed to implement the MPC controller becomes a serious drawback.

Feasibility. If the constraints are too restrictive on the optimization problem, then the problem may become infeasible. This difficulty is typically encountered when output constraints are imposed on the system. Although imposing constraints on the water level errors in the pools sounds like a good idea, this type of constraint usually makes the optimization problem infeasible. One way to overcome this drawback is to formulate the output constraints as soft constraints (i.e., a large penalty is added to the objective function every time the output constraints are violated). In this case, the infeasibility issues are avoided, and MPC tries to minimize the amount of time that the output constraints are violated.

Minimum Gate Movement Constraints. Minimum gate movement constraints can be thought of as a deadzone on the gates (i.e., if the computed gate movements are less than a certain amount, then the gate is not moved). This type of constraint is defined as a deadzone type II by Chow and Clarke (1994). The feasible regions for deadzone type II constraints are nonconvex, which makes the optimization problem difficult to solve (Chow and Clarke, 1994). In addition, persistent oscillations are inevitable in a closed-loop system that features deadzone type II constraints (Chow and Clarke, 1994).

State-Space Formulation of Model Predictive Control. For this work, a state-space formulation of MPC was utilized. A summary of this formulation of MPC follows. Most of the equations in this section are based on the work done by Lee et al. (1994) and Maciejowski (1998).

Process Model. The ID model can be placed into state-space form. To take advantage of the feedforward routine built into MPC, the state-space formulation presented earlier must be modified to include a vector of known disturbances $\Delta\mathbf{d}(k)$. The incremental, discrete-time state-space representation of the ID model including known disturbances is given by

$$\begin{bmatrix}\Delta\mathbf{x}(k+1)\\ \mathbf{y}(k+1)\end{bmatrix} = \begin{bmatrix}\mathbf{\Phi} & \mathbf{0}\\ \mathbf{C\Phi} & \mathbf{I}\end{bmatrix}\begin{bmatrix}\Delta\mathbf{x}(k)\\ \mathbf{y}(k)\end{bmatrix} + \begin{bmatrix}\mathbf{\Gamma} & \mathbf{\Gamma}_{\mathbf{d}}\\ \mathbf{C\Gamma} & \mathbf{C\Gamma}_{\mathbf{d}}\end{bmatrix}\begin{bmatrix}\Delta\mathbf{u}(k)\\ \Delta\mathbf{d}(k)\end{bmatrix} \tag{6.45}$$

$$\mathbf{y}(k) = [\mathbf{0}\ \mathbf{I}]\begin{bmatrix}\Delta\mathbf{x}(k)\\ \mathbf{y}(k)]\end{bmatrix} \tag{6.46}$$

where $\Delta\mathbf{x}(k)$ = state vector
$\mathbf{y}(k)$ = output vector
$\Delta\mathbf{u}(k)$ = control action vector

$\Delta\mathbf{d}(k)$ = known disturbance vector
$\mathbf{\Phi}$ = system matrix
$\mathbf{\Gamma}$ = control input matrix
$\mathbf{\Gamma}_d$ = known disturbance input matrix
$\mathbf{C}$ = output matrix
k = time index

The state vector $\Delta\mathbf{x}(k)$ consists of the changes in water level errors and the previous control actions. The output vector $\mathbf{y}(k)$ consists of the water level errors. The control action vector $\Delta\mathbf{u}(k)$ consists of changes in flow rate at the check structures, and the known disturbance vector $\Delta\mathbf{d}(k)$ consists of known changes in flow rate at the turnouts.

Prediction. The first step in MPC is to make a prediction into the future using the process model. In the prediction process, the predicted output of the system, $\hat{\mathbf{y}}(k+i \mid k)$, is determined from the current state vector $\Delta\mathbf{x}(k)$ and the future control actions $\Delta\mathbf{u}(k+i \mid k)$. Because the ID model is used as the underlying process model, no state estimation techniques are needed for the predictions. The predicted values for the state and output vectors one time step into the future are expressed as

$$\Delta\hat{\mathbf{x}}(k+1 \mid k) = \mathbf{\Phi}\,\Delta\mathbf{x}(k) + \mathbf{\Gamma}\,\Delta\mathbf{u}(k) + \mathbf{\Gamma}_d\,\Delta\mathbf{d}(k) \tag{6.47}$$

$$\hat{\mathbf{y}}(k+1 \mid) k) = \mathbf{C\Phi}\,\Delta\mathbf{x}(k) + \mathbf{I}\mathbf{y}(k) + \mathbf{C}(\mathbf{\Gamma}\,\Delta\mathbf{u}(k) + \mathbf{\Gamma}_d\,\Delta\mathbf{d}(k)) \tag{6.48}$$

The prediction for the state vector two time steps into the future is

$$\Delta\hat{\mathbf{x}}(k+2 \mid k) = \mathbf{\Phi}\,\Delta\hat{\mathbf{x}}(k+1 \mid k) + \mathbf{\Gamma}\,\Delta\mathbf{u}(k+1) + \mathbf{\Gamma}_d\,\Delta\mathbf{d}(k+1) \tag{6.49}$$

Inserting Eq. (6.47) into Eq. (6.49) yields

$$\Delta\hat{\mathbf{x}}(k+2 \mid k) = \mathbf{\Phi}^2\,\Delta\mathbf{x}(k) + \mathbf{\Phi}(\mathbf{\Gamma}\,\Delta\mathbf{u}(k) + \mathbf{\Gamma}_d\,\Delta\mathbf{d}(k)) + \mathbf{\Gamma}\,\Delta\mathbf{u}(k+1) + \mathbf{\Gamma}_d\,\Delta\mathbf{d}(k+1) \tag{6.50}$$

Similarly, the prediction for the output vector two time steps into the future is

$$\hat{\mathbf{y}}(k+2 \mid k) = \mathbf{C\Phi}\,\Delta\hat{\mathbf{x}}(k+1 \mid k) + \mathbf{I}\hat{\mathbf{y}}(k+1 \mid k) + \mathbf{C}(\mathbf{\Gamma}\,\Delta\mathbf{u}(k+1) + \mathbf{\Gamma}_d\,\Delta\mathbf{d}(k+1)) \tag{6.51}$$

Substituting Eqs. (6.47) and (6.48) into Eq. (6.51) yields

$$\hat{\mathbf{y}}(k+2 \mid k) = \mathbf{C}(\mathbf{\Phi} + \mathbf{\Phi}^2)\,\Delta\mathbf{x}(k) + \mathbf{I}\mathbf{y}(k) + \mathbf{C}(\mathbf{\Gamma} + \mathbf{\Phi\Gamma})\,\Delta\mathbf{u}(k) + \mathbf{C\Gamma}\,\Delta\mathbf{u}(k+1) \tag{6.52}$$

This recursive process continues until the end of the control horizon m is reached. At that point, the output prediction is expressed as

$$\begin{aligned}\hat{\mathbf{y}}(k+m \mid k) = {} & \mathbf{C}(\mathbf{\Phi} + \mathbf{\Phi}^2 + \cdots + \mathbf{\Phi}^m)\,\Delta\mathbf{x}(k) + \mathbf{I}\mathbf{y}(k) + \mathbf{C}(\mathbf{\Gamma} + \mathbf{\Phi\Gamma} + \cdots + \mathbf{\Phi}^{m-1}\mathbf{\Gamma})\,\Delta\mathbf{u}(k) \\ & + \mathbf{C}(\mathbf{\Gamma} + \mathbf{\Phi\Gamma} + \cdots + \mathbf{\Phi}^{m-2}\mathbf{\Gamma})\,\Delta\mathbf{u}(k+1) + \cdots + \mathbf{C\Gamma}\,\Delta\mathbf{u}(k+m-1) \\ & + \mathbf{C}(\mathbf{\Gamma}_d + \mathbf{\Phi\Gamma}_d + \cdots + \mathbf{\Phi}^{m-1}\mathbf{\Gamma}_d)\,\Delta\mathbf{d}(k) + \mathbf{C}(\mathbf{\Gamma}_d + \mathbf{\Phi\Gamma}_d + \cdots + \mathbf{\Phi}^{m-2}\,\mathbf{\Gamma}_d)\,\Delta\mathbf{d}(k+1) \\ & + \cdots + \mathbf{C\Gamma}_d\,\Delta\mathbf{d}(k+m-1)\end{aligned} \tag{6.53}$$

After the control horizon has passed, the remaining output predictions are based on the free response only (i.e., $\Delta\mathbf{u}(k+i) = 0$ and $\Delta\mathbf{d}(k+i) = 0$ for $i \geq m$). When the end of the prediction horizon p is reached, the predicted output is expressed as

$$
\begin{aligned}
\hat{\mathbf{y}}(k+p\mid k) = {} & \mathbf{C}(\mathbf{\Phi} + \mathbf{\Phi}^2 + \cdots + \mathbf{\Phi}^p)\,\Delta\mathbf{x}(k) + \mathbf{I}\mathbf{y}(k) + \mathbf{C}(\mathbf{\Gamma} + \mathbf{\Phi\Gamma} + \cdots + \mathbf{\Phi}^{p-1}\mathbf{\Gamma})\,\Delta\mathbf{u}(k) \\
& + \mathbf{C}\,(\mathbf{\Gamma} + \mathbf{\Phi\Gamma} + \cdots + \mathbf{\Phi}^{p-2}\mathbf{\Gamma})\,\Delta\mathbf{u}(k+1) + \cdots \\
& + \mathbf{C}(\mathbf{\Gamma} + \mathbf{\Phi\Gamma} + \cdots + \mathbf{\Phi}^{p-m-1}\,\mathbf{\Gamma})\,\Delta\mathbf{u}(k+m-1) \\
& + \mathbf{C}\,(\mathbf{\Gamma}_{\mathbf{d}} + \mathbf{\Phi\Gamma}_{\mathbf{d}} + \cdots + \mathbf{\Phi}^{p-1}\mathbf{\Gamma}_{\mathbf{d}})\,\Delta\mathbf{d}(k) \\
& + \mathbf{C}(\mathbf{\Gamma}_{\mathbf{d}} + \mathbf{\Phi\Gamma}_{\mathbf{d}} + \cdots + \mathbf{\Phi}^{p-2}\,\mathbf{\Gamma}_{\mathbf{d}})\,\Delta\mathbf{d}(k+1) \\
& + \cdots + \mathbf{C}(\mathbf{\Gamma}_{\mathbf{d}} + \mathbf{\Phi\Gamma}_{\mathbf{d}} + \cdots + \mathbf{\Phi}^{p-m-1}\,\mathbf{\Gamma}_{\mathbf{d}})\,\Delta\mathbf{d}(k+m-1)
\end{aligned}
$$

These output predictions can be put into matrix form by defining the following matrices.

$$
\mathbf{Y}(k+1\mid k) = \begin{bmatrix} \hat{\mathbf{y}}(k+1\mid k) \\ \hat{\mathbf{y}}(k+2\mid k) \\ \vdots \\ \hat{\mathbf{y}}(k+p\mid k) \end{bmatrix} \tag{6.55}
$$

$$
\Delta\mathbf{U}(k) = \begin{bmatrix} \Delta\mathbf{u}(k) \\ \Delta\mathbf{u}(k+1) \\ \vdots \\ \Delta\mathbf{u}(k+m-1 \end{bmatrix} \tag{6.56}
$$

$$
\Delta\mathbf{D}(k) = \begin{bmatrix} \Delta\mathbf{d}(k) \\ \Delta\mathbf{d}(k+1) \\ \vdots \\ \Delta\mathbf{d}(k+m-1 \end{bmatrix} \tag{6.57}
$$

$$
\mathbf{1} = \left.\begin{bmatrix} \mathbf{I} \\ \mathbf{I} \\ \vdots \\ \mathbf{I} \end{bmatrix}\right\} p \tag{6.58}
$$

$$
\mathbf{S_x} = \begin{bmatrix} \mathbf{C\Phi} \\ \mathbf{C}\left(\mathbf{\Phi} + \mathbf{\Phi}^2\right) \\ \vdots \\ \sum_{j=1}^{p} \mathbf{C\Phi}^j \end{bmatrix} \tag{6.59}
$$

$$
\mathbf{S_u} = \overbrace{\begin{bmatrix} \mathbf{C\Gamma} & 0 & 0 & \cdots & 0 \\ \mathbf{C}\left(\mathbf{\Gamma} + \mathbf{\Phi\Gamma}\right) & \mathbf{C\Gamma} & 0 & \cdots & 0 \\ \vdots & \vdots & \vdots & \ddots & \vdots \\ \sum_{j=1}^{p} \mathbf{C\Phi}^{j-1}\mathbf{\Gamma} & \cdots & \cdots & \cdots & \sum_{j=1}^{p-m} \mathbf{C\Phi}^{j-1}\mathbf{\Gamma} \end{bmatrix}}^{m} \tag{6.60}
$$

$$\mathbf{S_u} = \left[\overbrace{\begin{matrix} \mathbf{C\Gamma_d} & 0 & 0 & \cdots & 0 \\ \mathbf{C(\Gamma_d + \Phi\Gamma_d)} & \mathbf{C\Gamma_d} & 0 & \cdots & 0 \\ \vdots & \vdots & \vdots & \ddots & \vdots \\ \sum_{j=1}^{p} \mathbf{C\Phi}^{j-1}\mathbf{\Gamma_d} & \cdots & \cdots & \cdots & \sum_{j=1}^{p-m} \mathbf{C\Phi}^{j-1}\mathbf{\Gamma_d} \end{matrix}}^{m} \right] \tag{6.61}$$

Using these matrix definitions, the output predictions are written as

$$\mathbf{Y}(k+1 \mid k) = \mathbf{S_x}\,\Delta\mathbf{x}(k) + \mathbf{1}\cdot\mathbf{y}(k) + \mathbf{S_u}\,\Delta\mathbf{U}(k) + \mathbf{S_d}\,\Delta\mathbf{D}(k) \tag{6.62}$$

Thus, Eq. (6.62) is an expression for the output of the system as a function of the past and current control actions [the first two terms on the right-hand side of Eq. (6.62)] as well as the unknown future control actions and the known disturbances. The matrices $\mathbf{S_x}$, $\mathbf{S_u}$, and $\mathbf{S_d}$ are expressed only in terms of the state-space model (i.e., $\mathbf{\Phi}$, $\mathbf{\Gamma}$, $\mathbf{\Gamma_d}$, and $\mathbf{C}$). These matrices can be calculated before the MPC controller performs any type of optimization.

Optimization Problem. Once Eq. (6.62) is used to predict the future output of the system, the best possible set of future changes in control actions is determined by choosing the set that minimizes an objective function. A typical objective function is the sum of some combination of the water level errors and control actions over the prediction and control horizon, respectively. For this study, the objective function J is expressed as

$$J = \sum_{j=1}^{p} \left\| \mathbf{Q}_j \left[\hat{\mathbf{y}}(k+j \mid k) - \mathbf{y_r}(k+j)\right] \right\|^2 + \sum_{j=1}^{m} \left\| \mathbf{R}_j \left[\Delta\mathbf{u}(k+j-1)\right] \right\|^2 \tag{6.63}$$

where J = objective function
p = prediction horizon
m = control horizon
$\hat{\mathbf{y}}$ = predicted output
$\mathbf{y_r}$ = reference trajectory
$\Delta\mathbf{u}$ = unknown future control actions
$\mathbf{Q}_j$ = jth cost weighting matrix on the water level errors
$\mathbf{R}_j$ = jth cost weighting matrix on the control actions
k = time index

In this form of the objective function, the first term on the right-hand side is designed to keep the water levels at the desired setpoint, and the second term is designed to suppress the number of gate movements that are made.

The objective function can be put into matrix form by using the above definitions of $\mathbf{Y}(k+1 \mid k)$ and $\Delta\mathbf{U}(k)$ as well as the following definitions.

$$\mathbf{Y_r}(k+1) = \begin{bmatrix} \mathbf{y_r}(k+1) \\ \mathbf{y_r}(k+2) \\ \vdots \\ \mathbf{y_r}(k+p) \end{bmatrix} \tag{6.64}$$

$$\mathbf{Q} = \text{diag}\{\mathbf{Q}_1, \mathbf{Q}_2, \cdots, \mathbf{Q}_p\} \tag{6.65}$$

$$\mathbf{R} = \text{diag}\{\mathbf{R}_1, \mathbf{R}_2, \cdots, \mathbf{R}_m\} \tag{6.66}$$

Using these definitions, the objective function is written as

$$J = \|\mathbf{Q}[\mathbf{Y}(k+1 \mid k) - \mathbf{Y}_r(k+1)]\|^2 + \|\mathbf{R}\,\Delta\mathbf{U}(k)\|^2 \tag{6.67}$$

The problem can be summarized as minimizing the objective function J by adjusting the future control actions $\Delta\mathbf{U}(k)$.

FORMULATION OF THE OPTIMIZATION PROBLEM AS A QUADRATIC PROGRAMMING PROBLEM. There are many convenient methods to solve this optimization problem if it is expressed as a quadratic programming (QP) problem. In a QP problem, the objective function takes on the following form

$$J = \frac{1}{2}\,\Delta\mathbf{U}(k)^T\mathbf{H}_u\,\Delta\mathbf{U}(k) - \mathbf{G}(k+1 \mid k)^T\,\Delta\mathbf{U}(k) \tag{6.68}$$

where $\mathbf{H}_u$ = Hessian of the QP
$\mathbf{G}(k+1 \mid k)$ = gradient vector of the QP
T = transpose operator

To convert Eq. (6.67) to the form of the QP problem in Eq. (6.68), substitute the prediction equation (6.62) into (6.67).

$$J = \|\mathbf{Q}[\mathbf{S}_x\,\Delta\mathbf{x}(k) + \mathbf{1}\cdot\mathbf{y}(k) + \mathbf{S}_u\,\Delta\mathbf{U}(k) + \mathbf{S}_d\,\Delta\mathbf{D}(k) - \mathbf{Y}_r\,(k+1)]\|^2 + \|\mathbf{R}\,\Delta\mathbf{U}(k)\|^2 \tag{6.69}$$

At this point, it is convenient to define the error projection vector, $\mathbf{E}(k+1 \mid k)$. The error projection vector is the difference between the future target trajectory and the free response of the system (Maciejowski, 1998).

$$\mathbf{E}(k+1 \mid k) = \mathbf{Y}_r\,(k+1) - \mathbf{S}_x\,\Delta\mathbf{x}(k) - \mathbf{1}\cdot\mathbf{y}(k) - \mathbf{S}_d\,\Delta\mathbf{D}(k) \tag{6.70}$$

Substituting this definition into Eq. (6.69) yields

$$\begin{aligned} J &= \|\mathbf{Q}[\mathbf{S}_u\,\Delta\mathbf{U}(k) - \mathbf{E}(k+1 \mid k)]\|^2 + \|\mathbf{R}\,\Delta\mathbf{U}(k)\|^2 \\ &= \Delta\mathbf{U}^T(k)((\mathbf{QS}_u)^T\,\mathbf{QS}_u + \mathbf{R}^T\mathbf{R})\,\Delta\mathbf{U}(k) - 2(\mathbf{QS}_u)^T\,\mathbf{QE}(k+1 \mid k)\,\Delta\mathbf{U}(k) \\ &\quad + (\mathbf{QE}\,(k+1 \mid k))^T\,\mathbf{QE}(k+1 \mid k) \end{aligned} \tag{6.71}$$

The last term on the right-hand side of the above equation can be dropped out of the optimization problem because it is not a function of the unknown future control actions $\Delta\mathbf{U}(k)$. The objective function is then defined as

$$J = \Delta\mathbf{U}^T(k)((\mathbf{QS}_u)^T\,\mathbf{QS}_u + \mathbf{R}^T\mathbf{R})\,\Delta\mathbf{U}(k) - 2(\mathbf{QS}_u)^T\,\mathbf{QE}\,(k+1 \mid k)\,\Delta\mathbf{U}(k) \tag{6.72}$$

If Eq. (6.72) is divided by two, then it is equivalent to Eq. (6.68) if $\mathbf{H}_u$ and $\mathbf{G}(k+1 \mid k)$ are defined as follows.

$$\mathbf{H}_u = (\mathbf{QS}_u)^T\,\mathbf{QS}_u + \mathbf{R}^T\mathbf{R} \tag{6.73}$$

$$\mathbf{G}\,(k+1 \mid k) = (\mathbf{QS}_u)^T\,\mathbf{QE}\,(k+1 \mid k) \tag{6.74}$$

UNCONSTRAINED MODEL PREDICTIVE CONTROL. If there are no constraints on the problem, then an analytical solution can be found that minimizes the objective function. To find this solution, take the derivative of Eq. (6.68) (with $\mathbf{H}_u$ and $\mathbf{G}(k+1 \mid k)$ given by Eqs. (6.73) and (6.74), respectively), set it equal to zero, and solve for $\Delta\mathbf{U}(k)$.

$$\frac{\partial J}{\partial(\Delta\mathbf{U}(k))} = \mathbf{H}_u\,\Delta\mathbf{U}(k) - \mathbf{G}\,(k+1 \mid k) = 0 \tag{6.75}$$

Solving this equation yields the optimal set of control moves $\mathbf{\Delta U}(k)_{\mathrm{unc}}$.

$$\mathbf{\Delta U}(k)_{\mathrm{unc}} = \mathbf{H}_{\mathrm{u}}^{-1}\mathbf{G}\,(k+1 \mid k) = ((\mathbf{QS}_{\mathrm{u}})^T\,\mathbf{QS}_{\mathrm{u}} + \mathbf{R}^T\mathbf{R})^{-1}\,((\mathbf{QS}_{\mathrm{u}})^T\,\mathbf{QE}(k+1 \mid k)) \tag{6.76}$$

Note that

$$\mathbf{\Delta U}(k)_{\mathrm{unc}} = \begin{bmatrix} \mathbf{\Delta u}(k)_{\mathrm{unc}} \\ \mathbf{\Delta u}(k+1)_{\mathrm{unc}} \\ \vdots \\ \mathbf{\Delta u}(k+m-1)_{\mathrm{unc}} \end{bmatrix} \tag{6.77}$$

INFINITE HORIZON MODEL PREDICTIVE CONTROL. If the prediction horizon and the control horizon both approach infinity and if there are no constraints on the system, the MPC problem becomes a linear quadratic regulator problem. The Riccati equation can be used to determine the unknown future control actions in this case. Infinite horizon MPC is not possible if there are constraints imposed on the system.

CONSTRAINED PROBLEM. The more interesting problem comes when there are constraints present in the MPC problem. Some typical constraints that exist in an irrigation system are as follows.

1. The check gates are not allowed to shut all the way
2. The check gates are not allowed to come out of the water
3. There are minimum gate movement constraints
4. The water level in a given pool cannot rise above or fall below a certain elevation

Some of these constraints (1 and 2) are easy to incorporate into the MPC structure. Others (3 and 4) are more difficult. As mentioned earlier, constraint 3 is difficult to implement because it has a non-convex solution region and constraint 4 is an output constraint that can cause infeasibilities in the optimization problem.

A QP with constraints can be expressed as

$$\min_{\mathbf{\Delta U}(k)} \frac{1}{2}\;\mathbf{\Delta U}(k)^T\,\mathbf{H}_{\mathrm{u}}\,\mathbf{\Delta U}(k) - \mathbf{G}(k+1 \mid k)^T\,\mathbf{\Delta U}(k) \tag{6.78}$$

subject to $\mathbf{A}\;\mathbf{\Delta U}(k) \leq \mathbf{b}$

where **A** and **B** are vectors that define the constraints. The constraints on the optimization problem make it impossible to obtain an analytical solution. Instead, algorithms such as those contained in the MATLAB function FMINCON can be used to optimize the QP problem.

CONSTRAINED PROBLEM WITH DEADZONE TYPE II CONSTRAINTS. The minimum gate movement constraints are deadzone type II constraints that are difficult to directly incorporate into the optimization problem (Chow and Clarke, 1994). One possible method to handle these types of constraints is to place them directly into the objective function and use mixed integer optimization to perform the optimization. Another method is to not incorporate them into the objective function, solve the optimization problem, and then "clip" any of the solutions that violate the minimum gate movement constraints.

Implementation. Once the prediction and optimization routines are completed, the first set of future control actions is implemented on the system. The remaining calculated future control actions are discarded. At this point the system is updated and the entire procedure is repeated: prediction, optimization, and implementation.

Tuning. There is no good tuning procedure for MPC. Many times, tuning is done through trial-and-error techniques. In MPC, there are four tuning parameters: the prediction horizon, the control horizon, the cost weighting matrix for water level errors, and the cost-weighting matrix for the control actions.

The prediction horizon p should be long enough to include all the necessary dynamics of the system. The control horizon m must be less than or equal to the prediction horizon. A short control horizon usually provides smoother control. A control horizon of half the prediction horizon is a good place to start. As with the LQR controllers, the cost weighting matrices $\mathbf{Q}_j$ and $\mathbf{R}_j$ must be determined using a trial-and-error technique.

6.9 OVERALL CONTROL SCHEMES FOR IRRIGATION WATER DELIVERY SYSTEMS

As mentioned earlier, an overall automatic control scheme for irrigation canals should consist of a feedback controller, a feedforward controller, and a flow controller. The feedback controller is designed to eliminate unknown disturbances and account for any errors in the system. The feedforward controller is designed to route known flow changes through the system. The flow rate controller is designed to maintain the desired flow rate at each check structure. All three of these controllers are needed to effectively automatically control an irrigation water delivery system. Feedback control alone is generally not enough to effectively control an irrigation canal. A feedforward routine is needed to route the known flow disturbances through the canal and a flow controller is needed to account for the complex gate hydraulics of the check structures.

In this chapter, two different feedback controllers were examined in detail: the LQR controller and MPC. When an LQR controller is used, any feedforward method such as the volume compensation method or gate-stroking can be selected. When MPC is used as the feedback controller, the feedforward routine automatically built into MPC can be used as the feedforward controller. In this case, the performance of the feedforward controller is heavily dependent on the process model. An accurate process model will lead to good performance of the feedforward portion of MPC whereas an inaccurate process model will lead to poor performance of the feedforward portion of MPC (Wahlin and Bautista, 2003). Alternatively, an external feedforward controller, such as the volume compensation method, could be used instead of the feedforward routine that is built into MPC.

6.9.1 Examples

Various researchers have performed simulation studies on LQR type controllers and MPC type controllers. Some of these studies are listed here for the reader's reference. Clemmens and Schuurmans (2004) explain in detail the type of LQR feedback controllers described in this chapter. Clemmens and Wahlin (2004) show some simulation results using remote downstream feedback controllers that were tuned using LQR theory on ASCE's test cases for canal control algorithms [see Clemmens et al. (1998b) for details on the ASCE test cases]. These simulations use the volume compensation method as the feedforward routine. Wahlin (2002) performed unsteady flow simulation studies using remote downstream feedback controllers that were designed using LQR theory to successfully control a branching canal network run by the Salt River Project (SRP) in central Arizona. Once again, volume compensation was used as the feedforward controller. Wahlin (2002) also uses MPC and the feedforward routine built into the algorithm to control SRP's branching canal network.

Many researchers have used MPC to simulate the automatic control of irrigation water delivery systems. Ellerbeck (1995) used an upstream MPC controller to control an eight-pool canal using the ID model as the process model. Malaterre and Rodellar (1997) presented simulation results for the control of a two-pool canal using a downstream MPC controller. The process model was a linearized version of the Saint Venant equations. Eklund and Tufvesson (2001) presented results for the control of a two-pool canal using a combination upstream/downstream MPC controller (i.e., control actions were determined based on water levels upstream and downstream of the check structure). Eklund and

Tufvesson (2001) used system identification to define their process model. As mentioned earlier, Wahlin (2002) applied MPC to a large portion of SRP's branching canal network using the ID model as the process model. Wahlin (2004) also applied MPC to ASCE test canal 1.

REFERENCES

Åström, K. J. and B. Wittenmark (1997). *Computer-Controlled Systems: Theory and Design,* 3d ed., Prentice Hall, Upper Saddle River, N.J.

Bautista, E. and A. J. Clemmens (1999). "Computerized Anticipatory Control of Irrigation Delivery Systems," *Proceedings of the USCID Workshop on Modernization of Irrigation Water Delivery Systems,* Phoenix, Ariz., October 17–21.

Bautista, E., A. J. Clemmens, and T. S. Strelkoff (2002). "Routing Demand Changes with Volume Compensation: An Update," *Proceedings of Energy, Climate, Environment, and Water—Issues and Opportunities for Irrigation and Drainage,* USCID/EWRI, San Luis Obispo, Calif., July 10–13, 367–376.

Bautista, E., T. S. Strelkoff, and A. J. Clemmens (2003). "General Characteristics of Solutions to the Open-Channel Flow Feedforward Control Problem," *Journal of Irrigation and Drainage Engineering,* ASCE, 129(2): 129–137.

Burt, C. M. (1987). "Overview of Canal Control Concepts," in *Planning, Operation, Rehabilitation and Automation of Irrigation Water Delivery Systems,* ASCE, Portland, Ore., July 28–30.

Burt, C. M. and H. L. Plusquellec (1990). "Water Delivery Control," in *Management of Farm Irrigation Systems,* G. J. Hoffman, T. A. Howell, and K. H. Solomon, eds., ASAE, St. Joseph, Mich., pp. 371–423.

Burt, C. M., R. S. Mills, R. D. Khalsa, and V. Ruiz-Carmona (1998). "Improved Proportional-Integral (PI) Logic for Canal Automation," *Journal of Irrigation and Drainage Engineering,* ASCE, 124(1): 53–57.

Buyalski, C. P., D. G. Ehler, H. T. Falvey, D. C. Rogers, and E. A. Serfozo (1991). *Canal Systems Automation Manual,* vol. 1, USBR, Denver, Colo.

Camacho, E. F. and C. Bordons (1999). *Model Predictive Control,* Springer-Verlag, London.

Chow, C. M. and D. W. Clarke (1994). "Actuator Nonlinearities in Predictive Control," in *Advances in Model-Based Predictive Control,* D. W. Clarke, ed., Oxford University Press, pp. 245–259.

Clarke, D. W. (1994). "Advances in Model-Based Predictive Control," in *Advances in Model-Based Predictive Control,* D. W. Clarke, ed., Oxford University Press, 3–21.

Clemmens, A. J. (1998). "Progress on Canal Automation for Water Delivery Modernization," *Proceedings of Contemporary Challenges for Irrigation and Drainage,* USCID, Phoenix, Ariz. June 3–6.

Clemmens, A. J. and J. Schuurmans (2004). "Simple Optimal Downstream Feedback Canal Controllers: Theory," *Journal of Irrigation and Drainage Engineering,* ASCE, 130(1): 26–34.

Clemmens, A. J. and B. T. Wahlin (2004). "Simple Optimal Downstream Feedback Canal Controllers: ASCE Test Case Results," *Journal of Irrigation and Drainage Engineering,* ASCE, 130(1): 26–34.

Clemmens, A. J., E. Bautista, and R. J. Strand (1997). *Canal Automation Pilot Project Phase I Report,* USDA, ARS, USWCL, WCL Report #22, prepared for Salt River Project.

Clemmens, A. J., L. Feuer, and R. J. Strand (1998a). "Plug and Play Canal Automation: Is It Possible?" *Proceedings of International Water Resources Engineering Conference,* ASCE, Memphis, Tenn., August 3–7.

Clemmens, A. J., T. F. Kacerek, B. Grawitz, and W. Schuurmans (1998b). "Test Cases for Canal Control Algorithms," *Journal of Irrigation and Drainage Engineering,* ASCE, 124(1): 23–30.

Clemmens, A. J., G. Sloan, amd J. Schuurmans (1994). "Canal Control Needs: Example," *Journal of Irrigation and Drainage Engineering,* ASCE, 120(6): 1067–1085.

Eklund, P. and M. Tufvesson (2001). "Predictive Control of Irrigation Channels," Masters thesis, Dept. of Automatic Control, Lund Institute of Technology, Lund, Sweden.

Ellerbeck, M. B. (1995). "Model Predictive Control for an Irrigation Canal," Masters thesis, Faculty of Civil Engineering, TU Delft, Delft, the Netherlands.

Franklin, G. F., J. D. Powell, and A. Emani-Naeini (1994). *Feedback Control of Dynamic Systems,* 3d ed., Addison-Wesley, Reading, Mass.

Johnston, W. R. and J. B. Robertson (1990). *Management, Operation and Maintenance of Irrigation and Drainage Systems,* Manual #57, ASCE, Reston, Va.

Lee, J. H., M. Morari, and C. E. Garcia (1994). "State-Space Interpretation of Model Predictive Control," *Automatica*, 30(4): 707–117.

Lewis, F. and V. Syrmos (1995). *Optimal Control*, 2d ed., John Wiley, New York.

Maciejowski, J. M. (1998). *Predictive Control: A Lecture Course Given in the Aerospace Engineering Faculty TU Delft*, Delft University Press, the Netherlands.

Malaterre, P. O. and J. P. Baume (1999). "Optimum Choice of Control Action Variables and Linked Algorithms: Comparison of Different Alternatives," *Proceedings of the USCID Workshop on Modernization of Irrigation Water Delivery Systems*, Phoenix, Ariz., October 17–21.

Malaterre, P. O. and J. Rodellar (1997). "Multivariable Predictive Control of Irrigation Canals. Design and Evaluation on a 2-Pool Model," *International Workshop on Regulation of Irrigation Canals: State of the Art of Research and Applications*, Marrakech, Morocco, April 22–24.

Malaterre, P. O., D. C. Rogers, and J. Schuurmans (1998). "Classification of Canal Control Algorithms," *Journal of Irrigation and Drainage Engineering*, ASCE, 124(1): 3–10.

Pongput, K. and G. P. Merkley (1997). "Comparison and Calibration of Canal Gate Automation Algorithms," *Journal of Irrigation and Drainage Engineering*, ASCE, 123(3): 222–225.

Postel, S. (1992). *Last Oasis: Facing Water Scarcity*, W. W. Norton, New York.

Prett, D. M. and E. M. Garcia (1988). *Fundamental Process Control*, Butterworths, Boston.

Qin, S. J. and T. A. Badgwell (1997). "An Overview of Industrial Model Predictive Control Technology," *Proceedings of the Fifth International Conference on Chemical Process Control*, CACHE, AIChE, pp. 232–256.

Rogers, D. C. (1999). "Forty Years of Canal Automation—the Bureau of Reclamation Experience," *Proceedings of the USCID Workshop on Modernization of Irrigation Water Delivery Systems*, Phoenix, Ariz. Oct. 17–21.

Rogers, D. C. and J. Goussard (1998). "Canal Control Algorithms Currently in Use," *Journal of Irrigation and Drainage Engineering*, ASCE, 124(1): 11–15.

Schuurmans, J. (1992). "Controller Design for a Regional Downstream Controlled Canal," Masters thesis, Faculty of Civil Engineering, TU Delft, Delft, the Netherlands.

Schuurmans, J., O. H. Bosgra, and R. Brouwer (1995). "Open-Channel Flow Model Approximation for Controller Design," *Applied Mathematical Modeling*, 19: 525–530.

Shinskey, F. G. (1996). *Process Control Systems: Application, Design, and Adjustment*, McGraw-Hill, 4th ed., New York.

Wahlin, B. T. (2002). *Remote Downstream Feedback Control of Branching Canal Networks*, Ph.D. dissertation, Department of Civil Engineering, Arizona State University, Tempe, Ariz.

Wahlin, B. T. (2004). "Performance of Model Predictive Control on ASCE Test Canal 1," *Journal Irrigation and Drainage Engineering*, ASCE, 130(3): 227–238.

Wahlin, B. T. and E. Bautista (2003). "Feedforward Control: Volume Compensation versus Model Predictive Control," *Proceedings of the USCID Second International Conference on Irrigation and Drainage*, Scottsdale, Ariz., May 12–15.

Wylie, E. B. (1969). "Control of Transient Free-Surface Flow," *Journal of Hydraulics Division*, ASCE, 95(1): 347–361.

CHAPTER 7
OPTIMAL LOCATION OF ISOLATION VALVES IN WATER DISTRIBUTION SYSTEMS: A RELIABILITY/OPTIMIZATION APPROACH

Sukru Ozger
Carollo Engineers
Phoenix, Arizona

Larry W. Mays
Department of Civil and Environmental Engineering
Arizona State University
Tempe, Arizona

7.1 INTRODUCTION

7.1.1 Before September 11, 2001

The cornerstone of any healthy population is access to safe drinking water. The goal of the United Nations International Drinking Water Supply and Sanitation Decade from 1981 to 1990 was safe drinking water for all. A substantial effort was made by the United Nations to provide drinking water and sanitation services to populations lacking those services. Unfortunately, however, the growth in developing country populations almost entirely wiped out the gains. In fact nearly as many people lack those services today as at the beginning of the 1980s (Gleick, 1993, 1998).

The lack of adequate water supply systems is due to both the deterioration of aging water supplies in older urbanized areas, and to the nonexistence of water supply systems in many areas that are undergoing rapid urbanization, such as in the southwestern United States. In other words, methods for evaluation of the nation's water supply services needs to consider not only rehabilitation of existing urban water supply systems but also the future development of new water supply systems to serve expanding population centers. Both the adaptation of existing technologies and the development of new innovative technologies will be required to improve the efficiency and cost-effectiveness of future and existing water supply systems and facilities necessary for industrial growth.

An EPA survey (Clark et al., 1982) of previous water supply projects concluded that the distribution facilities in water supply systems account for the largest cost item in future maintenance budgets. The aging deteriorating systems in many areas raise tremendous maintenance decision-making problems that are further complicated by the expansion of existing systems. Deterioration of the water distribution systems in many areas has translated into a high proportion of unaccounted-for water due to leakage, which is not only a loss of a valuable resource, but also raises concerns of safe drinking water by possible contamination through cracked pipes.

The reliability of the existing aging systems is continually decreasing. Only recently have many municipalities been willing or able to finance rehabilitation of deteriorating pipelines, and are still deferring needed maintenance and replacement of systems components until a catastrophe or the magnitude of leakage justifies the expense of repair. Water main failures have been extensive in many cities.

As a result of governmental regulations and consumer-oriented expectations, a major concern now is the transport and fate of dissolved substances in water distribution systems. The passage of the Safe Drinking Water Act in 1974 and its amendments in 1986 (SDWAA) changed the manner in which water is treated and delivered in the United States. The U.S. Environmental Protection Agency (USEPA) is required to establish maximum contaminant level (MCL) goals for each contaminant that may have an adverse effect on people's health. These goals are set to the values at which no known or expected adverse effects on health can occur. By allowing a margin of safety (Clark et al., 1987; Clark, 1987) previous regulatory concerns were focused on water as it leaves the treatment plant before entering the distribution system (Clark, 1987), disregarding the variations in water quality that occur in the water distribution systems.

7.1.2 After September 11, 2001

Water utilities are implemented to construct, operate, and maintain water supply systems. In the past the basic function of these water utilities was to obtain water from a source, treat the water to an acceptable quality, and deliver the desired quantity of water at the desired time. The events of September 11, 2001 and subsequent anthrax attacks have shown the vulnerability of basic civilian infrastructure, somewhat changing the function of water utilities to include protection from various threats. Drinking water supply systems around the world offer great opportunities for terrorism as these systems are extensive, relatively unprotected and accessible, and often isolated. The threats to these systems include cyber, physical, chemical, and biological threats.

The events of September 11, 2001 have significantly changed the approach to management of water utilities. Previously the consideration of terrorist threats to our nation's drinking water supply was minimal. New trends in the technical management of water supply systems will be heavily focused upon security issues as related to the design, operation, and management of these systems. Of particular interest will be the evaluation of risk and reliability issues of the various components, the subsystems, and the systems as a whole from the viewpoint of their susceptibility to terrorism.

Within a very short time after September 11, 2001 we began to see a concerted effort at all levels of government to begin addressing issues related to the threat of terrorist activities to the U.S. water supply. We saw a number of acts passed such as the "Security and Bioterrorism Preparedness and Response Act" and the "Homeland Security Act" that addressed the nation's water supply. These acts resulted in agencies such as the U.S. Environmental Protection Agency (EPA) developing new protocols to address their new responsibilities under these acts.

7.1.3 Water Distribution Systems: A Brief Description

Water distribution systems are composed of three major components: pumping stations, distribution storage, and distribution piping. These components may be further divided into subcomponents, which in turn can be divided into subsubcomponents. For example, the pumping station component consists of structural, electrical, piping, and pumping unit subcomponents. The pumping unit can be divided further into subsubcomponents: pump, driver, controls, and power transmission. The exact definition of components, subcomponents, and subsubcomponents depends on the level of detail of the required analysis and to a somewhat greater extent on the level of detail of available data. In fact, the concept of component-subcomponent- subsubcomponent merely defines a hierarchy of building blocks used to construct the water distribution system.

A water distribution system operates as a system of independent components. The hydraulics of each component is relatively straightforward; however, these components depend directly upon each other and as a result affect each other's performance. The purpose of design and analysis is to determine how the systems perform hydraulically under various demands and operating conditions. These analyses are used for the following situations:

- Design of a new distribution system
- Modification and expansion of an existing system

- Analysis of system malfunction such as pipe breaks, leakage, valve failure, and pump failure
- Evaluation of system reliability
- Preparation for maintenance
- System performance and operation optimization

Books on the subject of water supply/water distribution systems include Mays (1989, 2000, 2002, 2004a). The subject of security of these systems is covered in detail in *Water Supply Security Systems* by Mays (2004b).

Distribution systems, consisting of pipelines, pipes, pumps, storage tanks, and the appurtenances such as various types of valves, meters, and the like, offer the greatest opportunity for terrorism because they are extensive, relatively unprotected and accessible, and often isolated. The physical destruction of a water distribution system's assets or the disruption of the water supply could be more likely than contamination. A likely avenue for such an act of terrorism is a bomb, carried by car or truck, similar to recent events. Truck or car bombs require less preparation, skill, or manpower than complex attacks such as those of September 11, 2001. However, we must consider all possible threats no matter how remote we may think that they could be.

7.2 DEMAND-DRIVEN ANALYSIS VERSUS PRESSURE-DRIVEN ANALYSIS

7.2.1 Definitions

Hydraulics of a water distribution network can be approached from two different perspectives. The difference between them comes from the level of primacy given to nodal demands vs. nodal pressures. The first approach assumes that consumer demands are always satisfied regardless of the pressures throughout the system and formulates the constitutive equations accordingly to solve for the unknown nodal heads. This approach is called demand-driven analysis and is used by almost all the traditional network hydraulic solvers. Examples are EPANET by Rossman (1994) and KYPIPE by Wood (1980).

A water distribution network can be designed or operated using demand-driven analysis under normal operating conditions. This is usually done by adjusting such independent decision variables as simple or rule-based controls for pump operations, valve settings, reservoir/tank levels, and so on. If a proper design has been made for a given design demand loading, then nodal pressures throughout the system should be well above the minimum pressure required. On the other hand, if the same network were analyzed with a higher demand loading, it would not be surprising to see warning messages such as "Negative pressures at 6:00 hrs" from the EPANET model while nodal demands are fully satisfied at the same time period. Almost all demand-driven models possess a partial recognition of this weakness with similar warning messages when negative pressures are calculated from a network hydraulic analysis.

The major drawback of the demand-driven approach is that it fails to measure a partially failed network performance resulting from abnormalities in physical and nonphysical system components (e.g., pipe breaks, fire-fighting demands, etc.). In such cases, if demand-driven analysis is used, it may produce very unrealistic results; for example, negative pressures are calculated at some nodes in the system. It is not uncommon that a hydraulic analysis using a demand-driven approach yields large negative pressures at one or more junctions of a distribution network.

The second approach to a network hydraulic analysis is called head-driven (occasionally called pressure-driven or pressure-dependent) analysis. Here, the primacy is given to pressures. A node is supplied its full demand only if a minimum required supply pressure is satisfied at that node. If the minimum pressure requirement cannot be met, then only a fraction of the nodal demand can be satisfied. That fraction is determined by a predefined relationship between nodal head and nodal outflow.

As mentioned above, demand-driven models work well under normal operating conditions. Major applications of head-driven analysis, on the other hand, are for analysis of network performance under partially failed conditions. Because a network reliability assessment requires evaluation of partially failed conditions of a network, pressure-dependent analysis is much more suitable for reliability assessment of water distribution networks.

One of the major problems with the pressure-driven approach is that it requires extensive field data collection to determine the relationship between nodal heads and nodal flows. Such a relationship may be unique for each junction of a network inasmuch as it depends on such physical characteristics as type of service connection and the development type that the junction serves (e.g., residential, high-rise, schools, hospitals, etc.). Calibration of models using field data is also part of the extensive effort needed to be able to use pressure-dependent analysis.

The second major drawback of head-driven analysis is that it lacks robust methods for the computational solution of the constitutive equations (Tanyimboh et al., 2001). Almost all the traditional network hydraulic models are built upon the demand-driven approach. Therefore, it is very desirable that reliability analysis techniques based on interpretations or transformations of existing demand-driven analysis from a pressure-driven approach be developed.

7.2.2 Hydraulic Equations for a Network Simulation Problem

Two sets of equations are needed to solve a network simulation problem. First, conservation of flow must be satisfied at each network junction. The second set of equations describes a nonlinear relation between flow and headloss in each pipe, such as the Hazen-Williams or Darcy-Weisbach equation. These equations form a coupled set of nonlinear equations whenever the network contains loops or more than one fixed-head source (Rossman, 2000). The fact that most distribution systems are looped and/or have multiple sources requires the aid of a computer to solve the nonlinear set of equations using an iterative procedure. The basic equations in a network model are described by Rossman (2000) for a network with N junction nodes and NF fixed grade nodes (tanks and reservoirs).

Let the flow-headloss relation in a pipe between nodes i and j be given as:

$$H_i - H_j = h_{ij} = rQ_{if}^n + mQ_{if}^2 \tag{7.1}$$

where H = nodal head
h = headloss
r = resistance coefficient
Q = flow rate
n = flow exponent
m = minor loss coefficient

The value of the resistance coefficient and the corresponding flow exponent depend on which friction headloss formula is being used. Most network models use one of the following flow-headloss relationships: (1) Hazen-Williams, (2) Darcy-Weisbach, or (3) Chezy-Manning.

In addition to the first set of equations described above, conservation of flow around all nodes must be satisfied, which forms the second set of equations

$$\sum_j Q_{if} - D_i = 0 \qquad \text{for } i = 1,\ldots,\text{N} \tag{7.2}$$

where D_i = flow demand (or supply) at node i
j = set of nodes directly connected to node i
Q_{ij} = flow in pipe ij

By convention, flow is positive if it is into the subject node and negative, otherwise.

As previously mentioned the conventional network hydraulic solvers such as EPANET by Rossman (1994) and KYPIPE by Wood (1980) are demand-driven. Demand-driven analysis seeks a solution for all heads H_i and flows Q_{ij} that satisfy the two sets of equations, (7.1) and (7.2), for a set of known heads at the fixed grade nodes. The basic assumption in demand-driven analysis is that nodes are assigned demands that are assumed fully satisfied. For example, variables used in the solution algorithm of EPANET are divided into unknown and known variables (Rossman, 1994).

The unknown variables are

y_s = height of water stored at tank node s
Q_s = flow into storage tank node s
Q_{ij} = flow in pipe connecting nodes i and j
H_i = hydraulic grade line elevation at node i (elevation plus pressure head)

and the known variables (constants) are:

A_s = tank cross-section area
E_s = elevation of node s
D_i = demand (or supply) at node i

The major assumption of demand-driven analysis, that demands are fully satisfied, works well under normal operating conditions. However, it is not uncommon that under abnormal conditions a solution arises where pressures are negative or unacceptably low, because demands are satisfied regardless of validity of calculated pressures at network nodes (Bouchart and Goulter, 2000). A few examples of such abnormal conditions may occur when demand loading on the network exceeds the design demand loading, fire-fighting demands for which the network was not designed, or closure of a portion of a network because of contamination or pipe main breaks. In reality, a shortfall in the volume of water actually delivered to consumers begins as the pressure in the network drops below some threshold value. Naturally and ideally, a lower bound for this threshold value is zero gauge pressure.

Water distribution networks are almost always skeletonized to some degree, depending on the modeling purpose forcing demands from secondary networks to be lumped into nearby junctions. Because of various headlosses associated with secondary networks and different elevations of outlets in these networks, it is very difficult, if not impossible, to define a relationship between the pressure at a junction and the flow available from that junction into the secondary network it serves. A tedious way of doing this is field data collection and calibration. A typical value for threshold pressure is below the minimum of 20 m in network studies (Bouchart and Goulter, 2000). Tanyimboh et al. (1999) state that pressures of 15 to 25 m are the minimum acceptable standards with the use of network models. All this implies that demand-driven algorithms cannot cope with situations where pressures are less than satisfactory at some demand nodes (Tanyimboh et al., 2001) and thus fail to depict deficient network performances.

7.2.3 Service Pressures

Criteria for assessing acceptable service pressures in a distribution network may vary from system to system. Therefore, there are no universally acceptable pressure ranges. Chin (2000) lists the following considerations for assessing the adequacy of service pressures:

1. Flow is adequate for residential areas when pressures are above 240 kPa (35 psi);
2. The pressure required at street level for excellent flow to a 3-story building is about 290 kPa (42 psi);
3. Adequate flow to a 20-story building would require about 830 kPa (120 psi) of service pressure at street level, which is not desirable because of the associated leakage and waste. Thus, very tall buildings are usually served with their own pumping system;
4. In general, pressures in the range of 410 to 520 kPa (60 to 75 psi) would be adequate for
    - Supplying normal water delivery to buildings up to 10 stories
    - Providing adequate sprinkler service in buildings of 4 to 5 stories
    - Supplying water for fire protection
    - Handling fluctuations in pressure caused by clogged pipes and excessive length of service pipes.

Overall, pressures in most cases should be maintained above 207 kPa (30 psi) and below 689 kPa (100 psi) during normal operations. Pressures above 689 kPa tend to increase water waste through undetected leaks and may cause damage to residential and commercial plumbing systems or pipe breaks. The minimum pressure requirement may be lowered to 138 kPa (20 psi) during emergency conditions, such as a fire. A 138 kPa pressure is sufficient to supply the suction side of pumps on a fire pumper truck. According to Walski (2000), the threshold pressure value below which a failure to satisfy the full demands occurs is usually assumed to be below the minimum of 20 m (≈196 kPa or 28.4 psi).

The Office of Water Services in England specifies a minimum acceptable static pressure of 7 m (≈68.5 kPa or 10 psi) below which customers may be entitled to compensation for less than satisfactory service. On the other hand, it is not uncommon that minimum acceptable pressure is specified as high as 25 m (≈245 kPa or 35 psi), to allow for possible increases in demand (Tanyimboh et al., 1999). In general, nodal pressures of 15 to 25 m will guarantee satisfactory service at all related stop taps in a distribution system. Pressures that are lower may cause a shortfall in supply and failure to supply the full demands. Tanyimboh et al. (1999) also suggest that in the absence of field test data, a good approximation to critical pressures can be obtained by the equivalent static pressure to the height above ground level of rooftop water tanks in typical two-story residential areas.

Water-using devices in residential houses may also require certain minimum pressures to operate. For example, most dishwasher manufacturers specify minimum working pressures anywhere from 20 to 40 psi.

Another factor that should be considered while estimating the critical pressure is the degree of network skeletonization. Typically, smaller pipes in a network are excluded from the modeling. The result is that total demand from the consumers located along those smaller lines is lumped into the nearby junctions located on larger pipes. Thus, when choosing the critical pressure for a junction, such characteristics as outlet elevations and headlosses of the secondary network that the junction serves must also be taken into account.

7.2.4 Head-Driven Analysis

Demand-driven analysis explained in the previous section assumes that consumer demands in a distribution network are fully satisfied regardless of validity of calculated pressures. It can be argued that many modern water-using devices such as toilets, washing machines, and dishwashers are essentially pressure-independent provided a minimum threshold pressure is available, which supports the assumption of primacy of demands.

The philosophical basis of head-driven analysis—the counterapproach to demand-driven analysis—is explained by Rossman (2001) as follows.

> ...[A] distribution system can be thought of as a completely closed, pressurized system with a set of variable sized orifices that connect to various devices that are open to the atmosphere (e.g., sinks, toilets, water heaters, washing machines, pipe leaks, etc.). For a specified opening of these orifices, the flow out of them will be dependent on the pressure maintained in the system (which in turn is dependent on the setting of the demand orifices)....

Thus, unlike demand-driven analysis, the head-driven approach treats demands as random variables and recognizes the primacy of pressures by defining a relationship between the outflow and the pressure at each node. This relationship is given, in its general form, by (Tanyimboh et al., 1999)

$$H_j = H_j^{\min} + K_j Q_j^{n_j} \tag{7.3}$$

where H_j = head at node j corresponding to demand Q_j
K_j = flow resistance coefficient
n_j = exponent
$H_j^{\min}$ = minimum required nodal head below which outflow at node j is unsatisfactory or zero

Because demands are unknown, Eq. (7.3) can be rearranged:

$$Q_j = \left(\frac{H_j - H_j^{\min}}{K_j} \right)^{1/nj} \tag{7.4}$$

The head-driven analysis requires substitution of Q_j into Eq. (7.2) in the place of D_i. The solution algorithms for solving the resulting problem have been described by Gupta and Bhave (1996) and Tabesh (1998).

As explained in the previous section, demand-driven analysis may produce unrealistic results (i.e., pressures being negative or less than satisfactory) especially under partially failed conditions of a system. In such cases, head-driven analysis is superior to the demand-driven approach in that the former is able to determine the nodes with insufficient supply and the respective magnitudes of the shortfalls. On the other hand, performing a head-driven analysis requires extensive field data collection and calibration to determine K and n values, possibly for each node (Gupta and Bhave, 1997). In addition to these practical difficulties, lack of robust methods for computational solution of constitutive equations in head-driven analysis makes the demand-driven network models the primary choice for hydraulic analysis or reliability assessments.

7.2.5 Node Flow Analysis

In order to overcome the weaknesses of demand-driven analysis, Bhave (1991) developed a method called Node Flow Analysis (NFA) to predict performance when a distribution network is deficient. The method was based on the assumption that when a network is deficient and thus unable to satisfy the nodal demands, the network will try to meet the demands, as far as possible, under the given conditions. In other words, the flow in the network will adjust such that the total supply is maximized under deficient conditions. Thus, the network flow analysis problem is considered as an optimization problem in their node flow analysis subject to certain constraints that relate hydraulic gradeline at each junction to available flow at that junction.

The complete optimization problem is as follows:

$$\frac{\text{Maximize}}{\text{Total outflow}} = \sum_j q_j \tag{7.5a}$$

where q_j is the flow available at node j
Subject to:

$$H_m = H_{om} \qquad \text{for all } m \tag{7.5b}$$

where m is number of source nodes and H_{om} is the specified hydraulic grade line (HGL) value at source node m.

$$q_j = q_j^{\text{req}} \qquad \text{if } H_j \geq H_j^{\min} \tag{7.5c}$$

or

$$0 < q_j < q_j^{\text{req}} \qquad \text{if } H_j = H_j^{\min} \tag{7.5d}$$

or

$$q_j = 0 \qquad \text{if } H_j \leq H_j^{\min} \tag{7.5e}$$

$$\sum_i Q_{ij} - q_j = 0 \qquad \text{for all } j \tag{7.5f}$$

where i is the set of all nodes connected to node j and assuming flow is positive if it is into node j.

$$\sum_{\text{loop}} h_L = 0 \qquad \text{for all loops} \tag{7.5g}$$

The basic differences from the usual demand-driven formulation are the additional constraints defined by Eqs. (7.5c) to (7.5f). Unlike Eqs. (7.3) and (7.4), a discrete flow-head relationship is defined. The method still requires defining a minimum pressure threshold at each node.

The network-solvability rule known as the unknown-number rule requires that in order for a network problem to be feasible, the total number of unknowns must be equal to the total number of nodes in the network. Let the number of source nodes and junctions in a network be S and J, respectively. For source nodes, the nodal flows are the only unknowns. However, neither the flows nor the HGL values at junctions are known in the NFA described above. Depending on the number of demand junctions, the number of these unknowns (junction heads + junction flows) will be between $J + 1$ (the network must have at least one demand node) and $2J$ (all junctions are assigned nonzero demands). Thus, the total number of unknowns for the entire network lies between $M + N + 1$ and $M + 2N$ which is greater than the total number of nodes in the network, $M + N$.

In order to bring down the total number of unknowns, NFA analysis treats either the demand or the HGL as unknown at each demand node. In other words, if the constraint (7.5c) is chosen, then nodal flow is treated as known, being equal to the nodal demand, and the nodal head as unknown. On the other hand, choosing the constraint (7.5d) implies a known nodal head but an unknown nodal flow, which is also constrained to being less than the nodal demand. Finally, choosing constraint (7.5e) indicates a strictly zero nodal flow and an unknown nodal head subject to being less than the pressure threshold.

The solution procedure of the NFA described above begins by imposing constraint (7.5c) on all demand junctions. Note that this is equivalent to the usual demand-driven analysis. Once the available HGL values are determined at all demand nodes, it is determined which one of the constraints [Eqs. (7.5c) to (7.5e)] actually applies to each demand node and a new solution is obtained. The solution should normally be stopped here if all the nodal constraints are indeed applicable, which can be verified from the results of the new solution. However, as Tanyimboh and Tabesh (1997) have also found, when a network with locally insufficient heads is simulated using the demand-driven approach, the deficiency appears to be far more serious and widespread than it is in reality. The result is that there are almost always some demand junctions in the network that are pressure deficient from an initial demand-driven analysis but then violate the constraint given by Eq. (7.5d), because of $q_j > q_j^{req}$, at the end of the next solution. Thus the solution proceeds in an iterative manner until a set of demand nodes is found in the deficient network that is truly pressure deficient. In other words, when no discrepancies are found between each nodal head and nodal flow and the constraint assigned to that node, the final solution is achieved.

7.3 SEMI-PRESSURE-DRIVEN ANALYSIS (SPDA) FRAMEWORK

Ozger (2003) developed a Semi-Pressure-Driven Analysis (SPDA) framework, using EPANET Toolkit functions and the C++ programming language, to facilitate the application of NFA (Bhave 1991) in the EPANET network hydraulics solver. The method can be used to predict deficient performance, in terms of Available Demand Fraction (ADF), under partially failed conditions of a network that may include one or more of the following conditions:

1. Fire events.
2. One or more sources in the system are shut down.
3. A pump station is out of service.
4. A control valve that has logistic importance in terms of flow delivery is broken.
5. A transmission line is damaged and taken out of service for repair.
6. A given portion of the network has to be isolated because of contamination.

The input requirements for any what-if scenario are an accurate description of the equipment/network component that will be out of service, the network isolation valve scheme, and, if an EPS analysis is needed, the duration of the repair/replacement process or emergency (e.g., fire-fighting).

Once a semi-pressure-driven analysis is performed for any what-if scenario, the following questions can be answered using the results:

1. Are there enough isolation valves so that the isolation can be reduced to as small an area as possible?
2. How much of the demand can be supplied at each junction?
3. How much of the networkwide demand can be satisfied?
4. What is the temporal variation of nodal and networkwide ADFs?
5. What portions of the network are the most vulnerable to the given what-if scenario?
6. What operational changes could lessen the deficiency in network performance?
7. How long does it take for the network to fully recover, in terms of hydraulics and/or water quality, once the repair/replacement work is finished?

The terms *minimum required pressure, critical pressure,* and *threshold pressure* are used interchangeably in the following discussion. By definition, critical pressure at a junction is the pressure value below which the system fails to supply the full demands lumped into that junction.

SPDA starts with a hydraulic analysis of the deficient network using EPANET2. In other words, the network is first simulated from a demand-driven point of view to identify junctions of the network with pressures below the predefined critical pressure values. Threshold values within a network may vary from junction to junction depending on the characteristics of the service connection and the type of development served by that node. Thus, ideally, the critical pressure value is a unique number for each junction in the network and must be determined empirically by field testing. However, in the absence of such field data, approximate values of critical pressures can be determined using one or more of the guidelines discussed in Sec. 7.2.1.

Once the pressure-deficient nonzero demand junctions are identified from an initial demand-driven analysis, the next step is to determine the available flows at those nodes, presumably knowing that the remaining demand nodes are fully satisfactory in terms of both pressure and demand, and the zero demand nodes have pressures above the cavitation limit. For this purpose, the following modifications are made to each pressure-deficient nonzero demand node

1. {New node elevation} = {original node elevation} + {threshold pressure head}.
2. Set demand to zero.
3. Connect an artificial reservoir to the node by an infinitesimally short control valve (CV) pipe that allows flow only from the node to the reservoir.
4. {Artificial tank elevation} = {new node elevation}.

With these modifications, demand at each pressure-deficient junction in the algorithm is treated as an unknown while a pressure threshold is imposed. Figure 7.1 shows a flowchart of the semi-pressure-driven algorithm. Note that the algorithm proceeds in an iterative manner. That is, if one or more artificial reservoirs receives more water than the nodes demand, those artificial reservoirs are removed from the network and the original elevations and demands at the corresponding nodes are restored. A performance index, later to be used for network reliability/availability assessment, can be defined at each node for a steady-state analysis as

$$\mathrm{ADF}_j = \frac{Q_j^{\mathrm{avl}}}{D_j} \tag{7.6}$$

where ADF_j = available demand fraction at node j
Q_j^{avl} = available flow to node j
D_j = total consumer demand allocated at node j

A networkwide-available-demand fraction is then given by

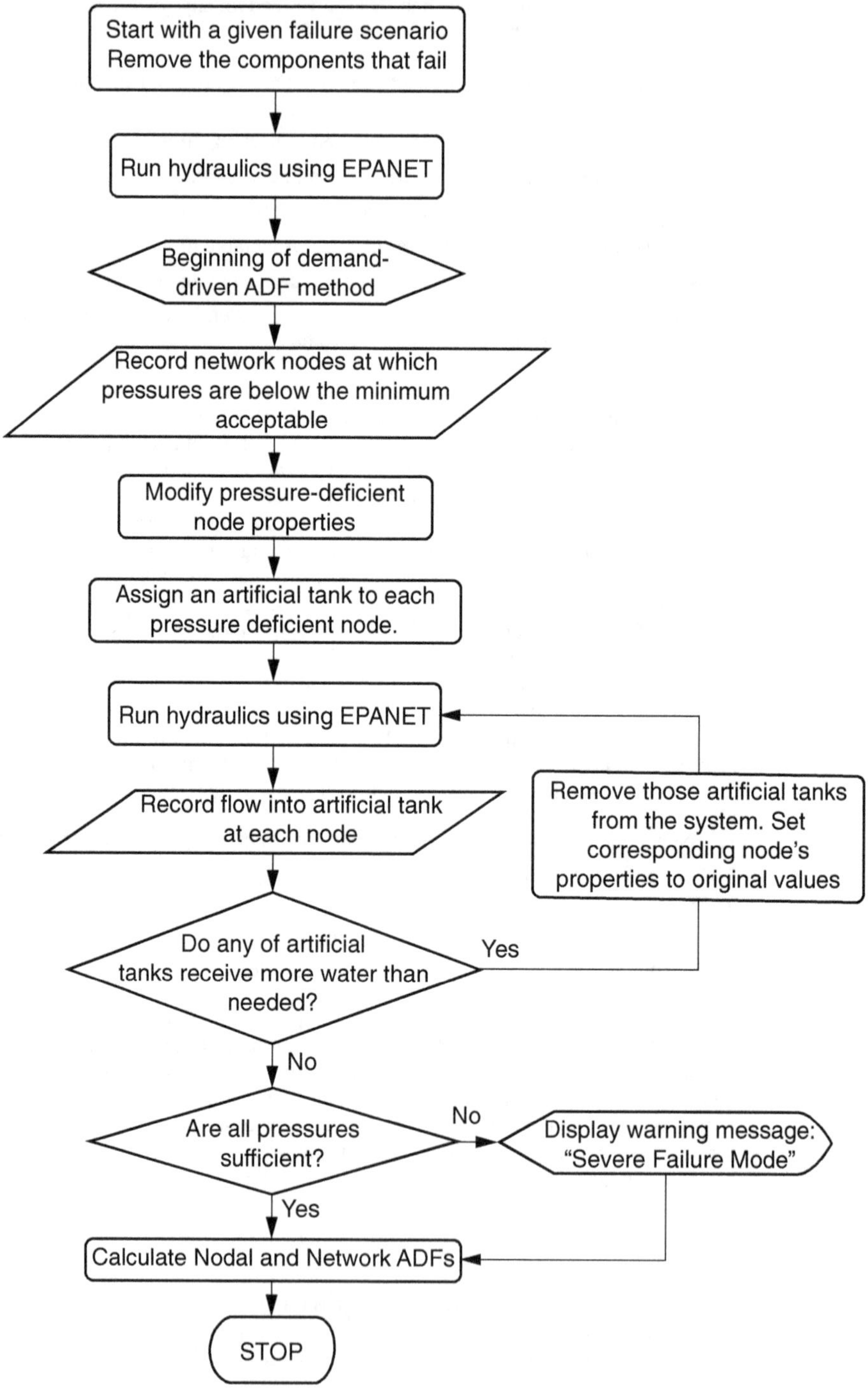

FIGURE 7.1 Flowchart of SPDA algorithm.

$$\mathrm{ADF}_{\mathrm{net}} = \frac{\sum_{\text{all nodes}} Q^{\mathrm{avl}}}{\sum_{\text{all nodes}} D_j} \tag{7.7}$$

Similarly, for an extended period simulation (EPS) analysis, nodal ADFs are given by

$$\mathrm{ADF}_j = \frac{\sum_{t=1}^{nt} Q_{jt}^{\mathrm{avl}}}{\sum_{t=1}^{nt} D_{jt}} \tag{7.8}$$

where t is the time step and nt is the total number of time steps from the time of component failure to when the network starts its normal operation following restoration of the malfunctioning equipment. Random pipe breaks typically take from several hours to a day to repair.

Systemwide ADF for an EPS analysis is then given by

$$\mathrm{ADF}_{\mathrm{net}} = \frac{\sum_{t=1}^{nt} \sum_{\text{all nodes}} Q_j^{\mathrm{avl}}}{\sum_{t=1}^{nt} \sum_{\text{all nodes}} D_j} \tag{7.9}$$

It can be shown, mathematically, that the systemwide ADF is equivalent to the demand weighted average of nodal ADFs. Thus, unlike conventional reliability models based on pressure deficiencies, there is a deterministic relationship between nodal reliabilities and the system reliability in this approach.

7.4 RELIABILITY MODELS

SPDA, developed by Ozger (2003) for predicting deficient performance of a network under partially failed conditions, can be used for network reliability assessment considering water pipe breaks. Water main failures can be modeled using the homogeneous Poisson process (HPP) if failure rates are stationary with respect to pipe ages. Alternatively, if deterioration causes an increase in pipe break rates with time, then the nonhomogeneous Poisson process (NHPP) can be used. Both models are applicable for extended period simulation as well as steady-state analysis for reliability assessment of a network.

7.4.1 HPP-Based Reliability Model

Pipe break rates in a distribution system can be determined from historical break/repair data. If such data indicate a monotonic break rate, the network is said to be nondeteriorating and consequently, reliability assessment of the network can be done using HPP for modeling the pipe failures. Two types of reliability measures are available if a network is found nondeteriorating. First, the reliability measure is based on individual pipe failure probabilities. Here, the probability of failure of an individual pipe is given by

$$p_i = 1 - e^{-\beta_i} \tag{7.10a}$$

and

$$\beta_i = \lambda_i L_i \tag{7.10b}$$

where β_i = expected number of failures per year for pipe i
λ_i = expected number of failures per year per unit length of pipe i
L_i = length of pipe i

Then, the first type of reliability measure based on individual pipe failure probabilities is given, following the work of Shinstine et al. (2002), by

$$R_{\text{net}} = 1 - \sum_{i=1}^{np} (1 - \text{ADF}_{\text{net}}{}^{i})\, p_i \tag{7.11}$$

where $\text{ADF}_{\text{net}}{}^{i}$ = network available demand fraction resulting from pipe i failure
p_i = probability of failure of pipe i
np = the number of pipes in the network

The reliability measure in Eq. (7.11) does not consider that pipes are repairable components. A more appropriate measure can be evaluated if the mean time between failure (MTBF) and the mean time to repair (MTTR) for individual pipes are known. This measure is called mechanical availability (ma) and given for individual pipes by (Ross, 1985; Cullinane, 1986)

$$\text{ma} = \frac{\text{MTBF}}{\text{MTBF} + \text{MTTR}} \tag{7.12}$$

where MTTR is the mean time to repair and MTBF is the mean time between failures and is given by

$$\text{MTBF} = \frac{1}{\lambda L} \tag{7.13}$$

where β_i is the expected number of failures per year per unit length of pipe i, and L is the pipe length. For simplicity, the constant value of 1 day can be assumed for MTTR.

Mechanical unavailability of a pipe is then given by

$$\text{mu} = 1 - \text{ma} = 1 - \frac{\text{MTBF}}{\text{MTBF} + \text{MTTR}} \tag{7.14}$$

The probability that all the pipes in a distribution network are operational is given by

$$\text{MA}_{\text{net}} = \prod_{i=1}^{np} \text{ma}_i \tag{7.15}$$

where np = total number of pipes. The probability of a failure of the ith pipe and all others remaining operational is given by:

$$u_i = \text{MA}_{\text{net}} \cdot \frac{\text{mu}_i}{\text{ma}_i} \tag{7.16}$$

The event of simultaneous failure of two pipes (the ith and kth pipes) has a probability of

$$u_{ik} = \text{MA}_{\text{net}} \cdot \frac{\text{mu}_i}{\text{ma}_i} \cdot \frac{\text{mu}_k}{\text{ma}_k} \tag{7.17}$$

Finally, using the zero-, first-, and second-order pipe failure probabilities above, network availability can be determined (Fujiwara and De Silva, 1990) from

$$A_{\text{net}} = \text{ADF}_{\text{net}}^{\ 0} \cdot \text{MA}_{\text{net}} + \sum_{i=1}^{np} \text{ADF}_{\text{net}}^{\ i} \cdot u_i + \sum_{i=1}^{np-1} \sum_{k=i}^{np} + 1\, \text{ADF}_{\text{net}}^{\ ik} \cdot u_{ik} \tag{7.18}$$

where $\text{ADF}_{\text{net}}^{\ 0}$ = available demand fraction of the network with a fully functional network
$\text{ADF}_{\text{net}}^{\ i}$ = available demand fraction of the network with the *i*th component removed
$\text{ADF}_{\text{net}}^{\ ik}$ = available demand fraction of the network with the *i*th and *k*th components removed

Individual nodal availabilities can be obtained using the ADF for the node of concern.

It should be noted that there are two alternatives for calculating the ADFs above. First, they can be determined with static demands at each node. In this case, ADFs can be calculated at each node using a steady-state analysis and the resulting network ADF can be found from Eq. (7.7). If, on the other hand, a repetitive demand pattern exists (e.g., a diurnal pattern with 24-h cycles), nodal and systemwide ADFs can be calculated using the results from an extended period analysis. Nodal ADFs are then given by

$$\text{ADF}_j = \frac{\sum_{t=1}^{nt} Q_{jt}^{\text{avl}}}{\sum_{t=1}^{nt} D_{jt}} \tag{7.19}$$

where t is the time step and nt is the demand pattern length. Similarly, systemwide ADF is given by

$$\text{ADF}_{\text{net}} = \frac{\sum_{t=1}^{nt} \sum_{\text{all nodes}} Q_j^{\text{avl}}}{\sum_{t=1}^{nt} \sum_{\text{all nodes}} D_j} \tag{7.20}$$

7.4.2 NHPP-Based Reliability Model

If historical break data indicate a deteriorating network, the reliability model should be based on pipe failure probabilities modeled using the nonhomogeneous Poisson process. The two most commonly applied, time-dependent models for the mean cumulative repair function are the power relation model and the exponential model (Tobias and Trindade, 1995).

The power relation model is given by

$$M(t) = at^b \tag{7.21}$$

where $M(t)$ is the expected number of failures between time zero and t, and b and c are empirically determined parameters. The corresponding intensity function is

$$\lambda(t) = \frac{dM(t)}{dt} = abt^{b-1} \tag{7.22}$$

Note that if the parameter b is between 0 and 1, the network is deteriorating. On the other hand, $b > 1$ indicates an improving network. The model parameters for the power relation model can be determined from either graphical or analytical analysis. Taking the natural logarithms of each side in Eq. (7.21),

$$\ln M(T) = \ln(a) + b\ln(t) \tag{7.23}$$

from which it can be seen that, if the power relation model is appropriate, the plot ln $M(t)$ versus ln t on linear by linear graph paper should approximate a straight line with intercept ln a and slope b.

For a nonrenewal process that can be modeled using a nonhomogeneous Poisson process, Crow (1974) has developed MLEs (Maximum Likelihood Estimates) for the parameters of the power relation model above (Tobias and Trindade, 1995).

Alternatively, the exponential model can be used to model the nonrenewal pipe failure processes. The intensity function then becomes

$$\lambda(t) = e^{c + bt} \tag{7.24}$$

The mean repair function can be found from simple integration of the intensity function

$$M(t) = \int_0^t \lambda(u)\,du = \frac{e^c}{b}\left(e^{bt} - 1\right) = \frac{a}{b}\left(e^{bt} - 1\right) \tag{7.25}$$

where $a = e^c$.

Similar to the power relation model, parameters of the exponential model can be determined from graphical or analytical analysis. Taking natural logarithms of both sides in Eq. (7.25) yields

$$\ln M(t) = \ln(a) + bt \tag{7.26}$$

Thus, the plot of ln $M(t)$ versus t should approximate a straight line with intercept ln a and slope b, if the exponential model is appropriate. Alternatively, analytical approaches can be used to estimate model parameters from MLEs.

If the NHPP model parameters (for either the power relation model or exponential model) are estimated from the historical break data, then the expected number of failures between time t_1 and time t_2 can be determined from the difference between the mean cumulative repair functions as

$$E(t_1, t_2) = M(t_2) - M(t_1) \tag{7.27}$$

where $M(t)$ is given by Eq. (7.21) or (7.25). What is more useful than the expected number of failures for reliability assessment is the failure rate as a function of time given by the intensity functions, that is, Eqs. (7.22) and (7.24). Pipe failure probabilities and pipe availabilities at time t can be calculated from Eqs. (7.10) to (7.18) using the failure rates at time t given by the appropriate intensity functions. Then, Eqs. (7.11) and (7.18) would yield the instantaneous network reliability and availability at the given time t.

The following derivation can be used if rather than an instantaneous measure, network availability is needed for a time horizon from a given base time. The expected volume of shortage at node j resulting from link l failures between time t_0 and t is given by

$$\text{VS}_{j,l}\,(t_0, t) = E_l\,(t_{0,\,t)(1\,-\,\text{ADF}j,l})Q_{j,l} \tag{7.28}$$

where $\text{ADF}_{j,l}$ and $Q_{j,l}$ are the available demand fraction and total demand at node j, respectively, during each repair work on link l. If a repetitive demand pattern exists for the duration of given time horizon at node j, the availability for that node between t_0 and t is then given by

$$A_j(t_0,t) = 1 - \frac{\sum\limits_{\text{all links}} E_l(t_0, t)(1 - \text{ADF}_{j,l})Q_{jl}}{n_c Q_{j,c}} \tag{7.29}$$

where n_c is the number of demand cycles from time t to t_0 and $Q_{j,c}$ is the total demand at node j per demand cycle. Note that the above formulation assumes single-order pipe failures only. In other words, it is assumed that no simultaneous pipe failures occur for the given time horizon. Equation (7.29) can be further simplified with the following three assumptions:

- Each failure occurs at the beginning of a demand cycle.
- Repair times are identical for all links (i.e., $Q_{j,l} = Q_j$).
- The duration of demand cycle and the repair time are equal (i.e., $Q_{j,l} = Q_{j,c}$).

Then Eq. (7.29) reduces to

$$A_j(t_0, t) = 1 - \frac{\sum_{\text{all links}} E_l(t_0, t)(1 - \text{ADF}_{j,l})}{n_c} \tag{7.30}$$

Systemwide availability can be determined by considering all the nodes in the system

$$A_s(t_0, t) = 1 - \frac{\sum_{\text{all links}} \sum_{\text{all nodes}} E_l(t_0, t)(1 - \text{ADF}_{j,l})Q_{j,l}}{\sum_{\text{all nodes}} n_c \cdot Q_{j,c}} \tag{7.31}$$

7.5 OPTIMIZATION MODEL

7.5.1 Effect of Valving on System Reliability

A valve is defined as a mechanical device by which the flow of fluid may be started, stopped, or regulated by a movable part that opens or obstructs passage. Valves are used in water distribution systems for a variety of purposes. Accordingly, different types of valves can be grouped into four (Ysusi, 2000): (1) isolation valves, (2) control valves, (3) blow-offs, and (4) air-release and vacuum-release valves.

Control valves are used to regulate flow or pressure at different points of the system by creating headloss or pressure differential between upstream and downstream sections. The mission of isolation valves is to isolate a portion of the system whenever system repair, inspection, or maintenance is required at that segment. The two most common types of isolation valves are gate valves and butterfly valves. Unlike the control valves, little attention has been paid to the layout and operation of isolation valves. Yet, from a reliability point of view, isolation valves are of great interest because they determine the extent of isolation should a portion of the system need repair, inspection, or maintenance. As an extreme example, a system with no valves would have to be shut down completely at the source for any maintenance. On the other hand, if there is sufficient valving in the system the outage or isolation can be limited to a small portion of the system.

Little is known about the reliability of isolation valves because such valves are fully opened under normal operating conditions and fully closed only when isolation of a portion of the system is needed. Valves that remain in either position for extended periods of time become difficult (or even impossible) to operate (Ysusi, 2000). Therefore, they should be exercised at least once a year or more often if the water is corrosive or dirty. According to Walski (2000), valve exercising is the single most important form of preventive maintenance for improving reliability.

Valves are typically located around junctions in a distribution system. Figure 7.2 shows a typical valve scheme for a typical city block. Their locations and numbers are usually determined using the rules of thumb that have evolved over the years. The most commonly used rule of thumb is to install a minimum of $n - 1$ valves around a junction to which n links (pipes) are connected. In other words, a minimum of three valves should be placed at every cross-intersection and a minimum of two valves is needed around every T-section.

Although regular exercising programs may increase their reliability, valves are not 100 percent reliable whenever they need to be closed for isolation of a distribution segment. Therefore, another consideration for valve placement is that no more than four valves should need to be operated when an isolation is needed. Consider, for example, pipe 1 in Fig 7.2. Should this pipe fail, six valves (valves a, b, c, d, e, and f) need to be closed in order to isolate the break from the remainder of the network. If the isolation is achieved through closing of all six valves, only those consumers whose service connection is linked to pipe segment 1 will be without water during the repair work. However, if there is a percent probability p that each valve can be operated (closed) successfully, the mission can only be successful with probability p^n. For example, for $p = 0.90$, the chance of operating all six is 53 percent ($0.90^6 = 0.53$). If valve c fails to operate among the six,* three additional valves (k, l, and m) need to

*The probability of that scenario occurring is $(1 - p)(p^5) = (1 - 0.90)(0.90^5) = 5.9$ percent.

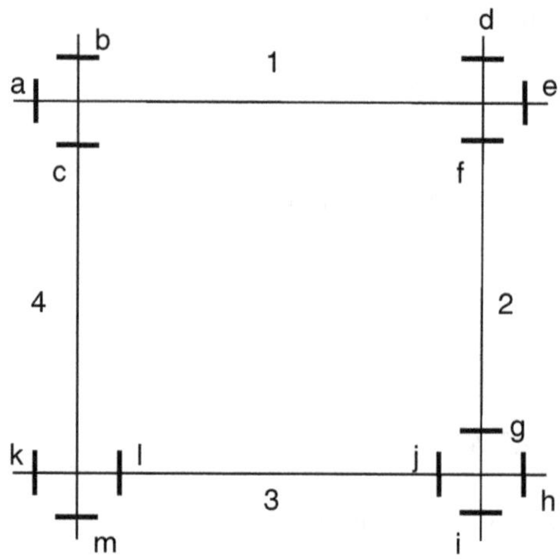

FIGURE 7.2 Example valve layout around a city block.

be closed for complete isolation. That means customers receiving water from pipe segment 4 will also be without water during the repair work on pipe segment 1.

The reliability of valves themselves has not been implicitly or explicitly incorporated in reliability assessments to date. Perhaps the main reason for this is that the human factor is the most important factor in determining their reliability. The more frequent the valve exercising programs, the greater the chance that they will operate when needed.

Walski (2000) gives a few more rules of thumb for locating isolation valves in distribution systems. The AWWA's Introduction to Water Distribution (1986) states that "Isolating valves in the distribution system should be located less than 500 ft apart (150 m) in business districts and less than 800 ft (240 m) apart in other parts of the system. It is a good practice to have valves located at the end of each block so that only one block will be without water during repair work." Similarly, The Ten States Standards state "Where systems serve widely scattered customers and where future development is not expected, the valve spacing should not exceed one mile." Finally, Walski (2000) points out a good practice, which requires placement of an isolation valve for every fire hydrant lateral.

The effect of valving in distribution system reliability has two aspects. First, some consumers will be out of service when a pipe break occurs in the system. The spatial extent of isolation is mostly determined by the locations of valves in the system. It is theoretically possible to determine the impact of isolation, induced by a link failure, by calculating the volume of water of which consumers located in the isolated portion of the system are deprived during the repair work. Let us call this the shut-off volume. Bouchart and Goulter (1991) studied this aspect of valving and expressed the reliability in terms of the expected volume of deficit along the isolated segment(s).

It should be noted, however, that network hydraulic models rarely possess the level of detail to determine the shut-off volume accurately. Typically, demands from city blocks are assigned to nearby junctions whereas in reality demands are withdrawn via the service connections located along the supply lines. The concept of lumping the continuous demands along the supply lines into single "point" loads at the network junctions using various demand allocation techniques is required in many cases (e.g., hydraulic simulation) for mathematical tractability (Bouchart et al., 1991).

The second implication of the valve scheme, besides determining the extent of direct shut-off (or volume of deficit), is the effect of isolation on system connectivity. The connectivity of a network is significantly affected by the valving scheme. For example, when pipe 1 fails in Figs. 7.3 and 7.4, connectivity among pipe segments 2, 3, and 4 would be preserved in Fig. 7.3 and it would be lost in the case of Fig. 7.4. The same goes for the links 5, 6, and 7.

Unlike the expected volume of deficit, the effect of the valve scheme on a network connectivity can only be assessed with hydraulic simulation of the deficient network.

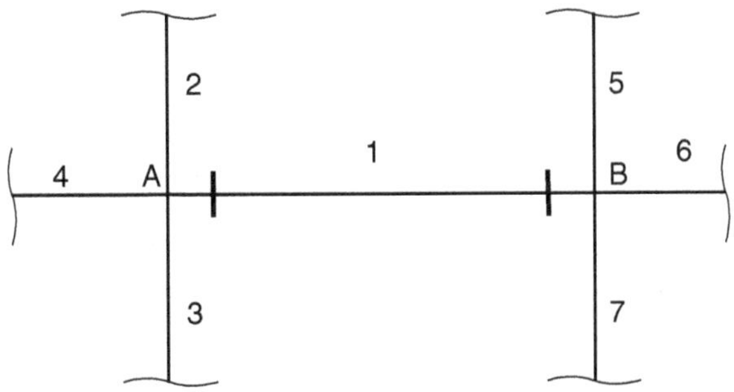

FIGURE 7.3 Valve scheme with pipe 1 fully valved.

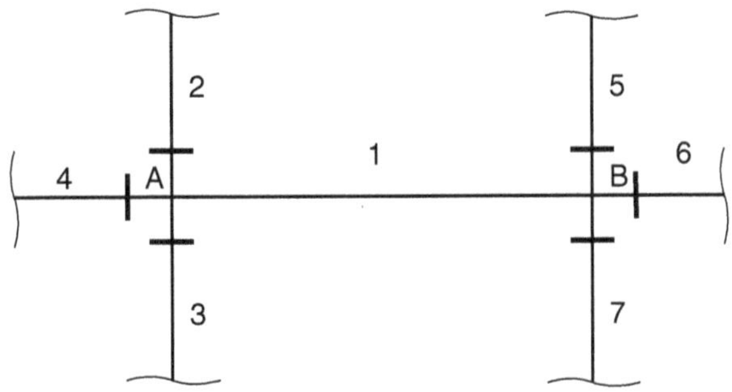

FIGURE 7.4 Valve scheme with no valves on pipe 1.

7.5.2 Optimal Valve Scheme Problem

Perhaps the single most important factor affecting the extent of isolation resulting from a pipe failure in a water distribution network is the valve layout in the vicinity of the break. A network with no isolation valves would have to be shut down at the source for any maintenance work. The more valves are situated in a network, the less the impact of a failure because the isolation can be restricted to a smaller portion of the system. Thus, only by knowing the valve layout in the vicinity of a pipe break is it possible to determine which links are to be closed and the deficient performance of the remaining network.

A fully valved network from a modeling approach is a network where each pipe in the network has isolation valves at both ends. Because demands are assumed lumped into junctions for mathematical tractability in conventional network modeling approaches, it would not make any difference if there were intermediate valves or not from a modeling point of view. In real life, on the other hand, demands are almost continuously distributed along a pipeline (withdrawn via service connections). Thus, the more intermediate valves are located on a pipe, the smaller the number of out-of-service consumers will be during the repair work following a pipe break.

In formulating the optimal valve location problem, a decision variable can be defined for each junction and the pipe that connects that junction to the rest of the system as

$x_{l,j}$ is the decision on whether to place a valve at the end of pipe l connected to junction j

$x_{l,j}$ is 1 if an isolation valve is placed and 0 otherwise

Thus the total number of decision variables is twice the number of pipes in a given network. Note that a solution to the problem is a binary string resulting in a combinatorial optimization problem. Due to the complex nature of the problem it is impossible to write the objective function explicitly in terms of the decision variables. It is well established that heuristic optimization techniques are very efficient in finding a near-optimal solution to such complex problems.

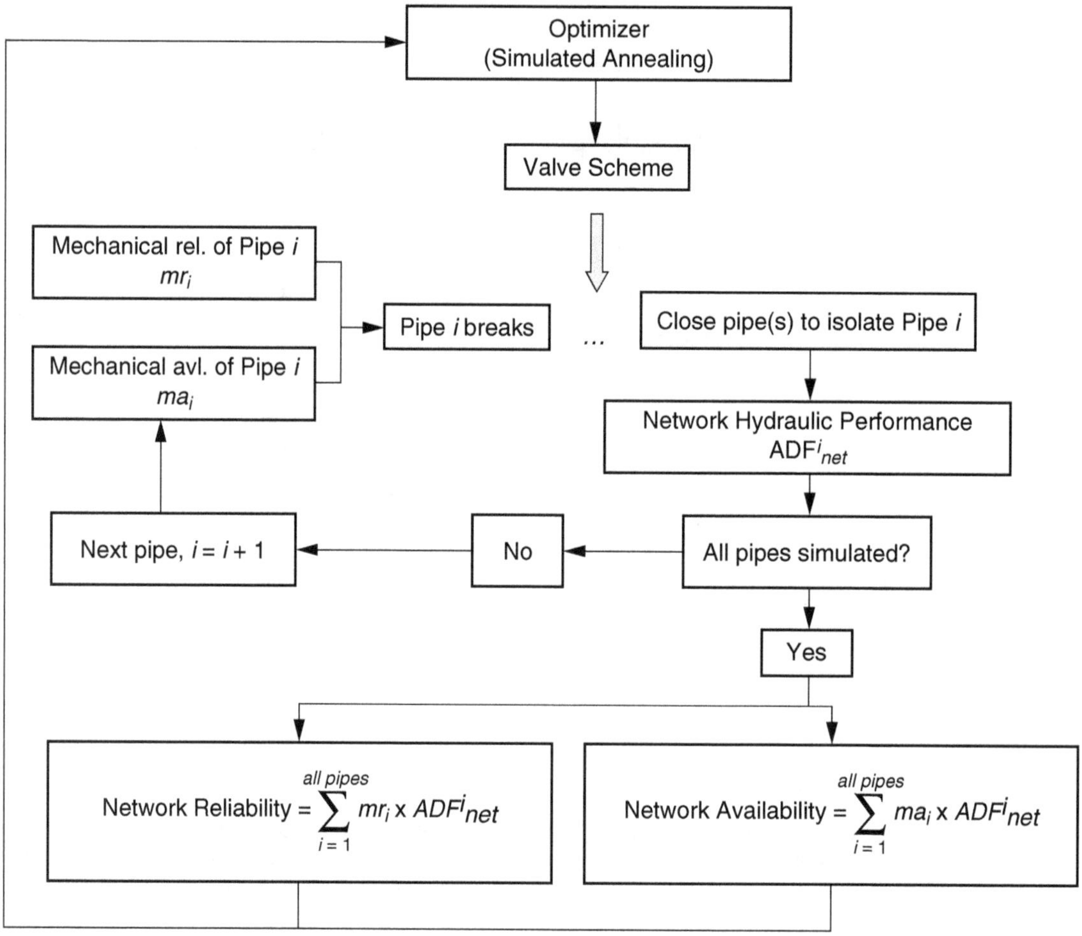

FIGURE 7.5 Flowchart of valve optimization problem.

Because the decision variables of our problem are binary, the solution space for a network with n pipes contains 2^{2n} solutions. The multiplier 2 of n comes from the fact that each pipe carries two decision variables, whether to place a valve at each end of a pipe. For a water distribution network with 50 pipes, there would be a total of $2^{100} = 1.26765 \times 10^{30}$ solutions. As the number of pipes reaches thousands (typical for even midsize networks), problem solution space and the time required for complete enumeration becomes even immeasurable.

Let us now add the following rule of thumb as a constraint to our problem: install $n - 1$ valves around a junction to which n links (pipes) are connected. For example, three valves should be placed at every cross-intersection and two valves are needed around every T-section. Mathematically, the "one less valve" constraint can be expressed as

$$\sum_{l}^{L_j} x_{l,j} = L_j - 1 \qquad \text{for } j = 1,\ldots, \text{NJ} \tag{7.32}$$

where $x_{l,j}$ = binary decision variable defined above
L_j = number of links around junction j
NJ = number of junctions in the network excluding dead-end nodes

It is assumed that this constraint does not apply to dead-end nodes where an isolation valve is always assigned unconditionally. The size of the solution space with the one less valve constraint is then given by

$$\mathrm{NS} = \prod_{j=1}^{\mathrm{NJ}} L_j \tag{7.33}$$

Thus, adding this constraint to the valve optimization problem would result in a significant reduction in the size of solution space. Note that the degree of this reduction depends largely on the level of looping in the network. Because total enumeration is impossible in many cases, local search heuristics such as simulated annealing or genetic algorithms can be used to find a near-optimal solution to the problem.

The objective function can be either the reliability or the availability of the given network

$$\text{Maximize } A_{\text{net}}$$

or

$$\text{Maximize } R_{\text{net}}$$

Note that the objective function does not include any cost term that depends largely on valve diameters in a given solution. The cost of a valve decreases as pipe diameter gets smaller. On the other hand, the probability of a pipe failure increases as the diameter gets smaller. This suggests that smaller diameter pipes are more likely to affect network reliability than larger diameter pipes and thus are more likely to have isolation valves in an optimal solution. However, this does not necessarily mean that isolation valves should be placed on all but the largest diameter pipe around each junction in an optimal valve scheme subject to the "one less valve" constraint, because there are other factors affecting a pipe's break rate such as a pipe's length. Connectivity also determines how important a pipe is from a reliability point of view. For example, a transmission main may have a low probability of failure but large supply deficiencies may result from its failure. Nevertheless, we expect that an optimal valve scheme subject to the "one less valve" constraint will also be one of the least-cost solutions for the reasons explained above and thus cost can be omitted from the objective function.

7.5.3 Simulated Annealing for Valve Optimization Problem

Simulated annealing (SA) has been termed a "biased random walk." The simulated annealing algorithm requires defining a local search neighborhood of the current solution in order to make its random moves. Although there are no strict rules for defining a neighborhood, a neighboring solution should be only slightly different from the current solution in order for the algorithm to be efficient. In order to maintain solution feasibility with respect to the "one less valve" constraint in the valve optimization problem, a random move to a neighboring solution is achieved as follows.

A junction in the network is chosen randomly and an isolation valve is added to the only valveless pipe around that junction. Next, a random pipe around the same junction is selected and the valve on that pipe is removed. However, this random pipe cannot be the same pipe as the valveless pipe, which is just assigned an isolation valve. This is done in order not to repeat the current solution. With this strategy, the "one less valve" constraint is maintained and the next solution is guaranteed feasible. The number of neighboring solutions NS with this random move approach is given by

$$\mathrm{NS} = \sum_{j=1}^{\mathrm{NJ}} (L_j - 1) \tag{7.34}$$

where L_j is the number of pipes connected to junction j and NJ is the number of junctions in the network excluding dead-end nodes.

Figure 7.5 shows the flowchart of valve optimization problem where ADF-based hydraulic performance is measured using the SPDA explained in previous sections.

The primary advantage of SA is the ability to escape from local optima. Thus, the ability to find the global optimum is not related to the initial conditions (i.e., the starting solution). Analogous to the physical process, simulated annealing slowly reduces the temperature T of the solution, thus gradually reducing the probability of accepting a poor move with time. The schedule by which T is reduced is called the cooling schedule and is critical to the success of SA. Two important parameters governing the cooling schedule are the step size for perturbation and T.

The major difficulty (art) in implementation of the simulated annealing algorithm is that there is no obvious analogy for the temperature T with respect to a free parameter in the combinatorial problem. If the initial temperature is too low or cooling is done insufficiently slowly the algorithm may trap in a local minimum. In order to avoid entrainment in local minima (quenching) the annealing schedule, the choice of initial temperature, number of iterations performed at each temperature, and how much the temperature is decremented at each step as cooling proceeds should be carefully planned.

The recommended parameter settings are step sizes and T values that allow approximately 80 percent of the poor moves to be accepted. The initial temperature for the SA algorithm can be determined to allow a certain fraction of poor moves to be accepted. One way of doing this is to search the entire neighborhood of the initial solution before the optimization starts and determine the average decrease Δ_{obj}^{-} of the objective function (assuming the objective is to maximize) with poor moves. Then the initial acceptance rate using the probability of acceptance for the Metropolis algorithm $e^{\Delta \text{obj}/T}$, is given by

$$P_{\text{init}} = e^{\Delta_{\text{obj}}^{-}/T_{\text{init}}} \tag{7.35}$$

where Δ_{obj}^{-} = average decrease of objective value from all poor moves
P_{init} = initial acceptance rate
T_{init} = user-specified initial temperature

Similarly, the final acceptance rate can be estimated using the final temperature T_{final} in the place of T_{init} in Eq. (7.35). Rearranging Eq. (7.35), initial temperature required for P_{init} is given by

$$T_{\text{init}} = \frac{\Delta_{\text{obj}}^{-}}{\ln(P_{\text{init}})} \tag{7.36}$$

For example, if approximately 60 percent of the poor moves are accepted at the beginning of SA, then Eq. (7.36) reduces to $T_{\text{init}} = -1.95762\Delta_{\text{obj}}^{-}$. Because this approach of setting the initial temperature might be sensitive to the initial solution the algorithm starts with, this procedure can be repeated for several different initial solutions and taking the average of the objective function decrease with poor moves.

The way in which the temperature is decreased in SA is known as the cooling schedule. The two most commonly used cooling schedules, in practice, are a linear cooling schedule ($T_{\text{next}} = T_{\text{previous}} - \Delta T$) and a geometric cooling schedule ($T_{\text{next}} = C \times T_{\text{previous}}$) where $C < 1.0$. The choice of cooling schedule is usually problem specific and should be considered as another tune-up parameter for an efficient optimization run.

There are several ways of determining when to stop an SA optimization. One way is to stop when the incumbent solution has not changed for a predetermined number of iterations. Another approach is halting the process when the total number of iterations reaches a predetermined limit. Another criterion is to stop when neighbors are being accepted at a rate of λ_f or less (e.g., $\lambda_f = 0.01$).

A less likely stopping criterion is when the incumbent value is sufficiently close to a good predetermined upper bound for the global optima. This is less likely because in most optimization problems, it is difficult, if not impossible, to find a good upper bound for the global optimal solution because the objective function is somewhat the combination of a problem-specific measure (e.g., reliability or availability) and the cost of the given solution. In the valve optimization problem, however, cost is eliminated from the objective function. Therefore, a good upper bound for the global optima can be obtained by determining the objective value (i.e., reliability or availability) of the fully valved network. Figure 7.6 shows the flowchart of valve optimization using simulated annealing.

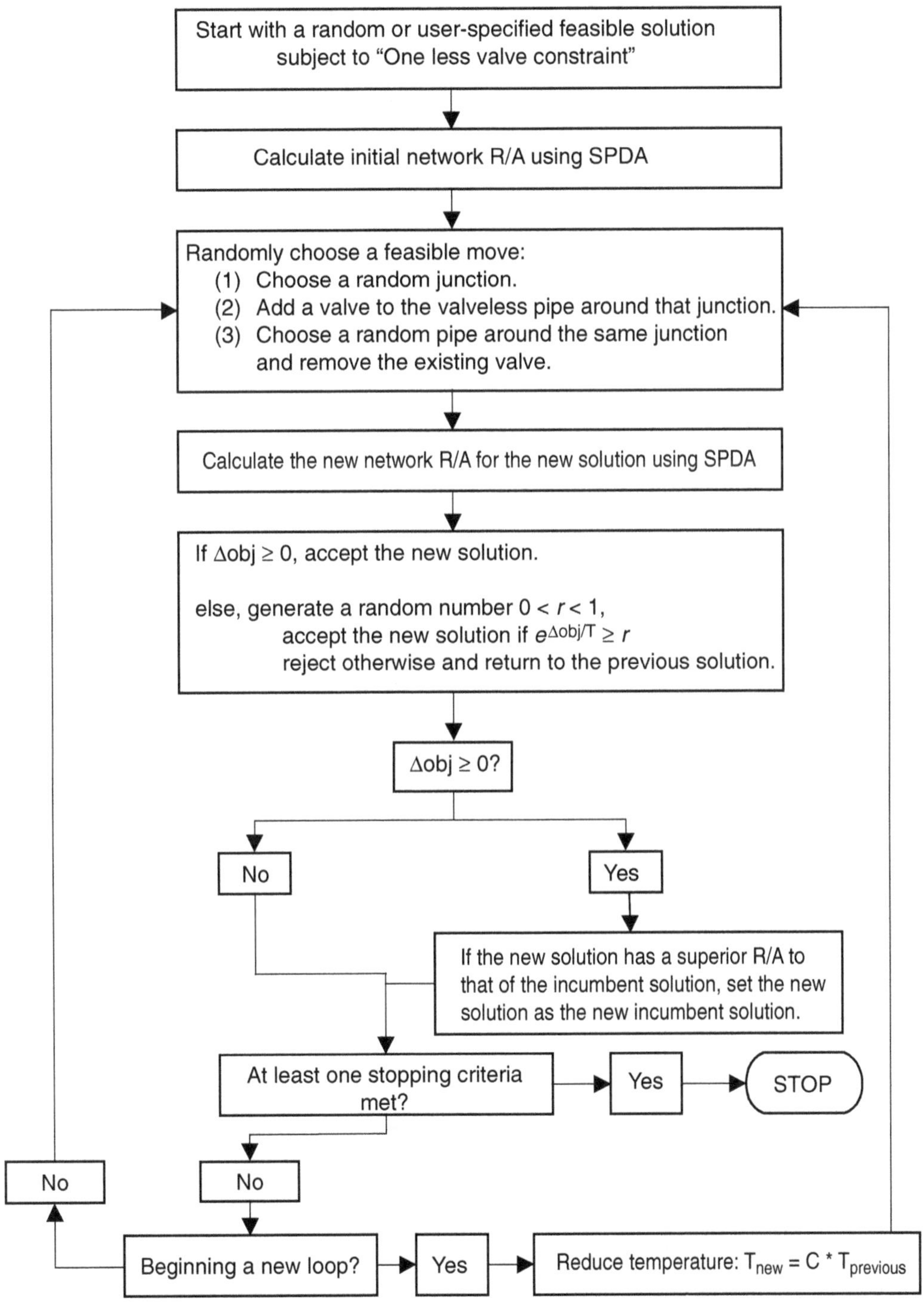

FIGURE 7.6 Flowchart of the simulated annealing search for valve optimization.

7.6 EXAMPLE APPLICATION OF RELIABILITY/OPTIMIZATION MODEL

The valve optimization using SA is illustrated using the example network shown in Fig. 7.7. The network consists of two fixed reservoirs, 21 pipes, and 13 junctions. Pipe characteristics and the junction properties are summarized in Tables 7.1 and Table 7.2, respectively. Demands are expressed in m^3/hour (CMH) and the calculated pressures are in meters. The total network demand for the steady-state simulation is 3146.4 cubic meters per hour.

A global value of 20 psi is assumed as threshold pressure for all nonzero demand junctions. Mean time to repair (MTTR) is assumed one day. The cooling schedule targets 1000 iterations with an initial temperature of 6, and 16 iterations per loop, and the temperature multiplier is 0.85. In other words, temperature is reduced by a factor of 0.85 every 16 iterations in the geometric cooling schedule. The initial solution is specified random and the optimization type is reliability-based.

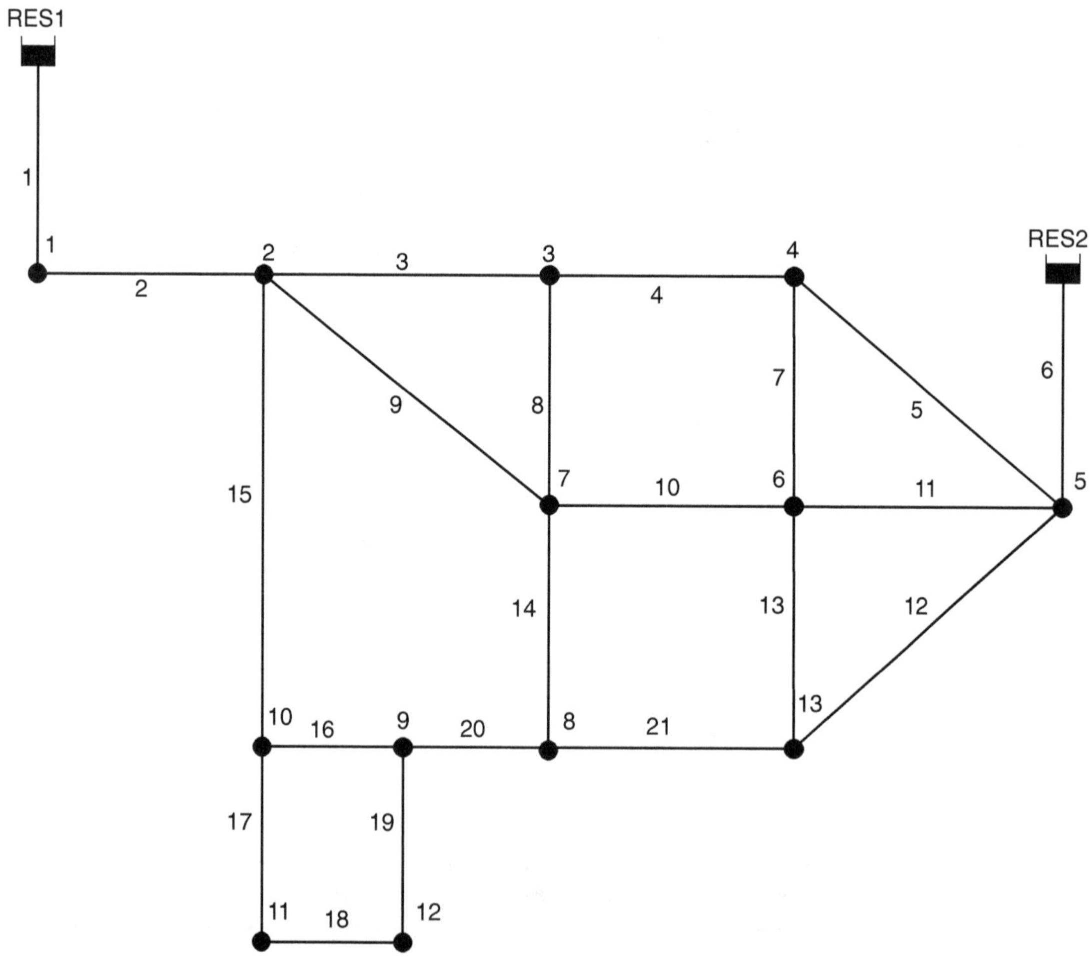

FIGURE 7.7 Application network.

TABLE 7.1 Pipe Characteristics

Pipe ID	Length, m	D, mm	C, H-W*
1	609.60	762	130
2	243.80	762	128
3	1524.00	609	126
4	1127.76	609	124
5	1188.72	406	122
6	640.08	406	120
7	762.00	254	118
8	944.88	254	116
9	1676.40	381	114
10	883.92	305	112
11	883.92	305	110
12	1371.60	381	108
13	762.00	254	106
14	822.96	254	104
15	944.88	305	102
16	579.00	305	100
17	487.68	203	98
18	457.20	152	96
19	502.92	203	94
20	883.92	203	92
21	944.88	305	90

*H-W = Hazen-Williams.

TABLE 7.2 Node Characteristics

Node ID	Elevation, m	Demand, CMH
1	27.43	0.0
2	33.53	212.4
3	28.96	212.4
4	32.00	640.8
5	30.48	212.4
6	31.39	684.0
7	29.56	640.8
8	31.39	327.6
9	32.61	0.0
10	34.14	0.0
11	35.05	108.0
12	36.58	108.0
13	33.53	0.0
RES1	60.96	N/A
RES2	60.96	N/A

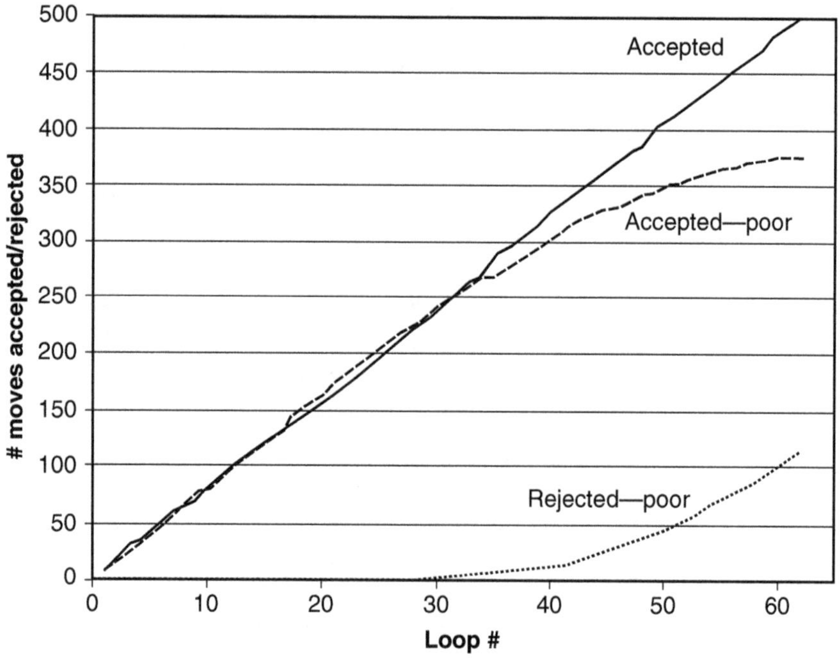

FIGURE 7.8 Cumulative accepted and rejected moves in SA search.

A test run involving 50 randomly selected initial solutions reveals that the initial and final acceptance rates among poor moves will be approximately 99.97 and 0.2 percent. This indicates that almost all the prospective moves will be accepted at the beginning of the simulation and the rate of acceptance decreases significantly at the end. The test run also determines the size of the solution space as 1,492,992 with the "one less valve" constraint and the neighborhood size as 27. The upper bound, corresponding to a fully valved network, is 0.9537.

The results indicate that there were 497 improving moves that were unconditionally accepted because they had a superior objective value to that of the current solution. Out of the remaining 503 nonimproving prospective moves, 403 were accepted although the objective function decreased with each, and 100 were rejected by the Metropolis algorithm. Figure 7.8 shows the cumulative number of accepted, accepted-poor, and rejected moves at the end of each loop. Although almost all the nonimproving moves were accepted at the beginning, the latter part of Fig. 7.8 shows the impact of reducing temperature T as the algorithm proceeds. The rejection rate starts going up at about the 30th loop. By the end of the simulation the acceptance rate among the poor-moves decreases to almost zero as also expected from the results of the test run. Overall, 80 percent of the poor-moves were accepted as targeted at the beginning.

The final incumbent value was 0.9492, a number within 0.5 percent of the upper bound found by assuming a fully valved network. Figure 7.9 shows the objective function value change with the solution progress. Note that about halfway through the simulation fluctuations in the objective value start diminishing, reflecting the effect of decreasing the temperature. In other words, fewer and fewer poor solutions are accepted as the algorithm proceeds. This consequence of the temperature effect can best be seen toward the end where the objective value variation has a much smaller range than the first two thirds of the simulation. This type of behavior is typical for SA. Gross features of the eventual solution appear at the higher temperatures with fine details developing at low temperatures (Kirkpatrick et al., 1983).The fact that the search escaped local optima many times and the final incumbent occurred toward the end of the simulation shows the effectiveness of simulated annealing.

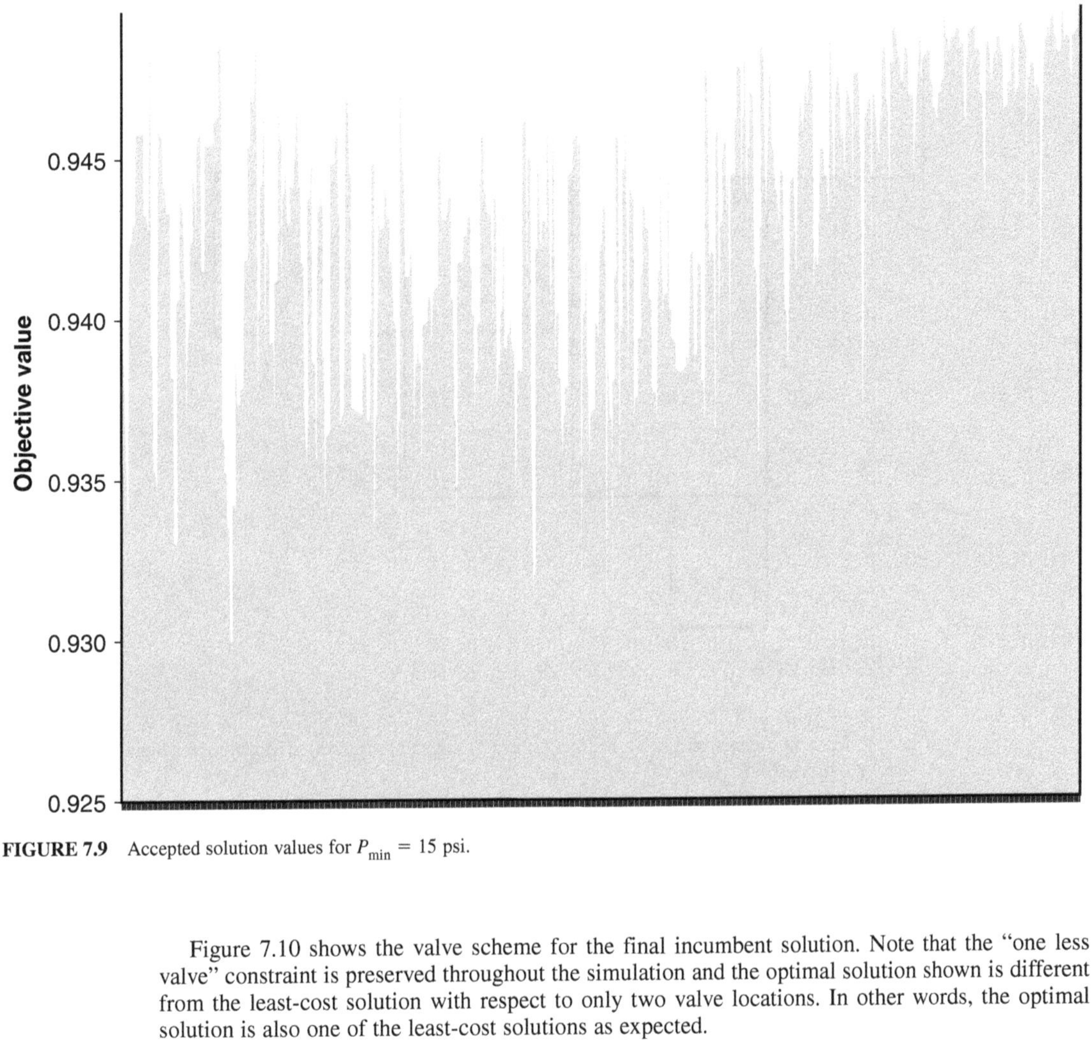

FIGURE 7.9 Accepted solution values for P_{min} = 15 psi.

Figure 7.10 shows the valve scheme for the final incumbent solution. Note that the "one less valve" constraint is preserved throughout the simulation and the optimal solution shown is different from the least-cost solution with respect to only two valve locations. In other words, the optimal solution is also one of the least-cost solutions as expected.

REFERENCES

AWWA (1986). *Introduction to Water Distribution,* American Water Works Association, Denver, Colo.

Bhave, P. R. (1991). *Analysis of Flow in Water Distribution Networks,* Technocomic, Lancaster, Pa.

Bouchart, F. and I. Goulter (1991). "Reliability Improvements in Design of Water Distribution Networks Recognizing Valve Location," *Water Resources Research,* American Geophysical Union, 27(12):3029–3040.

Bouchart, F. and I. Goulter (2000). "Chapter 18: Reliability Analysis for Design," in *Water Distributions Systems Handbook,* L. W. Mays, ed., McGraw-Hill, New York.

Chin, D. A. (2000). *Water Resources Engineering,* Prentice-Hall, Upper Saddle River, N.J.

Clark, R. M. (1987). "Applying Water Quality Models, Chapter 20," in *Computer Modeling of Free-Surface and Pressurized Flows,"* M. H. Chaudhry and L. W. Mays, eds. Kluwer Academic, Dordrecht, Netherlands, pp. 581—612.

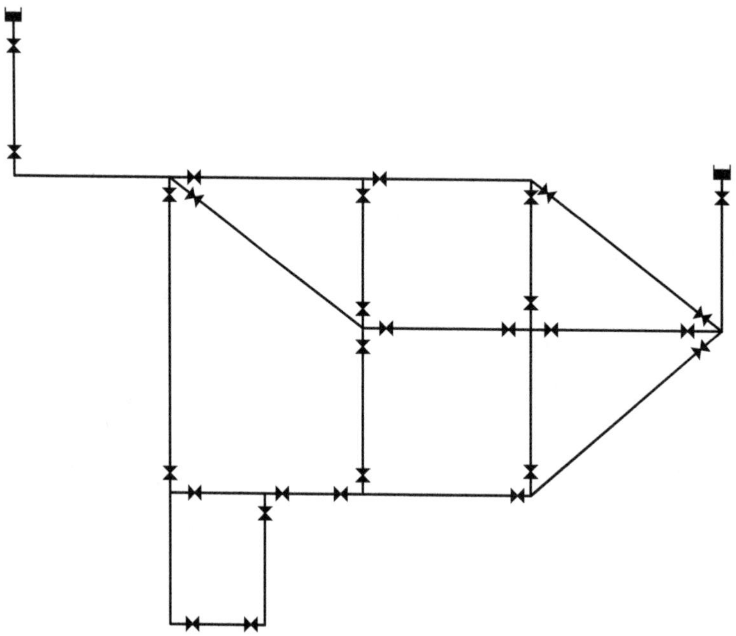

FIGURE 7.10 Optimal valve layout for example network I from SA.

Clark, R. M., J. Q. Adams, and R. M. Miltner (1987). "Cost and Performance Modeling for Regulatory Decision Making," *Water,* 28(3):26/24/040—27.

Clark, R. M., C. L. Stafford, and J. A. Goodrich (1982). "Water Distribution Systems: A Spatial and Cost Evaluation," *Journal of Water Resources Planning and Management,* ASCE, 108(8):243–256.

Crow, L. H. (1974). "Reliability Analysis for Complex, Repairable Systems in Reliability and Biometry." *SIAM,* F. Proschan and R. J. Serfling, eds. Philadelphia, 379–410.

Cullinane, M. J. (1986). "Hydraulic Reliability of Urban Water Distribution Systems," *Proceedings, Water Forum 1986: World Water Issues in Evolution,* ASCE, New York, pp. 1264–1271.

Fujiwara, O. and A. U. De Silva (1990). "Algorithm for Reliability-Based Optimal Design of Water Networks." *Journal of Environmental Engineering,* 116(3):575–587.

Gleick, P. H. (1993). *Water in Crisis, A Guide to the World's Freshwater Resources,* Oxford University Press, New York.

Gleick, P. H. (1998). *The World's Water 1998–1999, The Biennial Report on Freshwater Resources,* Island Press, Washington, D.C.

Gupta, R. and P. R. Bhave (1996). "Comparison of Methods for Predicting Deficient Network Performance," *Journal of Water Resources Planning and Management,* ASCE, 122(3):214–217.

Gupta, R. and P. R. Bhave (1997). "Closure of 'Comparison of Methods for Predicting Deficient Network Performance'," *Journal of Water Resources Planning and Management,* ASCE, 123(6):370.

Kirkpatrick, S., C. Gelatt, and M. Vecchi (1983). "Optimization by Simulated Annealing," *Science,* 220:671–680.

Mays, L. W., ed. (1989). *Reliability Analysis of Water Distribution Systems,* American Society of Civil Engineers, New York.

Mays, L. W., ed. (2000) *Water Distribution Systems Handbook,* McGraw-Hill, New York.

Mays, L. W., ed. (2002). *Urban Water Supply Handbook,* McGraw-Hill, New York.

Mays, L. W., ed. (2004a). *Urban Water Supply Management Tools,* McGraw-Hill, New York.

Mays, L. W., ed. (2004b). *Water Supply Systems Security,* McGraw-Hill, New York.

Ozger, S. (2003). "A Semi-Pressure-Driven Approach to Reliability Assessment of Water Distribution Networks," Ph.D. dissertation, Department of Civil and Environmental Engineering, Arizona State University, Tempe, Ariz.

Ross, S. M. (1985). *Introduction to Probability Models,* Academic, New York.

Rossman, L. A. (1994). "EPANET Users Manual," Drinking Water Research Division, Risk Reduction Engineering Laboratory, Office of Research and Development, U.S. Environmental Protection Agency, Cincinnati, Ohio.

Rossman, L. A. (2000). "EPANET Users Guide," Drinking Water Research Division, Risk Reduction Engineering Laboratory, Office of Research and Development, U.S. Environmental Protection Agency, Cincinnati, Ohio.

Rossman, L. A. (2001). "Thread: Modeling Demands," MIKE NET Support Forum. http://www.bossintl.com/forums.

Shinstine, D. S., I. Ahmed, and K. Lansey (2002). "Reliability/Availability Analysis of Municipal Water Distribution Networks: Case Studies," *Journal of Water Resources Planning and Management,* ASCE, 128(2):140–151.

Tabesh, M. (1998). "Implications of the Pressure Dependency of Outflows on Data Management, Mathematical Modelling and Reliability Assessment of Water Distribution Systems," PhD thesis, University of Liverpool, Liverpool, England.

Tanyimboh, T. T. and M. Tabesh (1997). "Discussion of 'Comparison of Methods for Predicting Deficient Network Performance,'" *Journal of Water Resources Planning and Management,* ASCE, 123(6):369–370.

Tanyimboh, T. T., R. Burd, R. Burrows, and M. Tabesh (1999). "Modelling and Reliability Analysis of Water Distribution Systems," *Water Science Technology,* 39(4):249–255.

Tanyimboh, T. T., M. Tabesh, and R. Burrows (2001). "Appraisal of Source Head Methods For Calculating Reliability of Water Distribution Networks," *Journal of Water Resources Planning and Management,* ASCE, 127(4):206–213.

Tobias, P. A. and D. C. Trindade (1995). *Applied Reliability,* 2d ed., Chapman & Hall/CRC, New York.

Walski, T. M. (2000). "Chapter 17: Maintenance and Rehabilitation/Replacement," *Water Distributions Systems Handbook,* L. W. Mays, ed., McGraw-Hill, New York.

Wood, D. J. (1980). "User's Manual—Computer Analysis of Flow in Pipe Networks Including Extended Period Simulation," Department of Civil Engineering, University of Kentucky, Lexington, Ky.

Ysusi, M. A. (2000). "Chapter 3: System Design: An Overview," *Water Distributions Systems Handbook,* L. W. Mays, ed., McGraw-Hill, New York.

CHAPTER 8
GROUNDWATER REMEDIATION DESIGN USING SIMULATED ANNEALING

Richard L. Skaggs
Pacific Northwest National Laboratory
Richland, Washington

Larry W. Mays
Department of Civil and Environmental Engineering
Arizona State University,
Tempe, Arizona

8.1 INTRODUCTION

8.1.1 What Is Simulated Annealing?

There has been an emergence in the use of combinatorial methods such as simulated annealing in groundwater management during the past 10 years. Although previous studies demonstrated the feasibility of using these methods, a general finding was that computational processing requirements were inordinately high relative to gradient-based methods. An enhanced annealing algorithm was developed and used to demonstrate the potential for greatly improving the computational efficiency of simulated annealing as an optimization method for groundwater management applications. The algorithm incorporates "directional search" and "memory" capabilities. Selecting search directions based on better understanding of the current neighborhood of the configuration space was shown to improve algorithm performance. Also, memory concepts derived from the tabu search method show particular promise for improving the rate and quality of convergence. Performance of the enhanced annealing method was evaluated and the resultant management method was demonstrated using an example from the literature.

In recent years, groundwater quality management models (consisting of groundwater flow and contaminant transport simulation models coupled with optimization) have been developed to aid in groundwater remediation design. The simulator allows hydrogeologists to realistically estimate the effects of water withdrawal, injection, and contaminant movement under alternative remediation conditions. The optimization algorithms used in these groundwater management models have included linear programming and nonlinear optimization techniques, and more recently less conventional combinatorial optimization methods including neural networks (Rogers and Dowla, 1994), genetic algorithms (Ritzel et al., 1994), and simulated annealing (Dougherty and Marryott, 1991; Marryott et al., 1993).

Building on these and other research efforts, the overall objective of the research presented here is the development of a methodology focused on evaluation of alternative groundwater contamination cleanup technologies and management strategies. The methodology couples flow and contaminant transport simulation and optimization to enhance decision-making effectiveness when considering multiple issues such as technical, environmental, regulatory, and financial risks. The focus of the work is on development of an optimization method that identifies near-global optima without exorbitant computational resource requirements. The optimization method included in the methodology is enhanced simulated annealing, simulated annealing modified to include the concepts of memory and directional search.

Simulated annealing has been shown to locate global or near-global optima with a high degree of reliability. However, an important offsetting disadvantage is the relatively large number of required objective function evaluations. This is a particularly acute problem in groundwater management when hundreds to thousands of objective function evaluations may be necessary to solve a given problem, and each objective function evaluation is based on a flow and transport simulation requiring up to several minutes of computer processing time. Thus, an enhanced annealing methodology was developed to significantly reduce the number of objective functions required in the optimization procedure.

8.1.2 Background

Since the late 1970s, there has been a considerable body of work directed at the development of management models to assist decision-makers in selecting the most reliable and cost-effective methods for remediation of contaminated groundwater. The general problem, as viewed by Wanakule et al. (1986), is one of discrete time optimal control where variables describing the aquifer system (e.g., heads, concentrations, etc.) are direct functions of the control variables (e.g., pumping rates). Management models generally consist of an optimization method linked to groundwater flow and solute transport models as subroutines to evaluate the cost and feasibility of alternative remediation designs. The majority of the optimizers used are linear, nonlinear, or differential dynamic programming optimal control methods. Recently, efforts have investigated the performance of other types of optimizers from the field of combinatorial optimization.

Dougherty and Marryott (1991) provide an overview of the simulated annealing algorithm and guidance on its application to groundwater management. They note the simplicity with which simulated annealing is implemented and coupled to flow and transport simulators. They apply their algorithm to several problems ranging from a highly idealized dewatering problem to one combining injection/extraction wells with a slurry trench. The dewatering problem is used to demonstrate relative performance of several variations of the algorithm and the iterative concept of *zooming* to reduce the number of objective function evaluations. Zooming, a refinement method in which a sequence of refined estimates of the objective function is solved, is found to significantly reduce the computation time but reduce the likelihood of locating a global minimum.

The pump-and-treat remedial design process is modeled by linking a simulated annealing algorithm with a flow-contaminant transport simulator. This application was used to demonstrate the advantages of problem-specific heuristics in reducing the objective function calls (e.g., screening alternatives based on different components of the objective function, and retaining objective function values for a limited number of previously attempted configurations). The authors also suggested the flow and transport problem could be solved only approximately in the early stages of the solution without significantly affecting results. With a fourth example they demonstrated the ease with which multiple technologies can be addressed by evaluating a pump-and-treat design coupled with a slurry trench. A comparison of simulated annealing with the MINOS (Murtagh and Saunders, 1980) algorithm suggested that CPU requirements for annealing increase at a slower rate than for the gradient-based method as the level of contaminant removal is increased.

In a sequel to the paper discussed above, Marryott et al. (1993) apply simulated annealing to an actual contaminated field site. Simulated annealing coupled with a flow and transport simulator is used to solve two problems. The first is a regulatory constraint formulation in which concentration levels in the aquifer are reduced below regulatory standards by the end of a designated planning period. The second is a velocity constraint formulation in which the goal is plume containment. Results from the evaluations show that the first formulation produces lower minimum costs and the second requires less computational effort. They indicate that although reducing the number of objective function evaluations at each temperature of the cooling schedule can greatly reduce the computational effort, it also decreases the probability of achieving the optimal solution.

The authors conclude that the computational effort required for simulated annealing, although large, is comparable to other nonlinear optimization techniques. And, although the simulated annealing method applied to groundwater management shows promise in reliably locating near-global optima for groundwater management problems, it is not yet mature and extensions and enhancements are needed.

8.1.3 Statement of the Problem

The solution process for simulated annealing is typically composed of several steps. Beginning at an elevated control parameter value, a completely stochastic search is conducted. Initially, sequential moves within the configuration space are large in magnitude as is the relative probability of making an uphill move (for a minimization problem). From initiation of the search to convergence, both the magnitude and direction of the move are blindly controlled by probabilistic functions. As the control parameter is decremented, the magnitude of the moves and the likelihood of moving uphill decrease exponentially at the same relative rate until convergence is achieved. In developing and selecting future moves, the simulated annealing algorithm contains no mechanisms that explicitly consider information about past moves (regions of relative success or failure). Nor can it consider the relative merit of one direction over another when evaluating candidate moves from a current location (e.g., gradients or projections). In other words, all moves are dictated by the current status of the annealing process and comparison of the current objective function value with that of proposed moves to randomly selected candidates.

The potential value in considering directional preferences and past moves in making search decisions can be deduced from a typical simulated annealing search pattern. Early, when the search range is large and the probability of moving uphill is high, large swings in objective function values are common. Almost all proposed moves, whether uphill or downhill, are taken. Later in the search, after only a few hundred iterations, the search has converged and narrowed to a smaller region providing only minor variations in objective function value. Thus, a significant percentage of the latter iterations are expended in a relatively small region of the configuration space. This is attributable to inherent features of the annealing process:

1. During the latter part of the search the potential change (i.e., search range) for all decision variables is uniformly small.
2. The likelihood of changing all variables is the same, regardless of which variables might have the greatest potential impact on the objective function value.

If the magnitude of movement could be increased for variables to which the objective function value is least sensitive (i.e., move into another region of the configuration space), the likelihood of locating a potential improvement and escaping a local minimum could be increased. The probability of wasting iterations on candidate moves representing little or no potential improvement could be reduced, thus increasing the efficiency of the search.

Also, near the end of the search a large number of iterations are expended in a small region of the configuration space and the likelihood of reencountering a given configuration becomes relatively high. When this occurs, no new information can be gained from reevaluating the objective function at these locations. In addition to individual sites, entire regions of the encountered configuration space may demonstrate little or no promise of containing a global or near-global optimum. It follows that the ability to remember recently encountered configurations or even sequences of iterations that proved unfruitful could, therefore, further reduce wasted (unfruitful) objective function evaluations.

8.2 ENHANCED ANNEALING

Simulated annealing is a stochastic computational method derived from statistical mechanics. As observed by Kirkpatrick et al. (1983), "there is a deep and useful connection between statistical mechanics (the behavior of systems with many degrees of freedom in thermal equilibrium at a finite temperature) and multivariate or combinatorial optimization (finding the minimum of a given function depending on many parameters)." Kirkpatrick et al. (1983) first demonstrated its applicability to large combinatorial optimization problems such as wire routing and component placement in computer design. Subsequent to its development, simulated annealing has proven effective in solving a multitude of hard optimization problems (i.e., locate near-global optima) containing multiple local optima, while being straightforward and efficient to implement (Davis and Steenstrup, 1987). In contrast to gradient-based methods, simulated annealing does not require calculation of derivatives, thereby eliminating the need for the objective function and constraints to be continuous and differentiable. Incorporating the concepts of directional search and memory into the simulated annealing algorithm results in an algorithm referred to as *enhanced annealing* (Skaggs, 1995; Skaggs et al., 2001a; Skaggs et al., 2001b).

Simulated annealing gets its name from and is based on the analogy between the simulation of the annealing of solids and solving large combinatorial optimization problems. A solid is annealed by increasing its temperature so that its molecules are randomly arranged in the liquid phase, followed by slow cooling to allow the molecules to arrange themselves in the low-energy ground state of a corresponding lattice. If the thermal ability is lost at a sufficiently slow rate, the molecules are able to align themselves into a completely ordered crystalline structure. This configuration is the state of minimum energy for the system. If the cooling occurs too rapidly, the system does not achieve the highly ordered state, but ends up in a meta-stable, higher-energy state.

Because the number of atoms in a given system is extremely large, statistical mechanics is used to stochastically analyze the aggregate properties of the material. Simulating the attainment of thermal equilibrium of a solid for a fixed temperature can be accomplished using a Monte Carlo method proposed by Metropolis et al. (1953). At each temperature value T, a system allowed to achieve thermal equilibrium has the probability of being in energy state E given by the *Boltzmann probability distribution*,

$$P\{\mathbf{E} = E\} = \frac{1}{Z(T)} \exp\left(-\frac{E}{k_B} T\right) \tag{8.1}$$

where $Z(T)$ = partition function depending on temperature T
k_B = Boltzmann constant
$\exp(-ET/k_B)$ = Boltzmann factor

Equation (8.1) specifies that a system in thermal equilibrium at temperature T has its energy probabilistically distributed among all different energy states, $E_I \in E$. Using this equation, it is possible to predict the thermodynamic behavior of an idealized natural system as it is annealed. At high temperatures, the system is constantly changing and the probability is relatively high for all energy states. As the system is cooled, the probability of changing to higher-energy states decreases. However, even at low temperatures, there is still a finite chance that the system will transform to a higher-energy state, and thus escape from a local energy minimum in favor of finding a better, more global one.

Computationally, given a current state of the solid characterized by the positions of the particles, a randomly generated small displacement is applied to a randomly selected particle. If the difference in energy ΔE, between the current state and the perturbed state is negative, the perturbed state is accepted as the new state. If $\Delta E \geq 0$, then the probability of accepting the new state is determined by Eq. (8.1). This acceptance rule (function) for new states is known as the Metropolis criterion. In applying this to combinatorial optimization, the configurations of the optimization problem are analogous to the states of the solid whereas the cost function takes the place of energy. Temperature simply becomes a control parameter for the algorithm.

Several steps are required in the implementation of simulated annealing. Initial values of the control parameter, c_0, and the cost function, C_0, are determined. A specified number L of Metropolis loops are then completed. Each loop consists of generating a new configuration, calculating its associated cost function value C_1, and applying the Metropolis criterion to determine whether the new configuration is accepted (i.e., C_0 is replaced by C_1) or rejected. After L loops have been completed the control parameter is adjusted and a new Metropolis loop is begun. Algorithms used for each of the key steps and the associated parameters are the following.

Initial value of the control parameter, c_0. For an idealized cooling schedule c_0 should be selected such that almost all transitions are selected. The approach by Vidal (1993) computes c_0 according to Eq. (8.2) where m_0 is a preset number of transitions for which m_1 produce $DC \leq 0$, m_2 produce $DC > 0$, $\Delta C^{(+)}$ is the average increase in C over the m_2 transitions, and χ is equal to a value close to 0.90.

$$c_0 = \overline{\Delta C^{(+)}} \left[\ln \left(\frac{m_2}{m_2 \chi - m_1(1 - \chi)} \right) \right]^{-1} \tag{8.2}$$

Decrements in the control parameter, c_0. A commonly used rule to ensure quasi-equilibrium is achieved from one Metropolis loop to the next is simply

$$c_{k+1} = \alpha_{k+1}\, c_k \qquad k = 0,1,2,\ldots, \tag{8.3}$$

where α ranges from 0.5 to 0.99 but generally is close to 1.0 (Laarhoven and Aarts, 1987). Chardaire and Lutton (1993) propose a slightly different approach that begins by presetting the number of control parameter decrements T such that Eq. (8.3) holds and

$$\alpha^T = 0.1 \tag{8.4}$$

which, for T equal to 25 would make α equal to about 0.91. However, it is suggested that T should be increased with decreasing values of the control parameter according to

$$T_{k+1} = T_k \omega \qquad \text{with } 1.0 < \omega < 1.05 \tag{8.5}$$

Markov chain length, L. Underlying the approach to selecting L is the notion that to achieve global minimization, quasi-equilibrium must be achieved at the completion of each Metropolis loop. Typically, the value of L is empirically determined, but Eq. (8.6), as recommended by Anderson and Vidal (1993), can serve as a guide where n is the number of decision variables.

$$L = 2(n - 1)n \tag{8.6}$$

Neighbor generation. The strategy used to make the transition from the current configuration to a neighboring configuration tends to be problem specific and dependent on how the solution space is configured. The major requirement is that transition from a given configuration to any other configuration should be possible in a finite number of moves. However, as suggested by Laarhoven and Aarts (1987), the neighbor generation mechanism should be biased toward "large" transitions at high values of the control parameter (i.e., produce large variations in the objective function value). And at low values of c, when large increases are often rejected, the bias should be toward small transitions. As a guide they suggest transitions should produce changes in the objective function approximately equal to the control parameter.

Dougherty and Marryott (1991) recommend an excursion limiting approach which simply varies the value of each of the decision variables by $\pm l$ levels from its current value. At the initial value of c, the magnitude of the excursion is defined by

$$\pm l = \pm s_l = \pm \frac{M}{2} \tag{8.7}$$

where s_l varies from $M/2$ to a lower limit equal defined as

$$s_l = \max \left[s_l \exp \left(\frac{C_0 - C_k}{C_k} \right), s_{l\min} \right] \tag{8.8}$$

and M is the number of alternatives available for each decision variable. Thus, a transition is accomplished by adjusting each decision variable by an amount equal to the integer nearest

$$l = s_l \text{ random}[0,1] \tag{8.9}$$

where C_0 and C_k are the initial and current values of the objective function, respectively. $s_{l\min}$ is equal to 0.51.

Convergence criteria. The annealing process is terminated when the rate of change in the average objective function values falls below a specified level as determined by

$$\overline{C_k} - k^{-1} \sum_{k-3}^{k} C_k \leq \epsilon \tag{8.10}$$

where e is a small positive number (say 0.05) and k equals 8.

8.3 ANNEALING WITH DIRECTIONAL SEARCH

An important aspect of traditional nonlinear optimization methods is the use of gradient operators at a point to determine the search direction. For a continuous and continuously differentiable function $f(\mathbf{x})$ at point x, the gradient is defined as:

$$\nabla f(\mathbf{x}) = \left[\frac{\partial f}{\partial \mathbf{x}} \right] = \left(\frac{\partial f}{\partial x_1}, \frac{\partial f}{\partial x_2}, \ldots, \frac{\partial f}{\partial x_n} \right)^T \tag{8.11}$$

and $\nabla f(\mathbf{x})$ represents the direction of steepest descent. In the combinatorial optimization problem addressed by simulated annealing, the conditions of continuousness and differentiability are not necessarily met; thus, selection of directions based on function gradients is not feasible.

However, in lieu of gradients, simulated annealing can be manipulated through its dependence on probabilities. Standard simulated annealing, when selecting a proposed transition from a given point, modifies all decision variables through the use of a probability function related to the current value of the control parameter c and a uniform probability distribution. The excursion length $s_{l\min}$ is equal for all the decision variables as is the exponential rate at which the excursion length decreases from its original value of $M/2$. Thus, the probability of a given magnitude dislocation at a particular stage of the annealing process is equal for all decision variables. An alternative approach would be, at strategic instances within the search, to adjust the excursion lengths of selected variables to increase the likelihood of moving to a potentially more desirable configuration.

It was observed by Ingber (1993) that in conducting annealing searches for many D-dimensional parameter space problems, due to physical considerations, different parameters might have different finite ranges and different annealing time-dependent sensitivities. This observation led to the consideration of independent annealing schedules for individual parameters in his adaptive simulated annealing method. Consequently, Ingber (1993) implemented a *reannealing* procedure whereby the range over which insensitive parameters are being searched is stretched out. In his algorithm the magnitude of search for a given parameter is a function of the annealing temperature wherein the annealing temperature is related to annealing time by

$$T_i(k_i) = T_{0i} \exp(-c_i k_i^{1/D}) \tag{8.12}$$

where k_i = annealing time step for parameter i
k = zero initially, incremented by 1 at completion of each temperature step
T_{0i} = initial annealing temperature for parameter i
c_i = tuning parameter
D = problem dimensionality

The reannealing is accomplished by periodically rescaling the annealing time for parameter α_i in terms of the sensitivities s_i calculated at the current value of the cost function L,

$$s_i = \frac{\partial L}{\partial \alpha^i} \tag{8.13}$$

Thus, the annealing time is rescaled according to

$$T'_{ik'} = T_{ik}\left(\frac{s_{\max}}{s_i}\right) \tag{8.14}$$

and

$$k'_i = \left(\frac{\ln(T_{i0}/T_{ik'})}{c_i}\right) \tag{8.15}$$

The effect of the reannealing is, for those parameters that are insensitive at the current configuration, to increase their annealing temperature and thereby increase the range of their search. Similarly, the components of the simulated annealing algorithm can be modified to incorporate the reannealing procedure.

The first step in incorporating reannealing into the simulated annealing algorithm is to include, along with the control parameter c, additional control parameters c_i where i varies from 1 to the number of decision variables. Thus, in the annealing process, the acceptance function is annealed according to control parameter c. And the decision variables are annealed independently according to respective values of c_i. At appropriate times during the annealing process, decision variable sensitivities are determined by

$$s_i = \frac{\partial C}{\partial Q_i} \cong \frac{C_{\text{scale}} - C_0}{Q_{\text{scale}} - Q_0} \qquad i = 1, \text{NW} \tag{8.16}$$

where s_i = sensitivity of objective function to decision variable i at current configuration
C_0 = current value of objective function computed for vector of decision variable values Q_0
C_{scale} = value of objective function computed for vector of decision variables Q_0 with Q_i incremented by DQ
NW = number of decision variables

Given the sensitivities from Eq. (8.16), the decision variable control parameters are then rescaled according to

$$c_i = \frac{s_{\max}}{s_i} \qquad i = 1, \text{NW} \tag{8.17}$$

and

$$c_i \leq c_0 \qquad i = 1, \text{NW} \tag{8.18}$$

where c_i = control parameter value for decision variable i
s_i = sensitivity for decision variable i
$s_{\max}$ = maximum sensitivity for all decision variables
c_0 = initial control parameter value
NW = number of decision variables

The annealing schedule for the simulated annealing algorithm has a parameter T that is analogous to the annealing time parameter k used by Ingber (1993). T varies throughout the annealing process and, along with w, controls the annealing rate. Thus, values of T_i must be determined for each of the rescaled c_i's according to the following equations:

$$k_i = \text{NINT}\left[\frac{\ln(c_i/c_0)}{\ln(0.1)^{(10/T_0 \cdot \omega)}}\right] - 1 \qquad i = 1, \text{NW} \tag{8.19}$$

and

$$T_i = T_0\, \omega^{ki} \qquad i = 1, \text{NW} \tag{8.20}$$

where k_i = nearest integer to rescale value of T_i relative to initial input value T_0; w = input annealing schedule parameter; and c_i, c_0, and NW are as previously defined.

Given the rescaled values of c_i and T_i, the decision variable control parameters are annealed as before. In addition, the excursion lengths are also rescaled by

$$s_{li} = s_{l0} \exp\left(\frac{c_i - c_0}{c_0}\right), \qquad i = 1, \text{NW} \tag{8.21}$$

where s_{li} = rescaled excursion length for decision variable i; s_{l0} = initial excursion length for each decision variable; and c_i, c_0, and NW are as previously defined.

Finally, at the completion of each Markov chain loop, s_{li} is adjusted based on the current value of c_i by replacing values of C_0 and C_k in Eq. (8.22) with c_0 and c_i, respectively, as follows:

$$s_{li} = \max\left[s_{li} \exp\left(\frac{c_0 - c_i}{c_i}\right), s_{l\min}\right] \qquad i = 1, \text{NW} \tag{8.22}$$

where s_{li}, c_0, c_i, $s_{l\min}$, and NW are as previously defined.

8.4 ANNEALING WITH MEMORY

It has been observed here and by other authors [e.g., Greene and Supowit (1986) and Ingber (1993)] that a significant portion of the total computational effort for standard simulated annealing is expended at low temperatures during the latter stages of the search. The bulk of this effort is spent evaluating proposed moves that are subsequently rejected. As the search proceeds the range of the search is reduced along with the likelihood of making an uphill move, and a significant chance arises of a given configuration being proposed multiple times. There is also no retention of information from previous moves. In other words, the move from a particular location is based solely on the relative values of the objective functions and the control parameter for the current and proposed move as dictated by the Boltzmann function. Finally, near the end of the search there can be a tendency to cycle over a limited number of moves, leaving and returning to the local minimum. It's noted that the use of rescaling, as was demonstrated in the previous section, increases the search range for at least a subset of the decision variables. A benefit of this is an increase in the number of accessible configurations (e.g., diversification of candidate configurations) from a given location and thereby reducing the chance of repeats. However, the literature related to the tabu search method (TS) suggests that the efficiency of local search procedures can be improved by memory processes.

The tabu search method was developed by Glover (1989, 1990, 1994b, 1995) and has been used to obtain improved solutions to problems in scheduling, sequencing. resource allocation, investment planning, telecommunications, and many other areas. However, there are no known applications of TS to groundwater management problems. Derived originally by introducing memory into the scatter search method (Glover, 1994a), TS is a combinatorial optimization approach for exploring solution space beyond local optimality. In contrast to memoryless methods (e.g., simulated annealing and genetic algorithms), TS is founded on the premise that intelligent problem solving must incorporate adaptive memory and responsive exploration. It is presumed that a bad strategic choice can contribute more to the search than a good random choice. The following highlights some of the key TS features. For a comprehensive discussion of the many alternative TS processes, the reader is referred to Glover (1995). An extensive anthology of TS applications and empirical experience is provided by Hammer (1993).

Implementation of TS begins similarly to other combinatorial search methods wherein some function $C(\mathbf{x})$ is optimized over a solution space $S = \{\mathbf{x}|\mathbf{x} = (x_1, x_2,\ldots,x_m)\}$ where $\mathbf{x}$ is the state vector. TS proceeds iteratively, moving from one solution to the next until the termination criteria are achieved. Each $\mathbf{x}$ has a neighborhood $N(\mathbf{x})\epsilon S$ and each solution $\mathbf{x}'\epsilon N(\mathbf{x})$. However, TS extends the search by modifying $N(\mathbf{x})$ as the search proceeds, replacing it by another neighborhood $N^*(\mathbf{x})$. A key element of TS is the use of memory to determine $N^*(\mathbf{x})$ and thereby organize the manner in which S is explored.

The solutions admitted to $N^*(\mathbf{x})$ are determined by any of several TS mechanisms. One, from which TS derives its name, labels solutions encountered as tabu and forbids their inclusion in $N^*(\mathbf{x})$ for a specified duration. Application of tabu status is intended to promote strategically aggressive search trajectories. In effect, original objective function evaluations are complemented by tabu evaluations to either encourage or discourage the selection of certain type solutions for inclusion in $N^*(\mathbf{x})$.

TS uses both explicit and attributive memory. Explicit memory retains complete solutions, typically elite or high-quality solutions encountered during the search or their neighbors. Attributive memory, on the other hand, records changes in solution attributes associated with moves. Both types of memory are applied as functions of event recency, event frequency, and differentiation between short and long term. They are applied in ways to concentrate focus on good regions or good solution features and exploring promising new regions.

The two basic types of TS memory are generally characterized as short term and long term. The most commonly used short-term memory is called recency-based memory and simply tracks solution attributes that have changed in the immediate past. Tabu status is applied to selected attributes that occur in recent solutions. Solutions that contain tabu attributes are thereby excluded from $N^*(\mathbf{x})$ and cannot be revisited while they retain their tabu status. The duration (i.e., number of iterations) that an attribute remains tabu-active is referred to as tabu tenure. Tabu tenure can be varied for different attributes or combinations of attributes as well as for different stages of the search. A flowchart describing the implementation of short-term memory is presented in Fig. 8.1. The procedure begins with an initial solution $\mathbf{x}$. The neighborhood $N^*(\mathbf{x})$ is generated, excluding solutions containing tabu attributes. From this list, the best solution $\mathbf{x}'$ is selected and a move is made. The tabu list is then updated and a check for termination is made. If the termination criteria are not met (e.g., exceed k_{max}, no solutions remaining in the neighborhood, etc.), a new neighborhood is generated around the new $\mathbf{x}$, and the process continues.

Variations of the above include using aspiration criteria that would overrule tabu status for high-quality solutions (e.g., best yet encountered) or graduated penalties that would allow for varying degrees of tabu status. Glover (1995) notes that the short-term TS memory components are sufficient to obtain high-quality solutions for many problems. However, he also states that TS can be made significantly stronger by introducing longer-term memory components. Long-term memory considerations are primarily based on the frequency of events. Transitional frequencies track how often selected attributes change, whereas residence frequencies track relative durations with which attributes occur in generated solutions. Two important long-term memory considerations are intensification strategies and diversification strategies. Intensification strategies are designed to influence choices toward move combinations and solution attributes that historically produce good solutions. An example would be, subsequent to convergence at a local minimum, use of restarting procedures that incorporate good attributes from previous solutions.

Diversification strategies, in contrast, drive the search into new regions. They are often based on restricting attribute choices to those infrequently used. A basic example is that referred to as strategic oscillation whereby the search is driven toward and away from selected boundaries of feasibility by manipulating the objective function (e.g., adjusting penalties and incentives). A strategy that links diversification and intensification is path re-linking. This approach retains elite solutions in

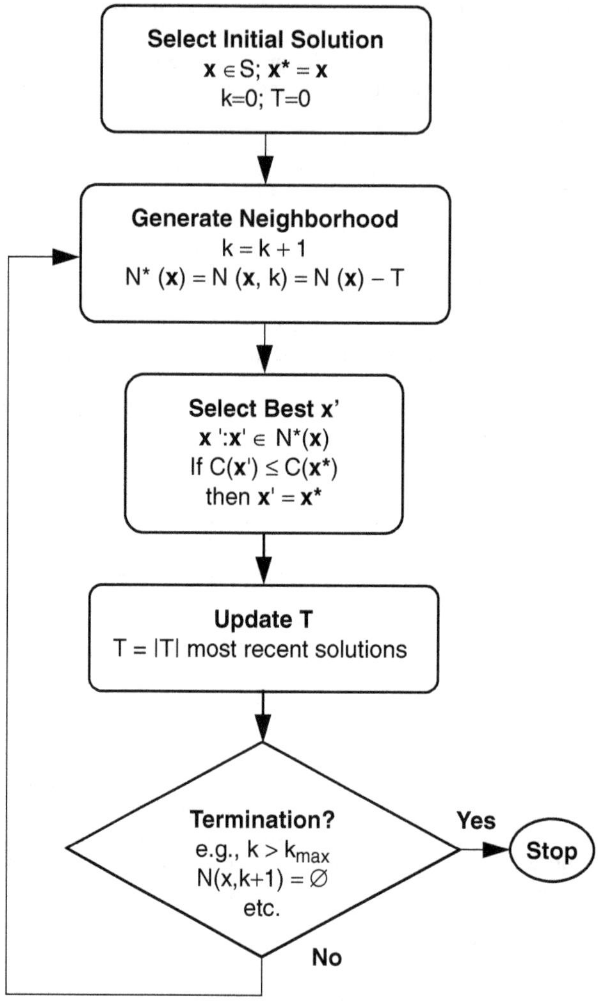

FIGURE 8.1 Flowchart of tabu search short-term memory process.

long-term memory. It is implemented by exploring trajectories connecting one or more of the elite solutions. Starting at an initiating solution, moves are selecting by introducing attributes from guiding solutions (also from the elite set) with the goal of creating a good attribute composition in the current solution.

From the description of TS, some conceptual parallels and differences between TS and simulated annealing are apparent. Although TS makes decisions relative to a neighborhood containing multiple alternative solutions, simulated annealing always makes decisions pertaining to a single proposed move. Both methods employ mechanisms for diversification and intensification. TS utilizes a variety of memory-based processes to balance diversification and intensification throughout the search, whereas simulated annealing tends to focus on diversification early (high uphill move probability and wide search range) in the annealing process, and more on intensification (low uphill move probability and limited search range) during the latter stages of the search. It is noted that the directional search capability discussed earlier does provide for continued diversification throughout the search.

The collection of similarities and differences suggests that simulated annealing could potentially be improved by incorporating elements of memory. Although it was beyond the scope of the effort presented here to exhaustively explore the numerous possibilities, some simpler TS memory processes were adapted for use by the enhanced annealing algorithm presented.

8.5 MEMORY COMPONENTS

Three memory components were selected for inclusion in the enhanced annealing algorithm:

1. Short-term recency memory of last NT1 moves
2. Short-term recency memory of an attribute for last NT2 iterations
3. Long-term record of all configurations visited after 1000 iterations

The first two components are inspired by similar TS processes. Retention of the last NT1 moves for a duration of NT1 iterations (i.e., a circular list wherein as one configuration is added to the list another is dropped) allows for the avoidance of cycling near the end of the annealing process. Although the first tabu list is in effect throughout the search, it only becomes functional near the end of the search when the search range is small and the likelihood of returning to a recently visited site is high. The expected benefits from this mechanism are twofold. First, the prevention of cycling should enhance the rate and/or quality of convergence by in turn preventing the expansion of iterations within loops that have little or no probability of containing a more optimal solution. And, computational resources are not wasted on reevaluating objective functions.

The second tabu list is intended as an intensification strategy. Subsequent to each move, the direction of change in total pumpage is recorded. For NT2 iterations subsequent to the move, configurations that would reverse the direction of change in total pumpage are tabu. In other words, if the last move was the result of an increase in pumpage, then for the next NT2 iterations, only those configurations reflecting an increase in pumpage would be considered. Again, dual benefits are expected from this strategy. The avoidance of reversals in changes in total pumpage will provide persistence in the search direction. If the search is heading toward a local optimum, it can't be derailed by a stochastically generated uphill move until the search has an opportunity to explore the current direction more deeply. In addition to accelerating the convergence process, for those proposed moves violating the tabu criteria, objective functions are not evaluated.

The third memory component, although not a TS-derived process, relies on long-term memory. It consists simply of retaining a condensed record of all configurations visited and their objective function values. When one of the configurations is reencountered, instead of reevaluating the objective function, it is simply located in memory. The tradeoff for this memory component is the savings in computational resources for the sites visited more than once (i.e., not calling the simulator), against the time required to search the record for every proposed configuration. The records are scanned using a binary search approach. The configurations are stored in a condensed form using a simple form of hashing as described by Woodruff and Zemel (1993). The hashing function used is

$$h_0 = \sum_{i=1}^{n} z_i x_i \tag{8.23}$$

where h_0 = hashed configuration
z_i = precomputed vector of pseudo-random integers in the range 1 to 4×10^9
x_i = configuration vector defining decision variables

Results of the hashing are stored as 4-byte characters. The probability of a collision (i.e., two different vectors being encountered with the same hash function value) is extremely remote. For example, for a construction site dewatering problem, given a scenario with 10 wells and 16 alternative flow rates for each well, there are 16^{10} configuration permutations. Therefore, each of the 4 billion hash function values represents approximately 275 different configurations. If all different encountered configurations were stored over 100 Monte Carlo solutions up to 100,000 configurations, the joint probability of encountering two of the 275 matches for a given hash function value is on the order of 6×10^{-10}.

8.6 OVERALL ENHANCED ANNEALING ALGORITHM

A flowchart depicting the enhanced annealing algorithm, having both directional search and memory capabilities, is shown in Fig. 8.2. As discussed above, modifications to the simulated annealing algorithm are as follows:

1. After each new configuration is generated, a check is made to determine if the configuration is tabu (i.e., recently visited or tabu total pumpage). If the configuration is tabu, another configuration is generated without computing the objective function. If the configuration is not tabu, a check is made to determine if it has been previously visited. If it has, the objective function value is retrieved from the record. Otherwise, the objective function is evaluated. The annealing process then proceeds as usual.

2. If the number of iterations is greater than NRSCL rescaling is implemented as described in Fig. 8.3. Otherwise, the control parameters for the acceptance function, individual decision variables, and the excursion length are adjusted. The excursion lengths are adjusted individually. In addition, they are adjusted as a function of the initial control parameter value and the current decision variable control parameter value.

8.7 APPLICATION OF ALGORITHM: N-SPRINGS SITE DESCRIPTION

8.7.1 Problem Description

Groundwater contamination from past and present industrial, agricultural, and commercial activities is one of the leading environmental challenges of our time. It is estimated that from 300,000 to 400,000 sites in the United States may contain contaminated groundwater. Past estimates suggest that the cost of cleanup for these sites over the next 30 years could amount to as much as $1 trillion (National Research Council, NRC, 1994). Resultant risks to public health and ecosystems, and the associated high costs of remediation make it essential that aquifer management strategies be developed in the most reliable and cost-effective manner achievable.

The enhanced annealing algorithm has been developed and used in the new methodology for groundwater remediation design. The flowchart of this new groundwater management methodology is illustrated in Fig. 8.4. Essentially the remediation design is a discrete-time optimal control problem that is solved by interfacing a flow and transport simulator with the simulated annealing optimizer.

N-Springs is located at Hanford, Washington and is the location of the 1301-N Liquid Waste Disposal Facility (LWDF) which was used for disposal of radiologically contaminated water associated with pass-through cooling, spent fuel storage, and other activities related to operation of the N Reactor (U.S. DOE, 1995). The study area, located adjacent to the Columbia River, is shown in Fig. 8.5. The LWDF is a long zigzagging trench running parallel to and about 800 ft from the Columbia River. From 1963 until 1985, the LWDF received discharges containing both radiological and nonradiological

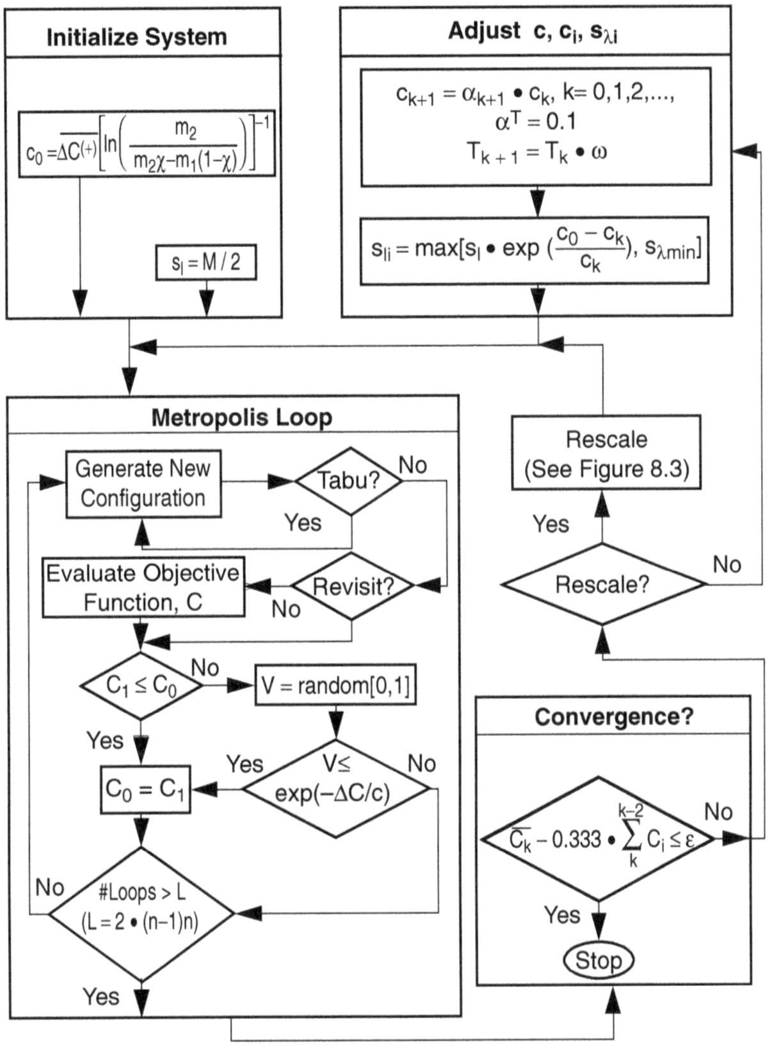

FIGURE 8.2 Enhanced annealing with memory and directional search.

contaminants at a rate of approximately 2100 gal/min. Almost immediately after receiving water, mobile contaminants from the LWDF were observed in springs along the banks of the Columbia River. Beginning in 1980, ^{90}Sr was detected in the groundwater near the river. It is estimated that 2997 Ci of ^{90}Sr were discharged to the facilities at N-Springs of which ~46 Ci were released to the Columbia River and 1085 Ci have decayed away. The majority of the radiological inventory remaining at the N-Springs area is located in the sediments underlying the LWDF including 75.5 Ci within the unconfined aquifer. Of this amount, 75 Ci are estimated to be adsorbed to the sediments and 0.5 Ci are dissolved in the groundwater.

Contours depicting the areal concentration distribution of ^{90}Sr in the groundwater at N-Springs are shown in Fig. 8.6. The plume covers approximately 55 acres having maximum concentrations on the order of 4000 pCi/l. The highest concentrations are shown to be immediately beneath and down-gradient (i.e., toward the river) from the LWDF.

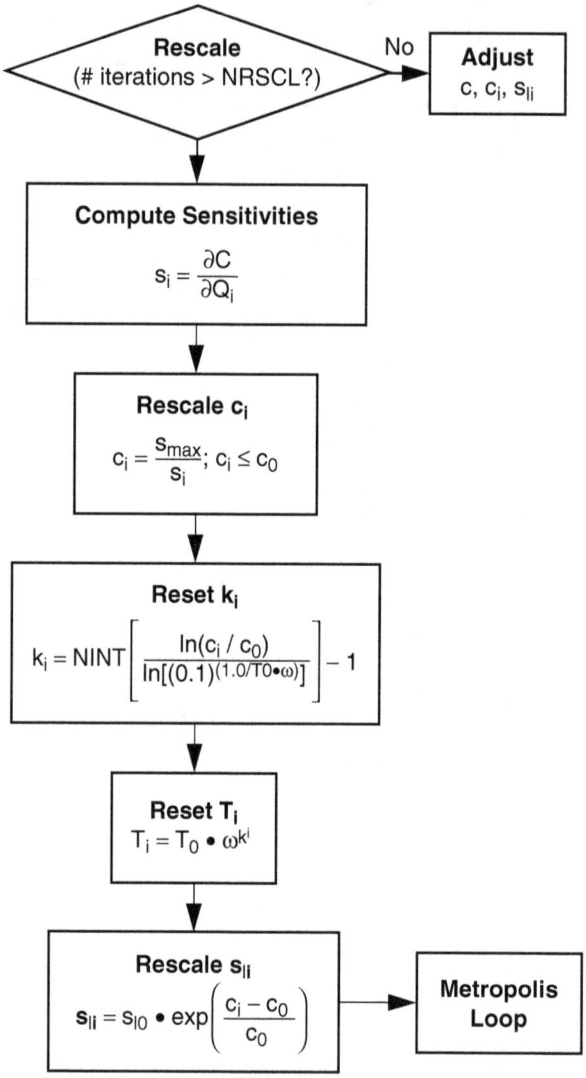

FIGURE 8.3 Flowchart of rescaling process (see Fig. 8.2).

8.7.2 Demonstration Scope

Alternative physical barrier and pump-and-treat designs were evaluated as part of an expedited response action to substantially reduce the seepage of ^{90}Sr into the Columbia River. The goal for the systems evaluated was to extend the travel path for the ^{90}Sr, taking advantage of its slow migration rate. The K_d for ^{90}Sr is 15.5 mL/g, resulting in a retardation factor of 100.2. Thus, increasing the path length to the river can dramatically enhance natural radioactive decay. Three barrier designs having lengths of 2000, 3000, and 4000 ft were considered. All fully penetrate the unconfined aquifer and are positioned approximately 30 ft from and parallel to the Columbia River. In conjunction with the barriers, the U.S. DOE study analyzed the performance of three different pump-and-treat systems.

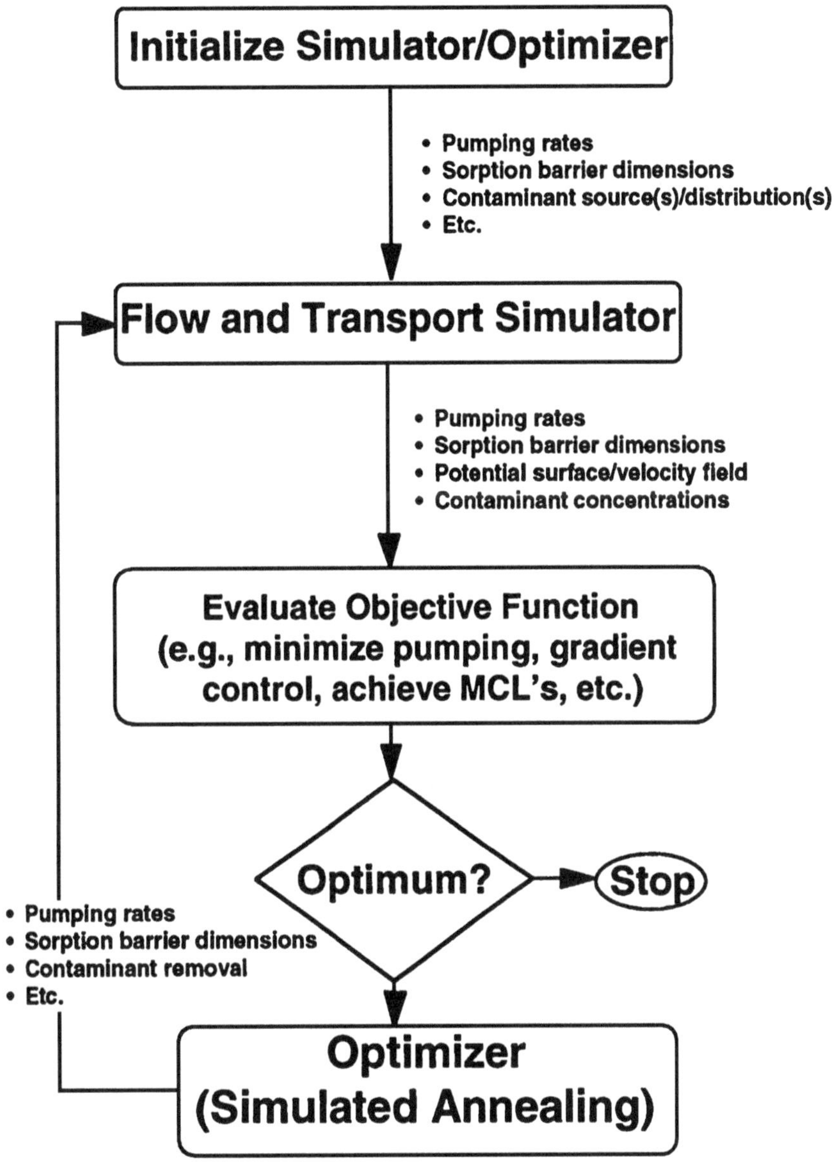

FIGURE 8.4 Flowchart of groundwater management methodology.

The combinations of barrier/pump-and-treat designs evaluated are summarized in Table 8.1. As indicated, the performance of 11 individual configurations was assessed. The 50-gal/min (gpm) system used one injection and one extraction well each operating at 50 gal/min. The 100-gpm system used two injection and two extraction wells operating at 50 gpm and the 180-gpm system used two injection wells operating at 90 gpm and three extraction wells operating at 60 gpm. The total rates of extraction and injection were equal such that there was no net withdrawal from the aquifer. Locations

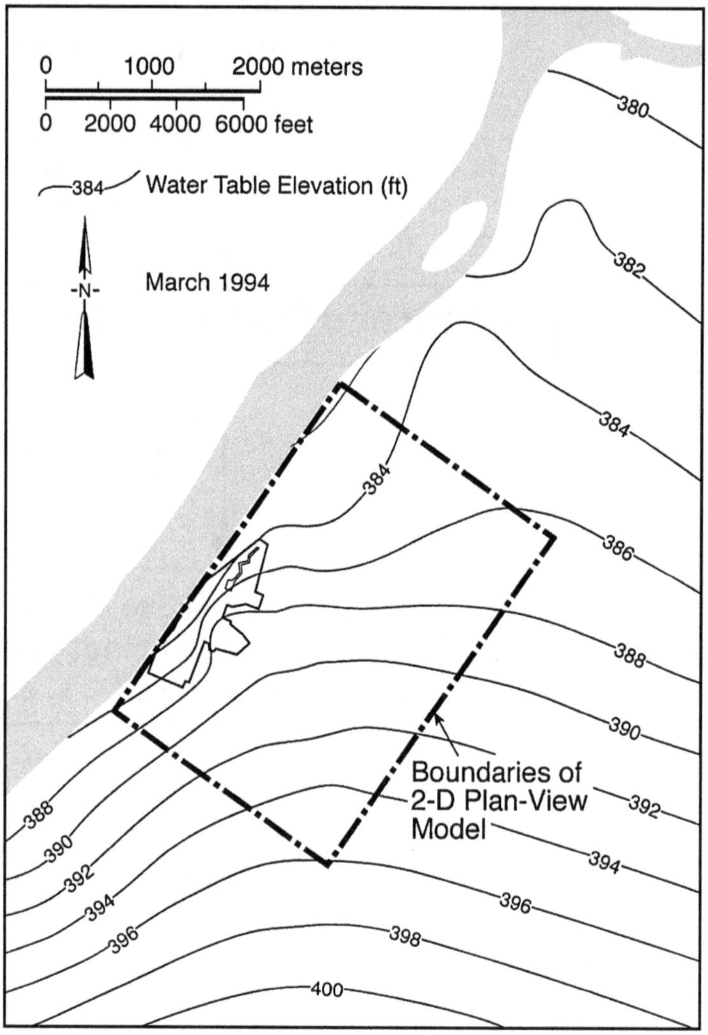

FIGURE 8.5 Plan view of N-springs study area. (*From U.S. DOE, 1995.*)

for the endpoints for the barrier designs and the extraction and injection wells analyzed in the U.S. DOE (1995) study are shown in Fig. 8.6.

The emphasis of the U.S. DOE (1995) study was in evaluating the performance of the barriers with and without pump-and-treat. The study did not report results for any systems using pump-and-treat alone. The findings from the study indicate that locating an extraction system in the area of highest ^{90}Sr produces only a minor decrease in total loading of ^{90}Sr to the Columbia River compared to loadings with a barrier. The barriers sufficiently increase the travel paths and times of the highly retarded ^{90}Sr, that release to the river is dramatically decreased. The findings of the U.S. DOE (1995) study are summarized in Table 8.2. The results indicate that without remediation, the releases to the river over a 10-yr period total 3.2 Ci. With 2000-ft, 3000-ft, and 3800-ft barriers in place, the releases decrease to 0.8 Ci, 0.1 Ci, and 0.0 Ci, respectively. With the well systems in place, the total for the

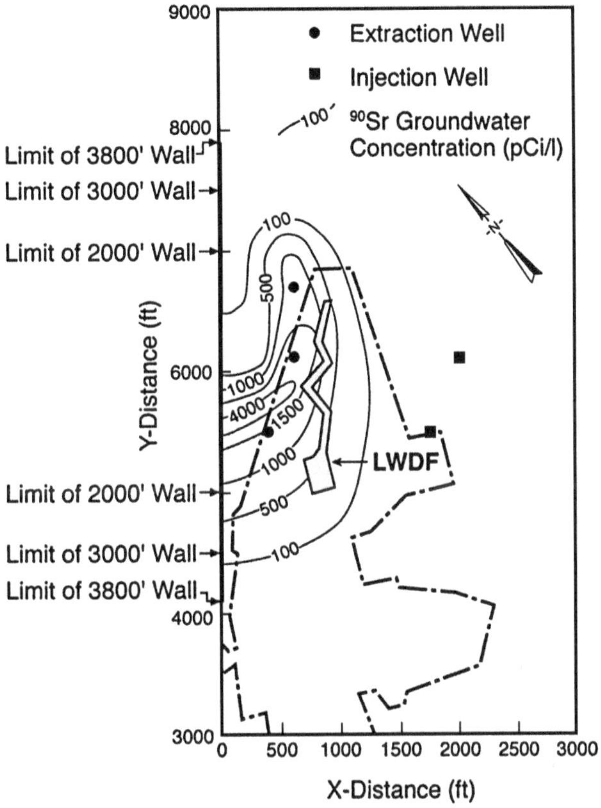

FIGURE 8.6 Areal distribution of ^{90}Sr in the groundwater at the N-Springs area.

TABLE 8.1 Summary of Barrier and Pump-and-Treat Designs Evaluated by U.S. DOE

Wall length, ft	Number of wells	50 gal/min	100 gal/min	180 gal/min
0	X	—	—	—
2000	X	X	—	X
3000	X	X	—	X
4000	X	X	X	X

Source: From U.S. DOE (1995).

2000-ft wall shows no change for a 50-gpm pumping system and a 25 percent reduction with a 180-gpm pumping system. Pumping had no apparent effects on the results for the 3000-ft and 3800-ft barrier releases.

These results strongly indicate that with a barrier in place, little if any incremental benefit is derived from adding the specified pump-and-treat configurations to the containment system. Therefore, the focus of the groundwater management methodology demonstration is on selecting an optimal design for eliminating ^{90}Sr releases to the Columbia River over a 10-yr period with or without a barrier in place. The possible configurations for the design will include no barrier or one of the

TABLE 8.2 Summary of 10-Year ^{90}Sr Releases (Ci) to the Columbia River for Barrier and Pump-and-Treat Designs by U.S. DOE

Design configuration	Estimated 10-year ^{90}Sr release, Ci
No barrier	3.2
2000-ft barrier	0.8
2000-ft barrier with wells (50 gpm)	0.8
2000-ft barrier with wells (180 gpm)	0.6
3000-ft barrier	0.1
3000-ft barrier with wells (50 gpm)	0.1
3000-ft barrier with wells (180 gpm)	0.1
3800-ft barrier	0.0
3800-ft barrier with wells (50 gpm)	0.0
3800-ft barrier with wells (100 gpm)	0.0
3800-ft barrier with wells 180 gpm)	0.0

Source: From U.S. DOE (1995).

three lengths analyzed by the U.S. DOE study, and one or more extraction or injection wells selected from 28 candidate locations. Details of the flow and transport models and formulation of the optimization problem are discussed below.

8.7.3 Flow and Transport Model

The model area for the demonstration is the same used by the U.S. DOE (1995) analysis as shown in Fig. 8.5. The model encompasses the 8000-ft by 12,000-ft study area. As shown, the prevailing hydraulic gradient is almost due north, from the lower right-hand corner of the model area to the upper left-hand corner. As groundwater approaches the river it turns northwesterly. Over the study area the gradient varies from 0.0005 to 0.003, producing average groundwater velocities between 0.1 to 2 ft/day in the vicinity of the LWDF. The uppermost stratigraphic unit at the site is the Hanford formation, which consists of uncemented gravels with sand and silt interbeds. The Hanford formation extends to just above the water table and transitions to the Ringold formation.

The U.S. DOE (1995) study was based on a two-dimensional finite-difference code with the 8000- by 12,000-ft study area discretized using a variable grid spacing consisting of 58 nodes in the y-direction and 38 nodes in the x- direction. The flow model assumed steady-state and transport was determined using particle tracking based on computed groundwater pathlines and accounting for retardation and radioactive decay. The site was discretized using a 35 by 68 element grid. In the vicinity of the ^{90}Sr plume (from 3000 to 9000 ft in the y-direction and 0 to 3000 ft in the x-direction) a subgrid, regularly spaced with element dimensions of 100 by 100 ft was used. Outside the plume area, element dimensions increased from 250 to 1500 ft. An overlay of the subgrid over the model area is shown in Fig. 8.7. The model parameters used for the demonstration were either those used in the U.S. DOE (1995) study or inferred from information provided. The parameters used are summarized in Table 8.3. The hydraulic conductivity value is 261 ft/day, the geometric mean of five pumping test results ranging from 100 to 500 ft/day. Barriers are represented in the model by simply reducing the hydraulic conductivity in selected elements to 0.0001 ft/day. The porosity used was 0.25. The aquifer bottom was assumed to be at elevation 365.0, which results in an average aquifer thickness of about 25.0 ft. The boundaries were all assumed to be fixed-head boundaries with values varying linearly between elevations estimated from the water table map in Fig. 8.5. Elevations at the northeast and southeast boundaries were 386.6 and 396.0 ft, respectively. Elevations along the river varied from 381.3 to 384.4 ft.

Initial conditions for the transport analyses were based on an element-by-element discretization of the plume shown in Fig. 8.7. Travel times to the river were determined considering advection only. The travel times were then adjusted for retardation based on the following equation

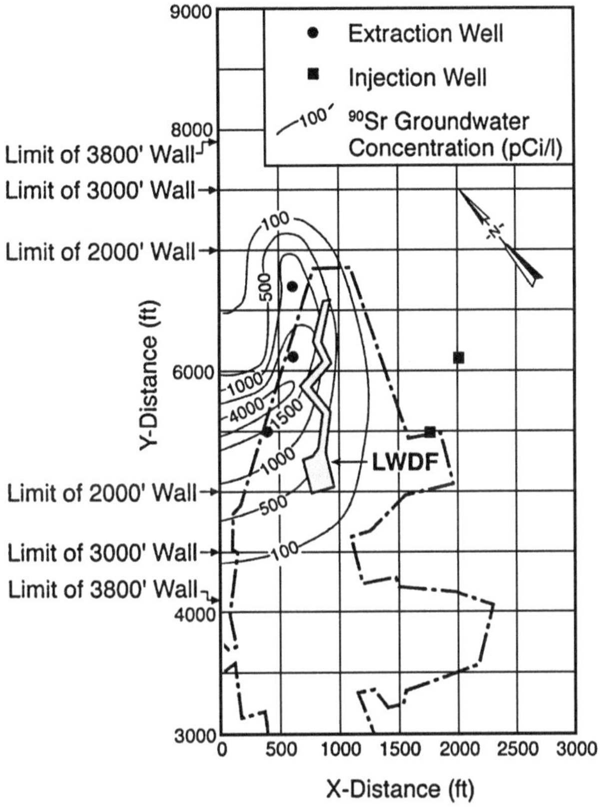

FIGURE 8.7 Overlay of 100 ft by 100 ft subgrid on the study area.

TABLE 8.3 Summary of Parameters for N-Springs Area Model

Parameter	Value
Hydraulic conductivity, K (ft/day)	261.0
Aquifer thickness, b (ft)	~25.0
Porosity, n	0.25
^{90}Sr distribution coefficient, K_d (mL/g)	15.5
Retardation factor	100.2
^{90}Sr half-life (yrs)	28

$$R_f = \frac{v}{v_c} = 1 + \frac{\rho_b}{\eta} K_d \tag{8.24}$$

where R_f = retardation factor
v and v_c = groundwater velocity and contaminant velocity, respectively
ρ_b = bulk distribution
η = porosity
K_d = distribution coefficient

Mass was then adjusted for radioactive decay using

$$\text{Mass} = \text{mass } 0.5^{(T/H_L)} \tag{8.25}$$

where T is the contaminant travel time and H_L is the half-life for ^{90}Sr.

The flow and transport models for the N-Springs area were validated based on comparisons with results obtained in the U.S. DOE (1995) study for flow and ^{90}Sr transport with no barrier and no wells. A particle released at x-y coordinates of (2000 ft, 6000 ft) had a computed groundwater travel time of 1579 days. The comparable value obtained from pathlines determined in the U.S. DOE study, is approximately 1500 days. Total release of ^{90}Sr to the Columbia River in 10 years was estimated to be 3.6 Ci. The 10-year release determined by the U.S DOE study was 3.2 Ci. These results suggest that the groundwater management methodology flow and transport models are reasonably simulating ^{90}Sr transport in the N- Springs area.

8.7.4 Optimization Problem Formulation

The management problem under consideration for the N-Springs area is selection of a least-cost containment system that will reliably prevent any release of ^{90}Sr into the Columbia River for a 10-year period. Factors evaluated include construction costs for a barrier and/or well(s), treatment costs for any extracted contaminated groundwater, plus maintenance and operating costs for both extraction and injection wells. In addition, the net groundwater withdrawal should be zero (i.e., total extraction rate equals total injection rate). The configuration space includes 28 candidate well locations as shown in Fig. 8.8. There are 12 possible flow rates for each of the wells: five injection rates (−125, −100, −75, −50, and −25 gpm), zero, and six extraction rates (25, 50, 75, 100, 125, and 150 gpm). As with the U.S. DOE study, four possible barrier configurations were considered: no barrier and candidate lengths of 2000, 3000, and 3800 ft. The endpoints for each of the barriers are noted in Fig. 8.9.

Approximate costs for the Hanford site are summarized in Table 8.4. In addition to the cost elements, other considerations are penalties associated with not achieving the stated objective of zero ^{90}Sr releases to the river, and not balancing the extraction and injection rates. The concerns about releasing radionucleides to the Columbia River are severe, so a high penalty is necessary to ensure that the search emphasize solutions that will achieve that objective. Also, the need for additional treatment or disposal of excess extracted water, or the need to develop a source for additional injection water, make the balancing of extraction and injection rates also very important.

Based on the above, the mathematical formulation for the optimal design of an N-Springs containment system is as follows:

$$\min \sum_{i=1}^{N_W} C_{Q+}\, Q_i^+ + \sum_{i=1}^{N_W} C_{Q-}\, Q_i^- + \sum_{i=1}^{N_W} I_i + C_B L_B + P_1\, \Delta Q + P_2 \text{ mass} \tag{8.26}$$

where C_{Q+} = cost per gpm for treating extracted groundwater for each extraction well i
Q^+_i = pumping rate (gpm) for extraction well i
C_{Q-} = cost per gpm for groundwater injection for each injection well i
Q_i^- = pumping rate (gpm) for injection well i
I_i = construction costs for well i if $Q_i \neq 0$
C_B = cost/ft of barrier construction
L_B = barrier length (ft)
P_1 = penalty for not achieving balanced extraction and injection rates
ΔQ = absolute value of difference in total injection and withdrawal rates (gpm)
P_2 = penalty for release of ^{90}Sr to the river
mass = computed mass of ^{90}Sr released to the river (Ci)

In addition to the estimated costs, another concern in selecting coefficient values for Eq. (8.26) is scaling the components of the cost function such that the desired emphasis is achieved, and there is adequate contrast in cost function values between good solutions and bad solutions. With this in mind,

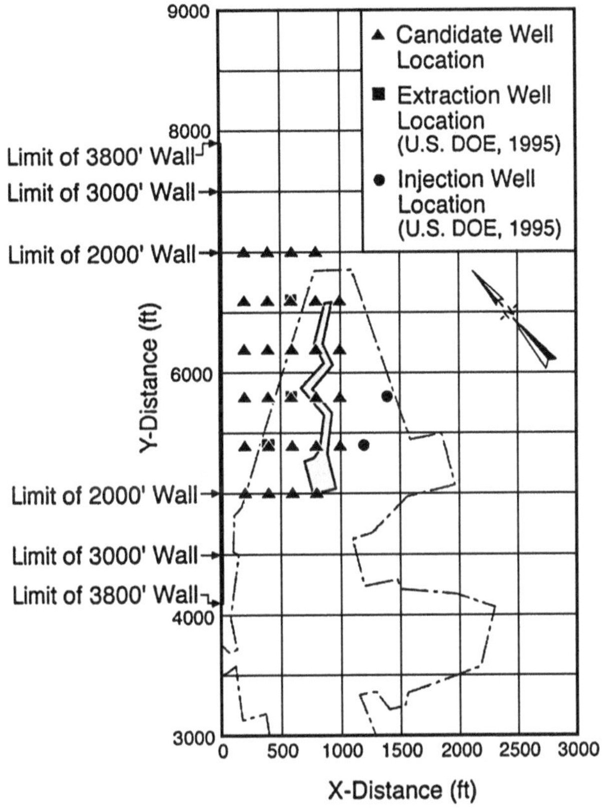

FIGURE 8.8 Well and barrier locations for N-Springs area.

the actual unit costs for well construction, groundwater treatment, and barrier construction were reduced by a factor of 1000. The coefficient for injection was set equal to 1.0, and the penalty factors $P1$ and $P2$ were both set equal to 5000. The resultant values are listed in Table 8.4.

The optimal solution obtained by the groundwater management methodology consists of a 3000-ft barrier with a single withdrawal well and a single injection well. The solution required 65,250 iterations, 30,883 objective function evaluations, and 1.5 CPU processing hours. The withdrawal well has a pumping rate of 25 gpm and is located at coordinates (6200 ft, 400 ft), on the downstream edge of the 4000 pCi/L contour of the ^{90}Sr plume. The injection well, also pumping at 25 gpm, is located almost immediately upgradient of the withdrawal well at coordinates (6600 ft, 800 ft). The water table produced by the design is shown in Fig. 8.9. The vertical scale is exaggerated, so the figure clearly depicts the disruption in the general gradient produced by the 3000-ft barrier at the river. This disruption significantly increases the travel time for ^{90}Sr to the river. In addition, the withdrawal well at the downstream edge of the plume is well positioned to intercept the highest concentration flows or at least incrementally impede their migration. The effect of the injection well immediately upgradient of the withdrawal well is likely to redirect the high concentration flows toward the withdrawal well.

Computed contaminant releases to the river are virtually zero until year 11 when the cumulative mass release is 0.001 Ci. The objective value cost elements associated with the design are summarized in Table 8.5. Given that the withdrawal and injection pumping rates are balanced, the only cost elements are installation costs for two wells, treatment costs for 25 gpm capacity, and construction of the 3000-ft barrier for a total of \$17.6 million. Applying the same cost factors to the U.S. DOE (1995) analysis of

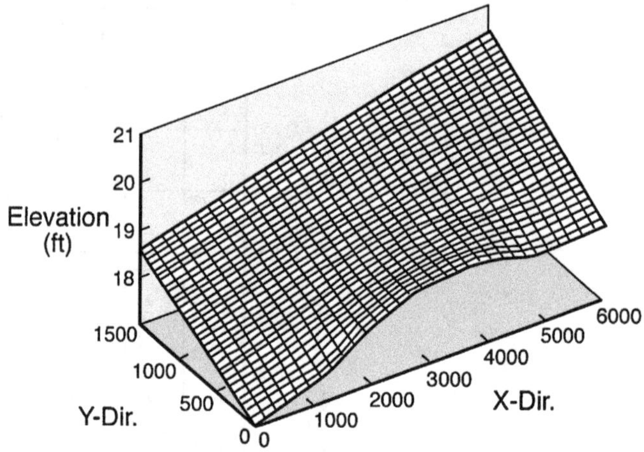

FIGURE 8.9 N-Springs area water table produced by optimal design.

TABLE 8.4 Summary of Estimated Construction and Operation Costs for ^{90}Sr at N-Springs

Cost element	Parameter	Value
Well construction: \$50,000/well	I_i	50.0
Groundwater treatment: \$220,000/gpm	C_{Q+}	220.0
Injection well operation	C_{Q-}	1.0
Barrier construction: \$4,000/ft	C_B	4.0
Penalty for excess extraction/withdrawal	P_1	5,000
Penalty for ^{90}Sr releases	P_2	5,000

TABLE 8.5 Summary of Optimal Design Costs

Cost element	Optimal design	U.S. DOE design
Well construction: \$50,000/well	\$0.1 million (2 wells)	\$0.1 million (2 wells)
Groundwater treatment: \$220,000/gpm	\$5.5 million (25 gpm)	\$11.0 million (50 gpm)
Barrier construction: \$4,000/ft	\$12.0 million (3,000 ft)	\$12.0 million (3,000)
Penalty for excess extraction/withdrawal	\$0.0	\$0.0
^{90}Sr releases	0.0 Ci	0.1 Ci
Total cost	\$17.6 million	\$23.1 million

the system consisting of a 3000-ft barrier and 50 gpm pumping (the lowest pumping rate analyzed) the comparable total cost would be \$23.1 million. In addition, the U.S.DOE design results in the release of 0.1 Ci of ^{90}Sr to the river within the 10-year study period. Although the wells are pumping at higher rates, their locations are less effective in intercepting or diverting the migration of ^{90}Sr.

8.8 SUMMARY AND CONCLUSIONS

The reliability of simulated annealing in locating global or near-global optima is a major advantage. Other noted advantages are its flexibility in considering complex, highly nonlinear or discontinuous objectives and/or constraints, and the ease with which it can be interfaced with application-specific simulators. However, the limited experience to date in applying simulated annealing to groundwater management problems highlights the large number of objective function evaluations required for convergence to a quality solution. Enhanced annealing departs from the purely random search approach inherent in simulated annealing by incorporating directional search and memory capabilities. These new capabilities compound the improvements in computational efficiency previously achieved, demonstrating the potential gains that accrue from combining concepts of related combinatorial optimization methods. Memory concepts derived from the tabu search method show particular promise for improving the rate and quality of convergence. Also, selecting search directions based on better understanding of the current neighborhood of the configuration space was shown to improve algorithm performance.

The capabilities of the groundwater management methodology with enhanced annealing were demonstrated by applying it to an actual contaminated groundwater site. The analysis considered optimal design of a contaminated groundwater containment system using pump-and-treat in conjunction with physical barriers. The flow and transport simulators proved to adequately represent important processes at the study site and optimal results were obtained that are consistent with previously reported findings. The application confirmed the ease with which simulated annealing can be adapted to new problem formulations and/or objective and constraint sets.

REFERENCES

Anderson, K. and R. V. V. Vidal (1993). "Solving the Quadratic Assignment Problem," in *Applied Simulated Annealing,* René V. V. Vidal, ed., Springer-Verlag, New York.

Chardaire, P. and J. L. Lutton (1993). "Using Simulated Annealing to Solve Concentrator Location Problems in Telecommunication Networks," in *Applied Simulated Annealing,* René V. V. Vidal, ed., Springer-Verlag, New York.

Davis, L. and M. Steenstrup (1987). "Genetic Algorithms and Simulated Annealing: An Overview," in *Genetic Algorithms and Simulated Annealing,* L. Davis, ed., Morgan Kaufman, Los Altos, Calif.

Dougherty, D. E. and R. A. Marryott (1991). "Optimal Groundwater Management, 1. Simulated Annealing," *Water Resources Research,* 27(10):2493–2508.

Glover, R. (1989). "Tabu Search—Part I," *ORSA Journal on Computing,* 1(3):190–206.

Glover, R. (1990). "Tabu Search—Part II," *ORSA Journal on Computing,* 2:4–32.

Glover, F. (1994a). "Genetic Algorithms and Scatter Search: Unsuspected Potentials," *Statistics and Computing,* 4:131–140.

Glover, F. (1994b). "Tabu Search for Nonlinear and Parametric Optimization (with Links to Genetic Algorithms)," *Discrete Applied Mathematics,* 49:231–255.

Glover, F. (1995). "Tabu Search Fundamentals and Uses," CAAI Research Report, University of Colorado.

Greene, J. W. and K. J. Supowit (1986). "Simulated Annealing Without Rejected Moves," *IEEE Transactions on Computer-Aided Design,* CAD-5(1):221–228.

Hammer, P. L., ed. (1993). "Tabu Search," *Annals of Operations Research,* 41.

Ingber, L. (1993). "Simulated Annealing: Practice Versus Theory," *Mathematical Computer Modeling,* 18(11):29–58. Kaufmann, Los Altos, Calif.

Kirkpatrick, S., C. D. Gelatt, Jr., and M. P. Vecchi (1983). "Optimization Simulated Annealing," *Science,* 220(4598):680.

Laarhoven, P. J. M. van (1988). *Theoretical and Computational Aspects of Simulated Annealing,* Centre for Mathematics and Computer Science, Amsterdam, the Netherlands.

Laarhoven, P. J. M. van and E. H. L. Aarts (1987). *Simulated Annealing: Theory and Applications,* Kluwer Academic, Norwell, Mass.

Marryott, R. A., D. E. Dougherty, and R. L. Stollar (1993). "Optimal Groundwater Management, 2. Application of Simulated Annealing to a Field-Scale Contamination Site," *Water Resource Research,* 29(4):847–860.

Metropolis, N., A. Rosenbluth, M. Rosenbluth, A. Teller, and E. Teller (1953). "Equation of State Calculations by Fast Computing Machines," *Journal of Chemical Physics,* 21:1087–1092.

Murtagh, B. A. and M. A. Saunders (1980). "MINOS/AUGMENTED User's Manual I." *Tech. Rep.* 80-74, 51 pp., Syst. Optimization Laboratory Department of Operations Research, Stanford University, Stanford, Calif.

National Research Council (1994). *Alternatives for Ground Water Cleanup,* National Academy Press, Washington D.C.

Ritzel, B. J., J. W. Eheart, and S. Ranjithan (1994). "Using Genetic Algorithms to Solve a Containment Problem," *Water Resource Research,* 30(5):1589–1603.

Rogers, L. L. and F. U. Dowla (1994). "Optimization of Groundwater Remediation Using Artificial Neural Networks with Parallel Solute Transport Modeling," *Water Resource Research,* 30(2):457–481.

Skaggs, R. L. (1995). "Enhanced Annealing Applied to Groundwater Restoration Management," Ph.D. Dissertation, 260 pp., Department of Civil Engineering, Arizona State University, Tempe, Ariz.

Skaggs, R. L., L. W. Mays, and L. W. Vail (2001a). "Simulated Annealing with Memory and Directional Search for Ground Water Remediation Design," *Journal of the American Water Resources Association*, 37(4): 853-866.

Skaggs, R. L., L. W. Mays, and L. W. Vail (2001b). "Application of Enhanced Annealing to Ground Water Remediation Design," *Journal of the American Water Resources Association*, 37(4): 867-875.

U.S. Department of Energy (1995). "Modeling Evaluation of N-Springs Barrier and Pump-and-Treat System," DOE/RL-94-132, Rev. 0, Richland, Wash.

Vidal, R. V. V. (1993). "Optimal Partition of an Interval—The Discrete Version," in *Applied Simulated Annealing.* R. V. V. Vidal, ed., Springer-Verlag, New York.

Wanakule et al. (1986).

Woodruff, D. L. and E. Zemel (1993). "Hashing Vectors for Tabu Search," *Annals of Operations Research,* 41.

INDEX

Index note: The *f* or *t* following a page number refers to a figure or table, respectively, on that page.

ABOUT THE EDITOR

LARRY. W. MAYS, PH.D., P.E., P.H., is Professor of Civil and Environmental Engineering at Arizona State University and former chair of the department. He is the former Director of the Center for Research in Water Resources at the University of Texas at Austin, where he also held an Engineering Foundation Endowed Professorship. A registered professional engineer in seven states and a registered professional hydrologist, he has served as a consultant to many organizations. Professor Mays is the author of *Water Resources Engineering* (John Wiley & Sons, Inc.) and *Optimal Control of Hydrosystems* (Marcel Dekker, Inc.), coauthor of *Applied Hydrology* and *Hydrosystems Engineering and Management* (both from McGraw-Hill), co-author of *Groundwater Hydrology* (John Wiley & Sons, Inc.) and editor-in-chief of the *Water Resources Handbook, Hydraulic Design Handbook, Water Distribution Systems Handbook, Stormwater Collection System Design Handbook, Urban Water Supply Handbook, Urban Water Supply Management Tools, Urban Stormwater Management Tools,* and *Water Supply Systems Security* (all from McGraw-Hill). He was also the editor-in-chief of *Reliability Analysis of Water Distribution Systems* (ASCE) and coeditor of *Computer Modeling of Free Surface and Pressurized Flow* (Kluwer Academic Publishers). The *Urban Water Supply Handbook* received the 2002 Honorable Mention in Engineering Award, given by the Association of American Publishers. Among his honors include a distinguished alumnus award from the University of Illinois at Urbana—Champaign in 1999. He lives in Scottsdale, Arizona and Pagosa Springs, Colorado.

www.ingramcontent.com/pod-product-compliance
Lightning Source LLC
LaVergne TN
LVHW062118200926
848484LV00007B/1712

* 9 7 8 0 0 7 1 4 4 3 8 1 4 *